T5-AFU-552

ISO 14000 Environmental Management Standards

ISO 14000 Environmental Management Standards

Engineering and Financial Aspects

ALAN S. MORRIS

Department of Automatic Control and Systems Engineering
University of Sheffield, UK

John Wiley & Sons, Ltd

Telephone (+44) 1243 779777

Email (for orders and customer service enquiries): cs-books@wiley.co.uk
Visit our Home Page on www.wileyeurope.com or www.wiley.com

Other Wiley Editorial Offices

John Wiley & Sons Inc., 111 River Street, Hoboken, NJ 07030, USA

Jossey-Bass, 989 Market Street, San Francisco, CA 94103-1741, USA

Wiley-VCH Verlag GmbH, Boschstr. 12, D-69469 Weinheim, Germany

John Wiley & Sons Australia Ltd, 33 Park Road, Milton, Queensland 4064, Australia

John Wiley & Sons (Asia) Pte Ltd, 2 Clementi Loop #02-01, Jin Xing Distripark, Singapore 129809

John Wiley & Sons Canada Ltd, 22 Worcester Road, Etobicoke, Ontario, Canada M9W 1L1

Wiley also publishes its books in a variety of electronic formats. Some content that appears in print may not be available in electronic books.

Library of Congress Cataloging-in-Publication Data

Morris, Alan S., 1948–
ISO 14000 environmental management standards: engineering and financial aspects / Alan S. Morris.
p. cm.
Includes bibliographical references and index.
ISBN 0-470-85128-7 (alk. paper)
1. ISO 14000 Series Standards. 2. Environmental protection – Standards. I. Title.
TS155.7.M64 2004
658.4′08 – dc22 2003058345

British Library Cataloguing in Publication Data

A catalogue record for this book is available from the British Library

ISBN 0-470-85128-7

Typeset in 10/12pt Times by SNP Best-set Typesetter Ltd., Hong Kong
Printed and bound in Great Britain by Antony Rowe Ltd, Chippenham, Wiltshire
This book is printed on acid-free paper responsibly manufactured from sustainable forestry in which at least two trees are planted for each one used for paper production.

Contents

Dedication

To the memory of Cyril, Joan and Glyn.

Preface

There is widespread concern about environmental matters in all developed countries around the world, and public interest in this is now so great that the implementation and operation of an efficient and effective environmental management system (EMS) is as important to the financial well-being of a company as it is to the environment that it is intended to protect. Apart from incurring financial penalties when environmental protection legislation is breached, a greater problem that businesses face is that poor environmental performance can lead to a boycott of a company's products and services by customers, with consequential serious damage to its financial health. In extreme cases, the general public may also take direct action that hinders or even shuts down a company's operations.

ISO 14000 is a descriptor for a set of standards that have been developed in response to this global concern about the environment. These standards represent a consensus agreement by national standards bodies around the world about the procedures that need to be followed in establishing an effective EMS. The primary standard amongst this set is ISO 14001, and the fundamental aim of this book is to cover the procedures that should be implemented by a company in order to satisfy the requirements laid down in this standard.

Environmental management to ISO 14001 standards is of similar importance to quality management to ISO 9001 standards in today's businesses, and the linkage between the two will be stressed in the text. Apart from the need to satisfy the stringent environmental control legislation that exists in most developed countries, the image of a company is damaged if pollution incidents occur, particularly if these are identified by environmental pressure groups, and this can have a severe impact on the marketability of products and services provided by the company. Conversely, ISO 14001 certification can have a very positive impact on a company's business, in view of the widespread public interest that now exists in environmental protection.

Whilst many texts are available that cover the management considerations in implementing environmental protection procedures that satisfy ISO 14001, these usually give little guidance about the necessary engineering procedures that are involved, or the associated financial implications. This text is intended to fill that gap, and its primary aim is therefore to provide a cross-disciplinary approach that bridges the management field and the engineering field. The book firstly presents the requirements of ISO 14001 environmental management systems, secondly summarises the company management schemes and procedures required for implementation of ISO

14001 systems, and thirdly discusses the engineering considerations and procedures necessary to ensure the successful operation of ISO 14001 systems. The relevant financial considerations are discussed throughout.

Chapter 1 provides an introduction to the ISO 14000 family of standards and summarises the main requirements of ISO 14001. Comparison with the requirements of the quality assurance standard ISO 9001 are also made, with guidance about how companies that already have expertise with ISO 9001 can use it beneficially in applying ISO 14001. Particular emphasis is given to the engineering considerations in applying ISO 14001, in terms of parameter measurement and recording, fault detection, waste reduction, equipment design and provision of emergency response procedures to minimise environmental damage when faults occur. Comment is also made about the need to tailor the EMS to the requirements of each situation and not to incur unnecessary costs in overspecifying the system.

Chapter 2 continues on from the brief introduction provided in Chapter 1 and explains the design and implementation of an ISO 14001 EMS in greater depth. The first section in this chapter covers the general design principles of an EMS. The following section then discusses the implementation of an EMS, and is subdivided into three subsections that cover respectively the general requirements, including the required level of documentation, the measurement and calibration requirements, and, finally, other engineering issues. The financial costs and benefits of operating an EMS are then discussed and evaluated in Section 2.3. Following this, the final three sections in this chapter cover, respectively, internal and external EMS auditing procedures, the procedure for getting ISO 14001 registration for the system, and advice about the need to publicise the environmental performance in order to maximise the financial benefits of operating the system.

The largest engineering contribution to the successful operation of an EMS is in the provision of systems that monitor system performance and measure environmental parameters. There is a particular requirement specified in ISO 14001 for the establishment of good measurement and calibration practices, so that the quality of measurements related to environmental management systems can be guaranteed, and proof established that the EMS is successfully ensuring that pollution does not exceed defined levels of acceptability. Because good measurement practice is so critical to the success of the EMS, two chapters in the book are devoted to the various aspects of this. The first of these (Chapter 3) describes the design of measurement systems, and covers the choice of instruments with appropriate characteristics, certified calibration procedures, documentation requirements and cost considerations. Following this, Chapter 4 considers measurement system errors, and describes procedures to ensure that measurements provided as part of an EMS are of adequate quality for their intended purpose.

The requirement to maintain accurate records of parameter measurements that are made as part of operating the EMS is also emphasised in ISO 14001, and Chapter 5 discusses the necessary mechanisms for this. This discussion starts with variable conversion elements, which are often necessary to convert sensors outputs to a recordable form. Discussion then continues with a review of the various data transmission mechanisms available, and the types of signal processing that are commonly necessary to maintain the accuracy and quality of data. Finally, the various means available for making permanent records of data are discussed.

The following block of chapters goes on to consider particular aspects of the control of environmental pollution. Firstly, Chapter 6 covers the sources and effects of air pollution and describes techniques for the quantification of air quality in terms of both particulate and gaseous pollutants. Chapter 7 provides a similar treatment for water pollution. Following this, Chapter 8 describes various ways of reducing air and water pollution by including appropriate features in plant design, and by designing, implementing and operating systems designed to control the emission of pollutants. Finally, Chapter 9 discusses other forms of environmental pollution, and considers ways of measuring and controlling noise, vibration and shocks. Several other engineering contributions to environmental protection are considered in the final four chapters, with emphasis on the financial considerations as well as the engineering aspects of each. Firstly, Chapter 10 discusses waste management and considers the various techniques that are available for reducing the amount of waste produced, including the application of mass-balance principles. Appropriate ways of disposing of waste that is produced are then considered, with the aim of minimising the environmental impact. In Chapter 11 that follows, procedures for assessing and quantifying the risk of pollution incidents when faults arise during the activities of a company are considered for both manufacturing operations and the provision of service functions. Various techniques to analyse and reduce risk are also discussed. In addition, this chapter covers reliability analysis, and describes various ways of assessing and improving the reliability of both normally operating plant and also special, emergency-response equipment. Chapter 12 continues on the theme of environmental protection through fault detection, by considering the technique of statistical process control and its role in detecting potentially pollution-creating situations at an early stage, thus allowing remedial action to be taken before serious pollution incidents have occurred. The principles of this, conditions for application and the main types of control chart used, are all covered. Finally, the accurate measurement of process variables in manufacturing systems is considered in Chapter 13, with separate sections covering temperature, pressure, fluid flow rate and level measurement. The importance of such measurements in ensuring that process plant operates as intended, without deviation of parameters from normal values that will cause a pollution risk, is emphasised.

Two appendices are provided. Appendix 1 summarises the main content of each standard within the ISO 14001 family. Appendix 2 provides a suggested layout for the EMS manual, which is a mandatory part of any EMS that conforms to the requirements of ISO 14001.

In terms of intended readership, the emphasis on the engineering aspects of environmental management systems, and the associated financial considerations, mean that the book is targeted primarily at company personnel who are concerned with developing, implementing, maintaining and modifying environmental management systems. However, it is anticipated that it will also be used by personnel at company management level who are directing environmental policy. The difficulties in discussing technical matters in language that is understandable to nontechnical personnel are well understood by the author. Therefore, all areas in the book are introduced in nontechnical language that is understandable to everyone, before going on to cover the technical detail that is necessary for personnel who are designing and implementing environmental management systems. This approach has been used successfully by the author previously (*Measurement and Calibration Requirements for Quality Assurance to ISO 9000*, John Wiley & Sons, 1997).

1

Introduction

There is now a considerable amount of public concern about the health of the environment in almost all developed countries of the world. As a consequence, the adoption by companies of procedures that minimise damage to the environment is becoming an important ingredient in their success, and is almost as important as the quality of the goods and services that they provide. Any actions of companies that lead to environmental pollution or damage, whether intended or not, cause widespread public anger that may lead to a boycott of the company's products or services, or even more direct action that interferes with its operations. In addition to this, environmental protection legislation is becoming increasingly stringent in most countries, and pollution incidents will, at best, lead to financial penalties and, at worst, result in orders to suspend operations until the cause of pollution is rectified. Thus, the implementation of an environmental management system (EMS) that minimises damage to the environment through a company's operations is becoming almost mandatory if the consequences of causing environmental damage are to be avoided.

'ISO 14000' is a global term for a set of standards that have been written in response to this need for environmental protection systems, in the same way that ISO 9000 standards were written to satisfy the need for quality assurance systems to control the goods produced and services supplied by companies. In fact, there is a very strong parallel between ISO 14000 standards and ISO 9000 standards, which will be discussed more fully later. Within the ISO 14000 series of standards, the fundamental standard that prescribes good practice in environmental management is ISO 14001[1]*. ISO 14001 specifies the various requirements that have to be satisfied

* Like many other ISO standards that have international recognition, ISO 14001 is often published in individual countries by national standards organisations written in identical words but with a slightly different prefix or code to achieve harmony with pre-existing national coding systems for standards. Thus, it is available from the British Standards Institute as BS.EN.ISO14001 and the European Committee for Standardisation (CEN) as EN.ISO14001.

ISO 14000 Environmental Management Standards: Engineering and Financial Aspects. Alan S. Morris.
 ISBN 0-470-85128-7

in setting up an effective EMS, such that the risks of pollution incidents and other forms of environmental damage through the operations and activities of a company are minimised.

The clauses in ISO 14001 are written in a general way so that the standard can be applied in a wide range of industries and in diverse geographical and social conditions. The guidelines specify the various procedures that need to be implemented in an EMS so that it successfully minimises environmental damage caused by the operations and activities of a company, but the standard recognises that every situation and application is different. Therefore, the standard is not prescriptive about how environmental protection procedures should be implemented in any particular situation, and does not set any particular emission targets, pollution levels or other parameters by which effects on the environment can be measured, except for specifying that at least the minimum environmental targets defined in legislation must be met. Some common examples of environmental targets set by legislation are limits on air and water pollution, waste management and waste reduction. Beyond such environmental targets set by legislation, and in respect of other environmental effects that are not subject to legislation, ISO 14001 recognises that the achievement of particular pollution-level and other environmental targets has to be balanced against the cost of achieving such targets and the economic well-being of the company. Thus, ISO 14000 prescribes only that the company sets environmental targets that are reasonably achievable at an acceptable economic cost. What is 'reasonably achievable' depends on the conditions in which a company operates and the type of industry and activities in which it is involved. Ultimately, what is 'reasonable' and what is 'achievable' is a matter for reasoned debate, and this is normally the subject of discussion with EMS auditors.

The ISO 14001 standard prescribes that a company shall establish an environmental policy that identifies all potential environmental effects arising out of its operations, and implements procedures designed to minimise these effects within the bounds of what can reasonably be achieved at an acceptable economic cost. As well as requiring the implementation of such an environmental management policy, ISO 14001 also prescribes that there must be a commitment to review the operation of the policy regularly and seek continual improvement in its performance in reducing environmental damage.

Documentation is a very important part of a company's EMS. First of all, this must record the assessment that is carried out initially to identify all operations carried out in the company and their environmental impact (if they have one). For each potential impact on the environment, appropriate procedures must be specified that are designed to minimise the environmental impact, and the manner in which such procedures are to be implemented and maintained must be defined. The documentation must also specify how the operation of the EMS is to be reviewed (and at what frequency) and must express a commitment on the part of the company to seek continual improvement in its performance with regard to reducing environmental damage. Such reviews must also be documented separately and kept with records of the company's environmental performance.

As well as the obvious benefit to the environment, the EMS implemented by a company in conformance with ISO 14001 usually leads to significant benefits to the company itself. A good record in environmental matters often makes it easier for a company to win new investment to finance technological developments and expansion. It is also usually well received by the customers of a company, and often leads to increased business. Furthermore, the commitment set by ISO 14000 to seek continual improvements in environmental matters that the company can influence is leading an increasing number of companies to require their suppliers to have ISO 14000 certification. Thus, failure to implement an ISO 14001 certified EMS is likely to lead to a serious decrease in business for many companies, and reduced inward investment. In addition to these benefits, procedures that are designed to prevent faults that can potentially cause environmental damage usually lead to greater efficiency in the company's operations. Hence, operating costs are reduced and profits rise, offsetting the cost of designing and implementing the EMS. Procedures implemented to minimise energy consumption, raw material usage, minimise waste and recycle the waste that is produced lead to further cost savings. Finally, improvements in the company's own working environment contribute towards increasing the job satisfaction of the company's employees. Job satisfaction of employees is clearly important in achieving the general quality goals of a company. If workers feel that the company looks after them, they are more likely to be inclined to work in the best interests of the company, and this will also enhance the company's economic performance.

Environmental Management Systems conforming to ISO 14001 can be implemented in one of two ways. The first approach is for the company to establish a system that is certified by a National Accrediting Body. The alternative second approach is to go for a self-certified system. Both approaches can lead to the successful implementation and operation of an EMS. However, the properly certified system, although being the more expensive approach, has a number of advantages. Firstly, many companies do not have personnel with sufficient knowledge and experience to develop an efficient and completely effective EMS, particularly within a reasonable timescale. If external consultants have to be employed, the cost of achieving a fully certified system is little greater than if only a self-certified system is implemented. A fully certified system is also more likely to win public and customer support, leading to the financial benefits noted earlier. Insurance costs for liability and litigation arising out of adverse environmental effects due to the company's operations will also normally be lower if the company operates a properly certified EMS. Finally, a certified EMS is a good defence if pollution incidents and other environmental effects occur that transgress environmental legislation. Provided that the company has not been negligent in the implementation of its EMS, penalties imposed for violations of environmental legislation are likely to be lower than would be the case if the company was not able to demonstrate a serious attempt to protect the environment via its EMS.

As alluded to earlier, the procedures required to satisfy ISO 14001 are very similar to those needed to satisfy ISO 9001[2], as demonstrated by the comparison given in Table 1.1. In fact, the aims of ISO 9001 and ISO 14001 are so similar that it is expected that a combined standard will be published in the near future that will set out the requirements necessary to establish a system satisfying both quality assurance and environmental protection objectives simultaneously. Although ISO 9001

Table 1.1 *Comparison of ISO 14001 and ISO 9001*

Feature	ISO 14001(2002)	ISO 9001(2000)
Aim	To identify all potential environmental effects of a company's operations, and to develop and implement a system that limits environmental effects to set targets	To identify all processes in a company that can affect the quality of the products and services provided, and to implement a system that assures the achievement of high quality
Audience	Customers of company plus general public	Customers of company
Documentation	Required for all aspects of environmental management system	Required for all aspects of quality management system
Required by legislation?	Partly	No
Overall responsibility	Executive management of company	Executive management of company
Day-to-day responsibility	Named persons in company	Named persons in company
Involvement of all personnel in company	Required	Required
Communication	Adequate communication channels must exist between all levels and functions of the company	Adequate communication channels must exist between all levels and functions of the company
Resourcing	Executive management of company must provide adequate resources for implementation and operation	Executive management of company must provide adequate resources for implementation and operation
Training	All personnel whose actions may affect the environment must receive appropriate training	All personnel whose actions may affect quality must receive appropriate training
Monitoring and measurement	Key parameters must be monitored and measured to demonstrate that the environmental protection objectives are being met	Key parameters must be monitored and measured to demonstrate that the quality assurance system objectives are being met
Accidents and emergency situations	Appropriate procedures in response must be established to minimise environmental damage	No equivalent clause
Regular review by executive company management to ensure system is operating as planned	Required	Required
Regular internal audits by person responsible for day-to-day operation	Required	Required
External audits at prescribed intervals	Required if system is certified	Required if system is certified
Records of reviews and audits	Required	Required
Continual development and improvement of system	Required	Required

and ISO 14001 remain separate for the present, the similarity in procedures between them means that, if a company already has an ISO 9001 certified system, the manager responsible for the quality system and the manager responsible for environmental policy will be able to work closely together, and in the case of small companies, one person might even fulfil both management roles. It also means that many of the features of the ISO 9001 quality system can be adapted for the EMS, which greatly simplifies the procedure of designing and implementing the EMS. However, this adaptation must not overlook the fact that, whilst there is much commonality, there are also some important differences that need to be considered. For example, whilst ISO 9001 is primarily concerned with satisfying the customers of a company, ISO 14001 has to satisfy the general public as well as its customers.

Companies that operate Total Quality Management (TQM) systems usually find that the philosophy engendered by TQM is a valuable aid in developing and implementing an EMS. TQM is a quality assurance buzzword that does not have a universally accepted standard definition. Different organisations and companies define and apply TQM in different ways. However, a global definition that encompasses most of the different interpretations is that TQM is an integrated approach to quality that operates in all parts of a company and encompasses a style of management that is aimed at achieving the long-term success of a company by linking quality with customer satisfaction. TQM requires that the quality of the company's product (whether the supply of manufactured goods or the provision of a service) should be the company's number-one priority, and demands an ongoing commitment to progressively increase quality still further. These aims are almost identical with the aims of an EMS in meeting environmental targets. This is why companies that have adopted the TQM philosophy find it relatively easy to apply similar principles to environmental protection.

1.1 General Approach to Developing an Environmental Management System

For an EMS to operate successfully, several conditions have to be satisfied. Firstly, the procedures instituted must be orientated towards preventing the occurrence of incidents that might cause environmental damage, rather than being mere fault detection systems that allow faults to be put right before someone complains about the environmental effects. Secondly, when introducing an EMS, it is very important that employees at all levels in a company are aware of the reason for it, understand fully how to operate it, and cooperate enthusiastically in implementing it. Thirdly, whilst it is necessary to appoint an EMS manager with designated authority for implementation and operation of the procedures designed to minimise the environmental impact of the company's operations, the responsibility for environmental protection must never be seen as being the responsibility of this one person alone. All personnel in a company must be encouraged to share in the duty of avoiding damage to the environment and to take pride in doing so.

It is also important that environmental management procedures should evolve and develop over a period of time. They must not be implemented and then stay the same for ever afterwards. Rather, regular review is necessary to ensure that the procedures continue to be efficient and remain the most appropriate as technological developments take place. A proper response must also be made promptly if new or modified environmental legislation is introduced.

Measurement is an essential ingredient in the operation of an EMS, and is one of the engineering aspects of environmental management that this book concentrates on. Firstly, measurement is necessary to ensure that process variables within a manufacturing process are maintained within acceptable limits, as large deviations may lead to undesirable environmental effects. Secondly, direct measurement of emission levels must be made to ensure that the target levels defined in the EMS are not exceeded. However, if the EMS is to operate satisfactorily, then the measurements themselves must be of high quality. There are several necessary conditions in achieving high-quality measurements. Firstly, only properly calibrated instruments and transducers must be used for making measurements, and appropriate calibration equipment must therefore be established and maintained. Secondly, all measurement errors must be identified, quantified and compensated for. Thirdly, only appropriate instruments must be used to make measurements. Fourthly, the operating principles and correct mode of usage of the measuring instruments used must be understood by the person making the measurements. Fifthly, data captured by measuring instruments must be transmitted from the point of measurement to the point of recording without deterioration in the quality of the data. Finally, suitable data-recording instruments must be used, so that the data can be included in records of the past performance of the EMS.

These general principles governing the establishment of an efficient EMS have been developed over a number of years, and national standards that existed prior to the publication of ISO 14001 (e.g. the British Standard BS 7750[3]) contained many of the clauses and recommended procedures that are now included in ISO 14001. Thus, ISO 14001 has resulted from the international community getting together under a technical committee set up by the International Standards Organisation and agreeing a common international standard that now supersedes the earlier national ones.

There are also initiatives to combine environmental management with other functions such as health and safety, as in the RC 14001 standard produced jointly by the American Registrar Accreditation Board and the American Chemical Council. RC 14001 combines ISO 14001 requirements with responsible care (RC) guidelines to safeguard health and safety.

1.2 Summary of Requirements of ISO 14001

As noted earlier, ISO 14001 is one standard within a set of standards known as the ISO 14000 series. However, it is ISO 14001 itself that actually sets down the requirements for achieving an efficient EMS. The other standards in the series are merely guides that give assistance in interpreting and implementing the various clauses written in ISO 14001. These other standards will be described briefly in the next section. However, before considering these, it is useful to first summarise what the main requirements of ISO 14001 are:

- The fundamental requirements of the EMS implemented are that it should:
  - Identify and assess the environmental impact of all of a company's operations, and repeat this on a regular basis.
  - Consider all of the company's operations and activities that are identified as having a potential or actual environmental impact, and set environmental protection targets that are appropriate to the scale and impact of the operations, but within the constraints of what is technically possible and economically affordable.
  - Irrespective of cost, ensure that the company at least complies with all relevant environmental legislation that its operations may be subject to in respect of their environmental impact.
  - Be continually reviewed and improved wherever possible.
  - Have all aspects of the policy written down in documentation that is available to the public.
- Everyone in a company must be fully committed to the EMS being operated.
- Appropriate communication paths must be established to ensure that the EMS operates efficiently.
- Responsibility for the implementation, operation and review of the EMS must be assigned to one designated person.
- The key characteristics of all operations that can have a significant effect on the environment must be regularly monitored and measured, and results must be documented.
- All instruments and equipment used to measure performance of the EMS must be used properly and calibrated regularly.
- All abnormal situations that might arise in the operations and activities of a company must be identified and their potential environmental impact must be assessed.
- Appropriate procedures must be established and documented for responding to abnormal situations that might cause environmental damage.
- The training needs of anyone in the company whose activities may impact on the environment must be identified and appropriate training provided.
- Regular audits must be carried out to ensure that the EMS is operating satisfactorily and meeting its target of protecting the environment in the way that is expected of it.
- The fundamental responsibility for implementation and successful operation of the EMS lies with the executive management of the company implementing it.
- The executive management must ensure that adequate resources are provided to support the EMS. These resources must include employees with the necessary skills as well as the financial resources necessary to buy whatever equipment is needed.
- The executive management themselves must regularly review the performance of the EMS. To do this, they should ask for performance reports from the person(s) with designated authority for operating the EMS and, having reviewed the reports, direct any necessary action to modify the EMS in order to improve the company's environmental performance. This review by executive management must be in addition to, and not instead of, the other internal and external performance audits that are carried out.

Whilst the above is an accurate summary of the main requirements of ISO 14001, copyright reasons prevent a verbatim reproduction of the exact phrases in the official ISO 14001 document as published. Hence, readers having direct involvement in planning and implementing an EMS that is to be certified under ISO 14001 are advised to actually read the official ISO document. This is not an onerous task, since the main part of the document only extends to some 10 pages.

1.3 Other ISO 14000 Standards

As mentioned earlier, ISO 14000 is not a standard in itself but rather the descriptor for a series of standards that have environmental management as the theme. The main standard, ISO 14001, sets out the requirements for achieving an efficient EMS, as described in the last section. The other standards do not set further requirements but merely offer guidance in satisfying the requirements set down in ISO 14001. The titles of these other standards are given below and a summary of the contents of each can be found in Appendix 1.

ISO 14004 (1996): *Environmental Management Systems: General Guidelines on Principles, Systems and Supporting Techniques.*
ISO 14010 (1996): *Guidelines for Environmental Auditing – General Principles.*
ISO 14011 (1996): *Guidelines for Environmental Auditing – Audit Procedures – Auditing of Environmental Management Systems.*
ISO 14012 (1996): *Guidelines for Environmental Auditing – Qualification Criteria for Environmental Auditors.*
ISO 14015 (2001): *Environmental Management Systems – Environmental Assessment of Sites and Organisations.*
ISO 14020 (2001): *Environmental Labels and Declarations – General Principles.*
ISO 14021 (2001): *Environmental Labels and Declarations – Self-declared Environmental Claims (Type II Environmental labelling).*
ISO 14024 (2001): *Environmental Labels and Declarations – Type I Environmental Labels – Principles and Procedures.*
ISO 14031 (2000): *Environmental Management – Environmental Performance Evaluation Guidelines.*
ISO 14032 (2000): *Environmental Management – Examples of Environmental Performance Evaluation.*
ISO 14040 (1997): *Environmental Management – Life Cycle Assessment – Principles and Framework.*
ISO 14041 (1998): *Environmental Management – Life Cycle Assessment – Goal and Scope Definition and Inventory Analysis.*
ISO 14042 (2000): *Environmental Management – Life Cycle Assessment – Life Cycle Impact Assessment.*
ISO 14043 (2000): *Environmental Management – Life Cycle Assessment – Life Cycle Interpretation.*
ISO 14048 (2002): *Environmental Management – Life Cycle Assessment – Data Documentation Format.*

ISO 14049 (2000): *Environmental Management – Life Cycle Assessment – Examples of Application of ISO 14041 to Goal and Scope Definition and Inventory Analysis.*
ISO 14050 (2002): *Environmental Management Vocabulary.*
ISO 14061 (1998): *Information to Assist Forestry Organisations in the use of Environmental Management System Standards ISO 14001 and ISO 14004.*
ISO 19011 (2002): *Guidelines for Quality and/or Environmental Management Systems Auditing.*

1.4 Engineering Aspects of ISO 14001 Requirements

Engineering input is implied specifically in the clause in ISO 14001 requiring monitoring and measurement of the key characteristics of operations that can have a significant impact on the environment. This requires design of sound measurement procedures and the use of measuring instruments that are properly calibrated. As it is usually necessary to record measurements for future reference, the recording process must be considered as well as the measurement process. The first problem often encountered is that the output of many measuring instruments is not in a form that can be directly input into a data-recording instrument, and signal conversion is therefore needed. It is also important to control the quality of signal transmission between the point of measurement and the point of data recording, and to ensure that suitably accurate recording equipment is used. However, in addition to this, further engineering input is required in the specification and implementation of procedures that are designed to prevent operations and activities of the company, and particularly malfunctions of personnel or equipment, from having an adverse effect on the environment.

1.4.1 Summary of measurement and calibration requirements

Whilst ISO 14001 specifies a requirement to establish documented procedures to monitor operations and activities that can have an impact on the environment and to maintain and calibrate measuring equipment properly, it does not give detailed guidance on how this requirement should be satisfied. However, since the measurement system design and maintenance requirements for ISO 14001 environmental management systems are similar to those specified in ISO 9001 for quality systems, the more detailed guidance given in ISO 9001 will be used as the basis for the measurement and calibration procedures described in this book, and in particular the procedures recommended in ISO 10012[4,5], which is a supplementary document to ISO 9001. Besides describing good measurement and calibration practice in detail, the ISO 10012 standard also gives some advice about implementation. Thus, the main measurement and calibration requirements appropriate to satisfying ISO 14001 can be interpreted as follows:

(1) All parameters relevant to the EMS must be measured with instruments that have adequate accuracy and other characteristics.
(2) Measurement procedures must be adequate for their purpose.

(3) The list of instruments used to make measurements relevant to the EMS must be documented.
(4) Responsibility for measurement and calibration procedures must be assigned to a named person.
(5) All personnel performing measurement and calibration duties must have appropriate qualifications and training.
(6) All measurements, whether for purposes of calibration or EMS performance assessment, must take into account all the errors and uncertainties in the measurement process.
(7) An effective system for the control and calibration of measuring equipment must be established and maintained. (Note: complete in-plant calibration is not essential if these services are provided by specialist calibration services companies that comply with the requirements of the standards.)
(8) Clear evidence must be available that the measurement system is effective.
(9) Calibration must be performed by equipment traceable to national standards.
(10) Calibration procedures must be documented.
(11) The calibration system must be periodically and systematically reviewed to ensure its continued effectiveness.
(12) A separate calibration record must be kept for each measuring instrument. Each record must demonstrate that the instrument is capable of performing measurements within the designated limits, and should contain at least the following information:
   - a description of the instrument and a unique identifier;
   - the calibration date;
   - the name of the person who performed the calibration;
   - the calibration results;
   - the calibration interval (plus the date when the next calibration is due).

   Also, depending on the type of instrument involved, some or all of the following information is also required:
   - the calibration procedure;
   - the permissible error bounds;
   - a statement of the cumulative effects of uncertainties in calibration data;
   - the environmental conditions (ambient temperature, etc.) required for calibration;
   - the source of calibration used to establish traceability;
   - details of any repairs or modifications that might affect the calibration status;
   - any use limitations of the instrument.
(13) Each instrument must be labelled to show its calibration status and any usage limitations (but only where it is practicable to do so).
(14) Any instrument that has failed or is suspected (or known) to be out of calibration must be withdrawn from use and labelled conspicuously to prevent accidental use.

1.4.2 Signal conversion, signal transmission and data recording requirements

Having assured the quality of measurements by using good measurement principles and properly calibrated measuring instruments, it is important to ensure that measurement quality is not lost before the measured data are captured in the data-recording system. As the output of many measuring instruments is not in a form that can be directly input into a data-recording instrument, signal conversion elements are often needed to translate the measurement signal into a more suitable form. For example, bridge circuits are commonly used to convert the varying-resistance output from instruments like resistance thermometers into a varying voltage. Following this, care must be taken to ensure that the quality does not degrade in transmission of the measurement data from the point of measurement to the point where it is recorded. Various signal-processing operations are often needed to preserve the quality of measurement data during transmission. These perform various functions like signal amplification and noise removal. Finally, recording equipment that can record the measurement data with sufficient accuracy must be chosen.

1.4.3 Other necessary engineering contributions to an EMS

Where malfunctions in manufacturing systems and other activities of a company may lead to environmental damage, procedures must be instituted to identify such malfunctions promptly and instigate remedial action. Risk analysis and reliability calculations are often useful tools in predicting the likelihood of faults that may lead to environmental damage. Reliability calculations are also necessary for the operation of systems that are designed to respond to faults and take remedial action. Various engineering techniques are available to respond to these requirements.

Engineering input is also required to satisfy other aspects of an EMS. Engineering improvements can often lead to a reduction in energy usage and a reduction in the raw materials used. Engineering input is also required to reduce the amount of waste generated and to achieve safe disposal of any waste that is produced.

1.5 What is Essential and What is Not When Implementing ISO 14001?

Many of the clauses in ISO 14001 are written using subjective words like 'appropriate', 'significant', and 'economically viable'. This leaves considerable room for argument about the environmental targets that a company needs to set and the procedures that it needs to institute to meet these. Clearly, any environmental targets that are set by legislation should be met. However, beyond this, no specific guidance can be given, as the requirements vary according to the size of the company, the industry in which it operates and what is commonly regarded as good environmental practice for companies in that industry. Often, experienced personnel in a company will be aware of what their competitors are achieving and will be able to design an EMS accordingly. Where such expertise is not available internally, many consultants are available who can give suitable advice.

The word 'appropriate' is particularly significant when considering the measurement procedures that are necessary. ISO 14001 specifies that measurement procedures must be adequate for their purpose. The cost in developing and implementing rigorous measurement procedures is not justified if less-rigorous procedures will be adequate for the particular purposes of the EMS operated. Thus, the procedures specified in Section 1.4.1 have to be interpreted intelligently according to the needs of the EMS. The procedures specified represent best practice and, if implemented fully, will produce measurements of the highest standards in terms of quality and consistency. Such high standards in measurement practices will be necessary to satisfy EMS requirements in some cases. However, in other cases, lesser standards of measurement will be satisfactory.

Ultimately, if the company seeks certification of its EMS, the team of auditors appointed to review it will be the final arbiters of whether the scope and operation of the EMS is satisfactory. One of the most important things that auditors will examine when a company submits an EMS for certification is commitment to the aim of protecting the environment. If the company has developed an EMS for the first time, the auditors will expect to see that it has made a reasonable attempt to satisfy the requirements prescribed in ISO 14001. However, they would not necessarily expect every aspect of the EMS to be operated at the highest level that is technically possible. For example, they would probably not expect measurement systems to be designed as rigorously as specified in Section 1.4.1 above. Nevertheless, the clause in ISO 14001 requiring regular review and enhancement of the operation of the EMS would require the company to make whatever improvements were economically viable, and therefore gradually institute more rigorous measurement and system monitoring procedures, as covered in Section 1.4.

References

1. *ISO14001, 1996, Environmental Management Systems: Specification with Guidance for Use* (International Organisation for Standards, Geneva) (also published as BS.EN.ISO14001 by the British Standards Institution, 1996).
2. *ISO 9001, 2000, Quality Management Systems – Requirements* (International Organisation for Standards, Geneva).
3. *BS7750, 1994, Environmental Management Systems* (British Standards Institution, London).
4. *ISO 10012-1, 1992, Quality Assurance Requirements for Measuring Equipment – Part 1: Metrological Confirmation System for Measuring Equipment* (International Organisation for Standards, Geneva) (also published as BS.EN.30012-1 by the British Standards Institution in 1994).
5. *ISO10012-2, 1997, Quality Assurance for Measuring Equipment – Part 2: Guidelines for Control of Measurement Processes* (International Organisation for Standards, Geneva).

2

Design and Implementation of ISO 14001 Environmental Management Systems

It is appropriate to start this chapter by repeating the guidance set out at the end of chapter one about the extent and sophistication of the environmental management system (EMS) adopted. There is much emphasis within ISO 14001 about the need to balance environmental protection needs with socioeconomic needs. The standard recognises that the cost of meeting some environmental targets could be almost unlimited if the targets were set too high. It is right that companies must have due regard to the cost of meeting environmental targets, and therefore they should not set targets where the financial burden of meeting them would require significant reduction in the number of people employed or might even threaten the survival of the company. Nevertheless, it is incumbent upon companies to implement all environment protection measures that can reasonably be achieved, and the reasons why higher targets are not set, whether for financial or other reasons, must be clearly stated in the EMS documentation.

In order to achieve its environmental goals, a company must develop administrative and technical systems to support its EMS and ensure that the required environmental targets are met and maintained. It is important that these administrative and technical systems identify all activities of a company that impact on the environment, whether in manufacturing operations or service provision, and combine to produce a cohesive and effective EMS. After such a system has been put into operation, it is also extremely important to monitor its continued effectiveness by generating and evaluating performance measurements.

The ISO 14001 standard prescribes a number of procedures, which, if applied correctly, successfully achieve these environmental protection goals. Sometimes, although unjustifiably, criticisms are directed at the standard, accusing it of being

ISO 14000 Environmental Management Standards: Engineering and Financial Aspects. Alan S. Morris.
 ISBN 0-470-85128-7

too bureaucratic, too costly to implement and requiring unnecessary documentation, especially when applied to small companies. In truth, the usual reason for such problems is that the consultant responsible for advising the company concerned about ISO 14001 certification has overspecified the requirements. The ISO 14001 standard has been drafted very carefully so as to leave room for intelligent interpretation according to the size of the company implementing it, the type of industry or activity involved and the circumstances prevailing. The only mandatory requirement is that the measures put in place are sufficient to satisfy legal requirements. Beyond that, the main requirement is that sufficient effort should be applied in implementing environmental protection procedures to satisfy both the company's customers and the general public. A good consultant will avoid unnecessary bureaucracy and cost as long as this main requirement is satisfied, but a bad consultant will overspecify the EMS and thus incur unnecessary cost. Thus, it is very important that companies adopting ISO 14001 choose their consultant with care and, if possible, seek recommendations about consultants from companies of a similar size who work in similar industries.

One matter of overriding importance in implementing an EMS is that of achieving the total involvement of all personnel in the company, and fostering a common commitment to environmental protection that is shared by everyone. Everyone must understand what is required of them and what effect their actions may have on the environment. The importance of engendering a commitment to the company's EMS cannot be overemphasised. Islands of good practice here and there are of no use at all: if workers in a few departments are achieving environmental protection targets but they see other departments who are not, they will quickly lose interest in working hard to achieve the targets themselves.

However, even when everyone in a company seems to be participating enthusiastically, one possible danger in operating an EMS is that complacency can creep in. Once this belief that 'our environmental performance is OK' has been adopted throughout a factory, a false sense of security can be generated. So the environmental performance is OK? But who has judged this to be so, and against what standards? Clearly, some independent confirmation that environmental performance is being maintained is necessary, and this is why regular audits of the EMS are essential.

2.1 Design of an Environmental Management System

Environmental policy must be more than a statement of intent: it must be manifest in a specific course of action designed to achieve the environmental objectives specified. To achieve this, the EMS must be designed and implemented according to a properly documented strategic plan. By using such a strategic plan, the required environmental targets will be met and maintained at minimum cost. The emphasis in the strategic plan should be on designing procedures that prevent environmental problems occurring, rather than on correcting defects after they have occurred. It is also important that a team approach is taken in developing the strategic plan. Whilst overall responsibility for the EMS must be in the hands of one person, there is much merit in involving a team of relevant personnel to discuss details of the environmental plan and the way in which it will be implemented. As a minimum, the team should include representatives of engineering, production and accounting functions in the company. This optimises the chance of gaining full commitment to the EMS by all parts of the company.

The essential activities in establishing a strategic environmental plan are shown schematically in Figure 2.1. The first step is to set the environmental targets. To do this, a careful assessment needs to be made of all operations and activities in the company, to identify and quantify their environmental impact. Writing down all operations and activities in the form of a flow chart is often a useful aid in this. It enables a systematic approach to be taken, where the environmental impact of each is considered in turn. In doing this, the environmental impact must be considered in its widest sense. Thus, environmental protection must go beyond limiting pollution of air and water in the environment, and should include other aspects such as waste minimisation and reducing the use of energy and raw materials.

In quantifying the effect of each environmental impact, due regard must be made not only to the magnitude of the effect, but also to the frequency or likelihood of

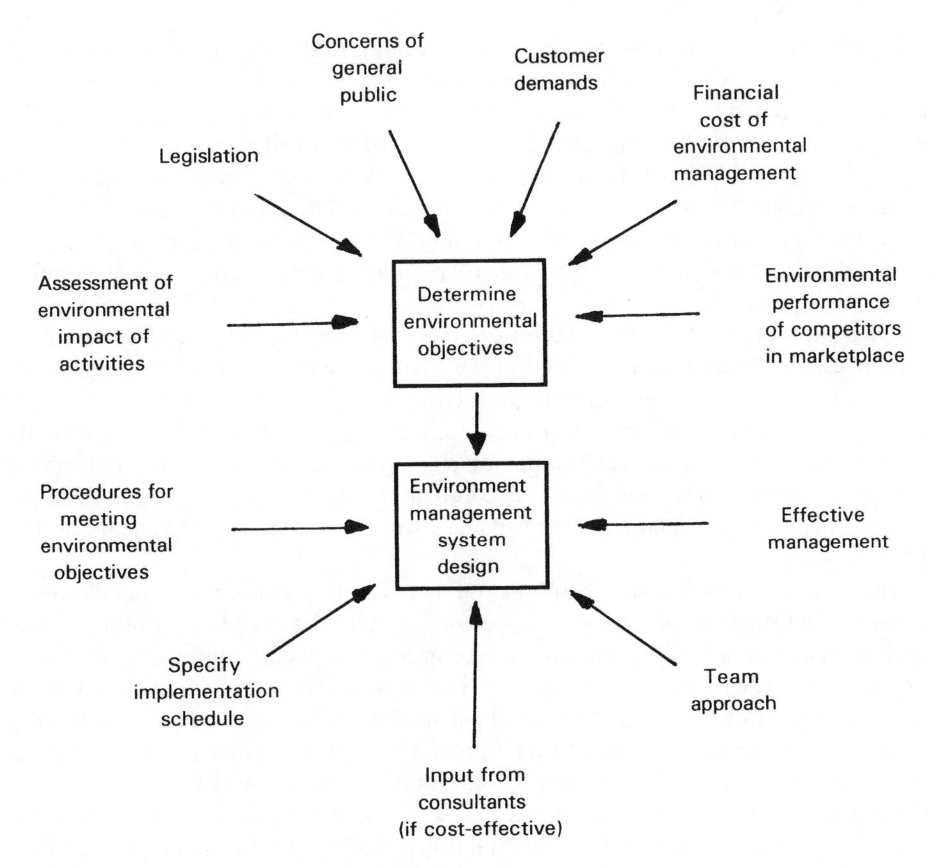

Figure 2.1 *Elements in a strategic environmental management plan.*

the impact occurring, and, if pollution is involved, the size and significance of the geographical area affected. Clearly, frequent environmental impacts are much more serious than infrequent ones, and pollution affecting areas with a large human population is more serious than pollution affecting areas with sparse or no human population. Further useful guidance on environmental impact assessment can be found in ISO 14015[1].

Once the impact of each operation and activity in the company has been quantified, the issues to be addressed by the EMS should be prioritised according to the environmental impact of each activity, and according to whether any legislative requirements have to be satisfied. Prioritisation of environmental issues to be addressed is necessary, because a company clearly cannot address every environmental issue simultaneously, due to the limitations of both staff and financial resources. Hence, the achievement of particular environmental targets has to be carefully balanced against the cost of meeting the targets, and due regard also has to be given to what the company's competitors are doing. The ISO 14001 standard recognises these limitations and allows a step-by-step approach, with the worst problems being addressed first. ISO 14001 does not neglect lesser environmental effects, but allows these to be addressed over a period of time, as part of the ongoing improvement of the EMS.

Legislation obviously has to be taken into account in shaping the EMS developed. Unfortunately, it is impossible to give specific guidance, since the statutory environmental requirements vary from country to country. However, due regard for these differences must be taken by any company involved in, or likely to become involved in, export markets. Whilst the fines imposed as a penalty for failing to comply with legislation might be seen as only a minor irritation that does not have any significant effect on the financial well-being of a company, a very much more serious penalty is involved in respect of the inevitable loss of reputation that a company would suffer when prosecuted.

Once appropriate environmental targets have been set, the next step is to examine each operation and activity included in the EMS, to determine how the targets can be achieved and maintained at minimum cost. To do this properly, each operation and activity must be broken down into separate elements. Each of these elements must have clearly defined environmental objectives and strategies that contribute towards a cohesive EMS for the whole company. Only when such a corporate strategy of environmental planning is instigated, will the full potential benefits of the EMS be realised.

The strategic aspects of environmental management cannot be considered in isolation, but must be part of a cohesive overall management plan. In other words, environmental management should not be considered as being fundamentally different from any other area of management, except in so far as the details of its practical implementation differ. The standard management procedures of planning, organising, directing and controlling should be applied equally in environmental management, as in any other management function. This allows environmental considerations to be managed in an effective way, whereby planning provides a basis for organising, which in turn enables the established EMS to be directed and controlled in an efficient manner.

As the EMS evolves over a period of time, the environmental performance of suppliers of materials to the company becomes a relevant matter for consideration. This is an area that a company often has control over, and this provides it with an opportunity to encourage better environmental consideration by its suppliers. Thus, as part of the continual improvement to the EMS demanded by ISO 14001, a company can move towards including ISO 14001 conformance as a requirement in the contract signed with suppliers. This requirement is most easily met if the supplier also operates a certified EMS. Indeed, in industries like automobile manufacture, it is becoming mandatory that all components are traceable to a chain of suppliers who are all certified to ISO 14001 or equivalent standards.

Mention must also be made of the need for the EMS to include specification of the planned programme of measurements designed to ensure that the environmental targets are being met. The details of all inspection procedures specified, their prescribed frequency, and the measurement techniques required, must also be fully documented.

Finally, one important concluding comment should be made about the design of the EMS and the development of a strategic plan to implement it. This comment is that the EMS must not be allowed to become stagnant. Technological change brings about continual improvements in the potential for reducing the environmental impact of a company's operations, and it is essential to take full advantage of these, as long as the economic cost is not too great. Also, the company's actual operations may change, with a consequent change in the environmental impact. Thus, all aspects of the EMS must be reviewed regularly, and changes made as necessary. This requirement to carry out reviews and seek regular enhancements to the system is enshrined in ISO 14001.

2.2 Environmental Management System Implementation

The EMS designed must be implemented such that the environmental targets set are met. It has already been noted that the targets set will not necessarily limit environmental impacts to the minimum possible according to the best technological capabilities, since ISO 14001 recognises that a company has to operate in an international market and must keep costs within strict financial constraints. These economic constraints may prevent the company from implementing very costly pollution-reduction technology. ISO 14001 only requires a company to strive to reduce environmental damage as far as possible, but within the constraints of what is economically reasonable and technically possible.

The key components in implementing an EMS that satisfies ISO 14001 (see Figure 2.2) are: commitment of company executives, cooperation of all company personnel, effective management, establishment of effective communication systems, appropriate training, design and implementation of supporting equipment, planning of emergency procedures, collection of data to monitor performance, regular system reviews and maintaining full documentation of the system. These components can be divided into general requirements, measurement and calibration requirements and other engineering considerations.

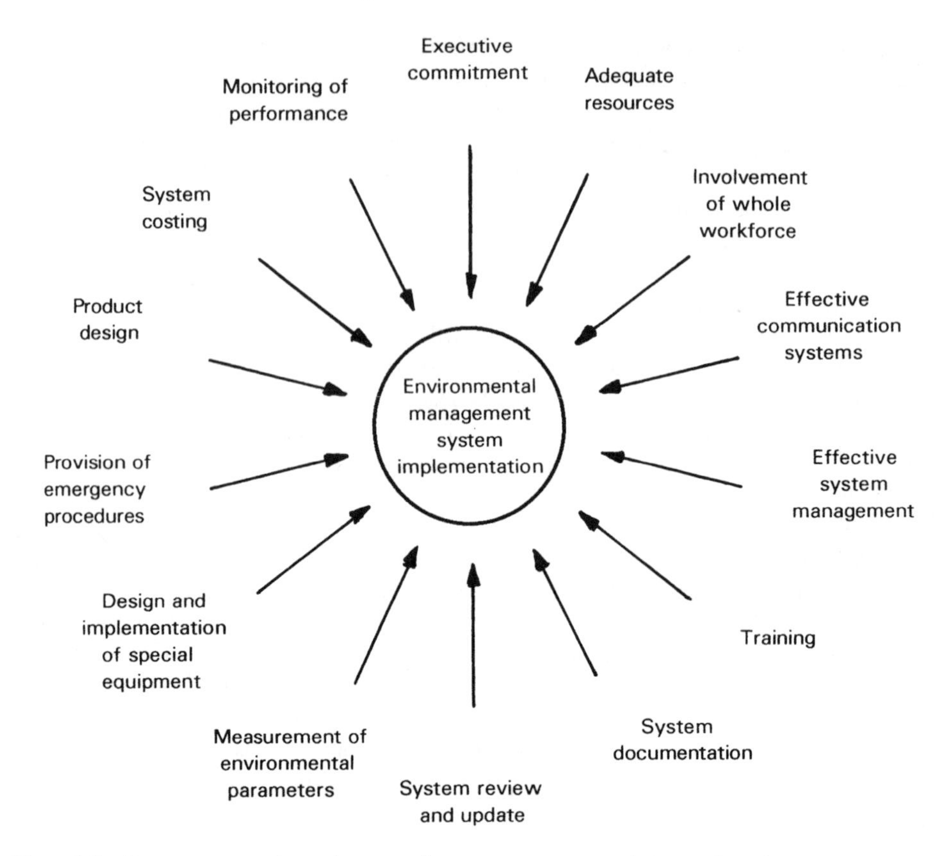

Figure 2.2 Key components in environmental management system design and implementation.

2.2.1 General requirements in implementing an ISO 14001-approved EMS

The general, nonengineering requirements specified by ISO 14001 to ensure that the EMS is implemented and maintained in an efficient and effective manner, include: executive commitment, total involvement of all company personnel, effective communication systems, managerial responsibility, adequate training, system documentation, and system review and update. These issues will be discussed in more detail in the following paragraphs.

Executive commitment

Full commitment by the executive management of a company is an essential requirement for successful implementation and operation of an EMS. They must fully support it by providing all financial resources necessary, and they must be the driving force behind the EMS. This commitment of the company executives to the EMS must also be highly visible, because any apparent lack of commitment on their part will spread like a plague through the rest of the company.

Executive management are responsible for ensuring that everyone in the company is sympathetic with the aims of the EMS and tries to play their part in making it successful. A spirit must be generated within the company that environmental protection is everyone's concern, not just that of the EMS director. 'Everyone' means all personnel, not just departmental managers. Effective communication of the company's environmental targets and details of the EMS operation are key to this, and executive management must ensure that this communication happens.

Whilst executive management have to accept overall responsibility for the EMS implemented, they should delegate day-to-day responsibility for its operation by appointing suitably qualified personnel to oversee EMS activities. Such personnel should be given the opportunity to decide for themselves how to tackle any problems in the operation of the EMS, since it is the people who are intimately acquainted with the workplace who will often come up with the optimum solutions to environmental problems. Company executives should only intervene if the delegated personnel fail to make the EMS work.

In order to fulfil their responsibility of ensuring that environmental targets are being met, executive management must conduct regular reviews of the EMS, and these reviews must be documented. To satisfy this requirement, they must seek detailed reports and data from the personnel with day-to-day responsibility for the EMS. If the review highlights any shortcomings in the performance of the EMS, executive management must direct appropriate remedial action and ensure that this action is taken.

Finally, executive management must ensure that the scope and design of the EMS are reviewed from time to time, and enhancements made if possible, to fulfil the requirements specified in ISO 14001 for continual development and improvement of the EMS. This review should also consider the effectiveness and need for existing EMS procedures. Thus, as a consequence of the review, 'improvement' might involve removing some existing procedures that are ineffective or unnecessary, as well as adding new procedures.

Total involvement

As discussed earlier, ISO 14001 requires that the responsibility for successfully operating the company's EMS should be shared by everyone in the company. Everyone has a contribution to make, even if it is just turning taps off to avoid wasting water and turning lights off to reduce energy consumption when they are not needed. Executive management are ultimately responsible for ensuring that this motivation towards environmental protection is generated in their employees.

The maintenance of such motivation in all employees to meet the standards set by the EMS is often assisted by the use of anonymous questionnaires from time to time to measure the morale of the workforce – to remove any irritants in the workplace and also assess the attitudes and degree of cooperation between different departments in the company. Schemes whereby workers are rewarded for making suggestions that lead to environmental improvements or elimination of environmental problems are also generally very useful. Display boards, where successes in meeting environmental targets are reported, can also be strong motivators in fostering the correct culture in environmental matters. Posters in the workplace encouraging things like reducing waste production and energy consumption can also be helpful.

Quality circles, which now exist in many companies as part of an ISO 9000 certified quality assurance system, can usually be extended to consider environmental matters, and are a good way of getting everyone in the company involved in the EMS. A quality circle consists of a group of people who collectively represent all functions within the company that can have an effect on quality. For instance, such a group must represent goods packing and delivery sections as well as shop-floor operatives from the production departments. The discussion that takes place at periodic meetings of the quality circle fulfils several functions. Firstly, it ensures that thoughts about quality maintain a high profile throughout the company. Secondly, it provides a feedback mechanism whereby breakdowns in or difficulties with the quality system can be reported and suggestions for improvement made. Thirdly, by giving everyone such a personal involvement in meeting quality targets, an atmosphere is generated where people have a pride in their work, understand the reasons for the quality control procedures implemented and are fully committed to their operation. These same benefits can usually be obtained for the company's EMS if this is discussed within the quality circle.

Communication

Whether achieved by quality circles or otherwise, communication paths are extremely important for achieving environmental targets, and they must exist from the shop floor right to the top of the company. Personnel on the shop floor involved in production operations are a valuable source of information with regard to production problems that may impinge on the environment and their likely causes, and mechanisms must exist for this information to be transmitted rapidly to the company management. Communication in the reverse direction is also very important, with management arranging meetings to explain environmental issues, describe improvement plans, discuss performance targets and report on progress made.

On top of this communication upward and downward in a company, sideways communication between different departments is also essential. It is not sufficient for each department to operate its own environmental protection scheme; the EMS must operate as a cohesive whole across the whole company. This can only be achieved by complete cooperation and constant dialogue between the different parts of a company.

At departmental level within a company, it can be very useful to ask every member of the department the following questions on an annual basis: 'What process are you involved in?'; 'What are the potential adverse effects of the process on the environment?'; 'How do you monitor performance of the process with respect to its environmental effects?' and 'What can you do to improve performance?' A further useful annual exercise is to construct a flowchart of the processes that each department is responsible for and ask, 'Why do you carry out each operation?'; 'Is each operation necessary?' and 'Can the process route be simplified to reduce adverse effects on the environment?'

Management of EMS

The responsibility for implementation, operation and review of an EMS must be assigned to one designated person. This person may subcontract responsibility for particular aspects of environmental policy in certain areas to other designated personnel, but a clear chain of persons with designated responsibility must exist, and the one named person with overall responsibility must be at the head of the chain. The existence of one named person in charge of the EMS is crucial. It is entirely insufficient to just assign responsibility to a department or group of people in an organisation, as this leads to people blaming each other when things go wrong.

As well as having one named individual with full responsibility for the EMS, it is also essential that this person has sufficient authority to do the job effectively. This is normally achieved by giving the person concerned a place on the management board of a company, usually with a title like Environmental Management System Director. This level of authority is necessary so that the person concerned can control the EMS system fully, and shut down manufacturing processes or other systems as necessary if there is a risk of serious environmental damage occurring, however much the managers of these processes or systems may protest.

Management of EMS implementation and operation has to fulfil several functions. The most obvious function is ensuring that all hardware and personnel involved in environmental protection activities operate efficiently and within the cost-estimates established at the design stage of the system. Ensuring that all staff involved have been properly trained is a necessary part of this. A second role is to make sure that the EMS interacts properly with all other general company management functions. To accomplish these two roles effectively requires that all the tasks involved are identified and carried out by assigning responsibility, delegating authority and creating accountability for each separate task.

Training

To satisfy ISO 14001, the documentation provided with the EMS must specify the training needs of everyone in the company whose activities may impact on the environment. The amount of training necessary will depend on the extent to which a person's activities can affect the environment. Clearly, personnel who are involved in designing, implementing, operating or maintaining an EMS will require in-depth training about particular aspects of the EMS, whereas other people will require much less training. However, the aim should be to ensure that the whole workforce in a company, including any contractors that are used, receives some training, even if the training for some only consists of a short half- or one-day course that acquaints them with the general environmental policy of the company, explains why the company operates an environmental policy, summarises the EMS, expounds the intended benefits of the EMS and emphasises the importance of everyone conforming with it. ISO 14001 also requires that, irrespective of the amount of training given to company personnel, all training undergone should be recorded in the EMS documentation.

Workers' attitude to training is almost as important as the training itself. Making people go on training courses is relatively easy, but ensuring that they assimilate the necessary knowledge is considerably more difficult. If the training courses are not managed properly, there is a strong likelihood that they will be treated as a welcome break from the normal working environment but their purpose will not be taken seriously. To prevent this happening, workers must see the need for courses and positively want to go on them. In some cases, this can be achieved by setting environmental targets and procedures to meet them that the workforce knows can only be achieved once they have obtained the necessary knowledge from a training course.

To be effective, training needs must be formulated and driven at departmental level within a company, and targeted towards meeting the department's environmental objectives. Figure 2.3 shows some necessary procedures in an effective departmental training plan. Training plans must be formulated individually for each member of a department so that he/she understands the relationship between their activities and the potential environmental effect. Appropriate people also need training about the proper response to make when unexpected events occur. This response may take various forms, such as implementing emergency procedures, taking remedial action to avoid pollution and plant shutdown procedures. Each training plan must be fully discussed with the person it is designed for, while obtaining the person's full agreement about the details of the training plan. It is also important that the effectiveness of the training is reviewed after it has taken place. Internal EMS audits can be a good time to do this, by asking employees questions about the environmental impact of what they do and about the actions that they take to avoid environmental damage. Their answers will determine whether training has been effective or whether further

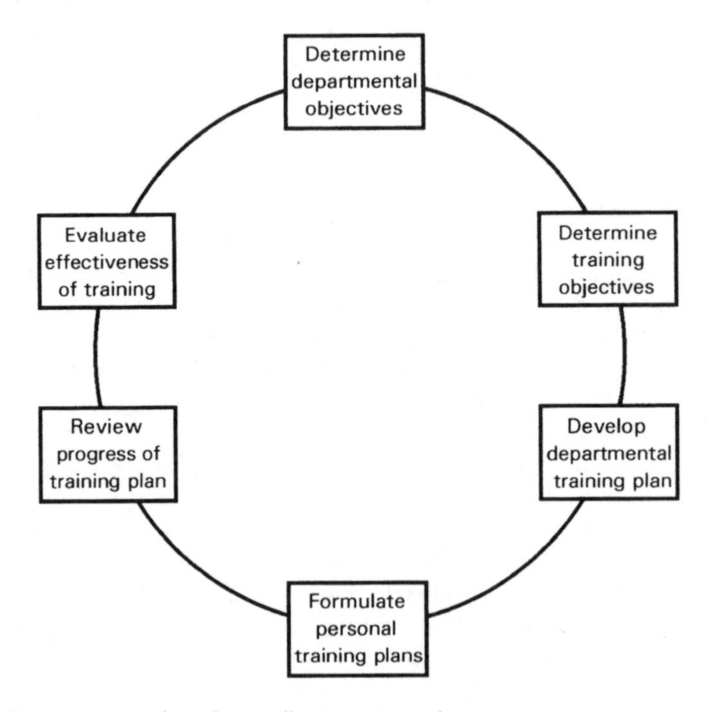

Figure 2.3 *Necessary procedures for an effective training plan.*

training is needed. Finally, because production methods change over time and the EMS has to evolve, training schemes need to be periodically reviewed, since the competence of the workforce may also need to improve through further training. It is also often beneficial to extend training to customers. By offering courses to users of the equipment that companies produce, the incidence of misuse of equipment by the customer can be greatly reduced and consequential adverse environmental effects avoided.

System documentation

Full documentation of all aspects of an EMS is essential. This should take the form of a manual, which can be in either electronic or paper form, in which every EMS procedure is carefully set out and the whole philosophy and purpose of the system is described. The environment targets and their justification in terms of balancing the cost/benefit equation is a very necessary part of this. Whilst there is no particular format defined for the manual, the information must be presented in a clear, systematic and orderly way. In the case of a small company, all EMS documentation would typically be bound within one manual. However, for larger companies, it is often more appropriate to maintain separate corporate and divisional EMS manuals. A suggested structure for the EMS manual is given in Appendix 2.

Each identified element in the EMS should be documented such that the details of hardware designs (with full drawings), operational instructions, purpose, running cost and interaction with other elements are all expressed clearly. All documentation should be dated and have the version number clearly marked. The documentation must include details of any elements of the system that involve the activities of suppliers and customers, and explain how the environmental performance of such outside bodies is monitored and influenced. Lists of approved suppliers must also be maintained. Review procedures for the EMS should also be defined in the documentation, and the required frequency for measuring its performance should be given. Such inspection, testing, auditing and costing of EMS procedures must be included within the manual as environmental performance records.

Documentation should identify clearly the organisational structure and mode of operation of the EMS, including a description of the associated training procedures. Assignment of responsibility for each elemental part of the system is particularly important, and care should be taken to see that activities on the boundaries between functions are managed adequately. The person responsible for modifying and controlling documentation must also be identified. To summarise, the system must be foolproof, without any loopholes!

The EMS manual must be readily accessible to all relevant personnel, and it is therefore important that copies are kept at all locations where operations are carried out that can affect the environment. However, as it is essential that all the copies issued are changed whenever revisions are made to the manual, the number issued and distributed around the company should not be greater than the minimum necessary to provide ready accessibility. To assist with document control, it is sensible to maintain a distribution list that shows where each copy is located and who is responsible for it. Of course, the problems of needing multiple copies and making sure each one is the latest version are avoided if the manual is produced in electronic form. This allows a single copy to be accessed from all locations, via the company's computer network.

As implied above, modifications and additions have to be made to the manual from time to time following periodic reviews of EMS operation, and it is essential that efficient, documented procedures exist for effecting such changes. To draw attention to the changes made, it is beneficial to highlight them in some way in the documentation. This can be achieved in various ways. One way would be to put a box around revised procedures, prefaced by the words, 'Revised procedure' in bold type. To ensure that there is no ambiguity about which is the latest version of the manual, it is essential that all pages within the manual are marked with a date and revision number. This should ensure that only the correct version of the manual can be used. Normal practice would always be to dispose of immediately any parts of the documentation that have become redundant or outdated. However, it is sometimes necessary to keep obsolete documentation for legal purposes, usually to demonstrate the extent of environmental protection efforts at some past date. This is acceptable as long as a system is in place to clearly mark such documentation as obsolete, so that it is not used accidentally in place of the current documentation.

System review and update

One factor complicating the design and operation of an EMS is that the environmental targets change continually under the influence of technological developments, market forces and new/revised legislation. This requires the EMS to be updated at various points in time in order to meet the changed requirements. Modifications to the system also become necessary if the monitoring exercise about costs and performance shows a deviation from the target cost and performance goals. A regular review of the EMS is therefore required to determine whether changes to the system are necessary. Such reviews should be additional to, and not instead of, the regular system audits specified by ISO 14001.

Even if reviews do not identify improved technological procedures for avoiding environmental damage or changes in legislation, ISO14001 still requires that the EMS should evolve and be continually improved. When a company implements an EMS for the first time, it is accepted that it has to prioritise its efforts towards reducing the worst environmental impacts of its operations first. However, once the worst problems have been dealt with, ISO 14001 expects future system reviews to identify lesser problems that can be given attention to further improve the company's environmental performance. At this stage, attention can also be given to other things that the company can influence, such as the environmental performance of its suppliers and customers. Increasingly, in order to attain higher standards in their own EMS, companies are putting suppliers under contractual obligation to implement an EMS. In many cases, suppliers are actually required to obtain certification of their EMS, to confirm that it conforms to ISO 14001. Companies can also influence the environmental performance of customers by giving them advice and training in the use of their products and also advice about safe disposal when the products come to the end of their life.

2.2.2 Measurement and calibration requirements in applying ISO 14001

ISO 14001 sets out a clear requirement that all parameters associated with the operation of the EMS must be monitored and recorded to ensure that changes do not occur that may lead to environmental damage. Measurements for an EMS fulfil two main functions. Firstly, the environmental consequences of a company's operations in terms of air pollution, water pollution, noise and vibration need to be measured and compared with the environmental targets set. Secondly, the value of parameters in manufacturing processes and other systems need to be monitored if significant variation in the parameters may lead to faults and a risk of increased environmental pollution. Ideally, all parameters should be monitored continuously, but it is often either impractical or too expensive to achieve this in practice. If continuous measurement is not possible, a suitable programme of sampling the parameters at given time intervals needs to be established.

The quality of parameter measurements must be assured, and appropriate measurement and calibration procedures to achieve this were summarised in Section 1.4.1. As emphasised in Section 1.5, these procedures represent best practice, but will not be necessary in their entirety in every EMS. However, for the present, it is sensible to look at how all the procedures specified in Section 1.4.1 can be implemented. Thereafter, the rigour with which the procedures should be applied requires intelligent interpretation according to the needs of each individual EMS considered.

Whatever the purpose of the measurements, the EMS must specify the quality required for each parameter measurement. Quality is usually defined in terms of measurement accuracy, and the measurement procedures instituted must ensure that the measurement errors do not deviate outside stated error bounds. The error bounds that are set depend on the nature of the parameter measured and the needs of the EMS. However, the minimum quality acceptable in measurements is an accuracy level that will satisfactorily demonstrate that the environmental targets are being met to the satisfaction of both the company's customers and the general public.

Once the company has set appropriate error bounds, it must then institute a system that guarantees that measurement errors do not exceed these bounds. Section 1.4.1 has already set out the general requirements for achieving quality measurements, and appropriate procedures that can be implemented to satisfy these requirements are summarised in Figure 2.4, and are discussed in more detail in the paragraphs below.

Appropriate choice of measuring instrument

To ensure measurement quality, it is essential that the instrument chosen to measure each parameter is suitable. Knowledge of the characteristics of different kinds of instrument constitutes necessary background knowledge for this, and the appropriate details are covered in Chapter 3. As well as having necessary characteristics in terms of things like measurement accuracy, sensitivity and resolution, instruments must also have a range appropriate to the values of the parameter to be measured. In addition, the instruments used must be suitable for use in the intended measurement environment, so that their characteristics are not changed significantly and they are not damaged in any way.

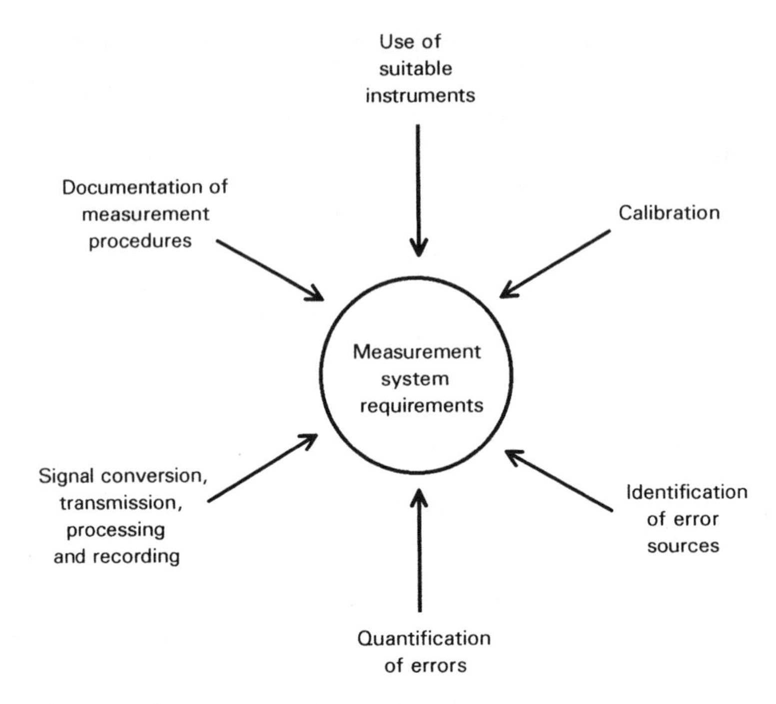

Figure 2.4 *Summary of measurement system requirements.*

Calibration

To ensure accuracy in the measurements made, all instruments used must be calibrated at appropriate intervals using procedures explained in Chapter 3. Instrument calibration ensures that the measuring accuracy of each instrument involved in the measurement process is known over its whole measurement range, when it is used under specified environmental conditions. This knowledge is gained by comparing the output of the instrument under test against the output of an instrument of known accuracy when the same input is applied to both instruments.

Assessment of measurement errors

Even when properly calibrated measuring instruments are used, errors occur in measurements for a number of reasons. Such errors must be quantified as far as possible, and corrections applied to the measurements so that the quality of measurements used in the EMS is assured.

The first source of error in measurements is the variation in their characteristics that occurs when they are used in environmental conditions that differ from those in which they were calibrated. In most circumstances, this error is unavoidable, since it is impractical to replicate the calibration environmental conditions in normal usage of a measuring instrument. Therefore, the environmental conditions that apply during usage should be measured and the consequent change in instrument characteristics should be quantified. Such error quantification requires knowledge of the way in which environmental conditions affect particular instruments. However, once this knowledge has been acquired, it is usually possible to establish measurement procedures that minimise the effect of environmental condition changes and thereby improve the quality of measurements. The necessary procedure can therefore be summarised as: firstly, using the calibration of the instrument under standard conditions as a baseline, and, secondly, making appropriate modifications to account for the variation in characteristics due to the differing environmental conditions existing in its normal place of usage. Having done this, it is then possible to quantify the quality of measurements made by calculating the bounds of measurement error of each instrument, as it is used in its normal operating environment. These matters will be covered in greater detail in Chapter 4.

The second source of error in measurements arises because the quality of measurements made by an instrument does not remain at a constant level for ever, but changes over a period of time as the characteristics of the instrument change. The characteristics of all instruments change in the long term, but the magnitude and the rate of this change are influenced by many factors that are difficult or impossible to predict theoretically. It is therefore necessary to apply practical experimentation to determine the rate of such changes in instrument characteristics. Once the maximum permissible measurement error has been defined, knowledge of the rate at which the characteristics of an instrument change allows a time interval to be calculated which defines the moment in time when an instrument will have reached the bounds of its acceptable performance level. The instrument must be recalibrated either at this time or earlier. This measurement error level that an instrument reaches just before recalibration is the error bound, which must be quoted in the documented specifications for the instrument.

A number of other measurement error sources exist, as discussed in Chapter 4. These must all be identified and quantified, and appropriate steps taken to minimise their effect on the quality of measurements used for the EMS. Finally, once all error sources have been identified and minimised as far as possible, the cumulative measurement error in a system needs to be calculated if the system has more than one measurement component obtained from different instruments. The appropriate mechanisms for doing this are also covered in Chapter 4.

Usage of measuring instruments

In order to ensure the quality of measurements obtained, all measuring instruments must be maintained and used by personnel who have appropriate qualifications and training. Also, only instruments of the type specified in the EMS manual should be used for making each measurement. Documented procedures must be followed in making measurements, and the instruments used must be handled carefully. All instruments should be properly calibrated, as confirmed by the calibration status that should be marked on them. Also, if instruments have adjustable elements that are set during calibration, seals must be applied to prevent tampering.

Signal conversion, transmission, processing and recording

The data captured by a measuring instrument are often not in a suitable form for input into a data recorder, such as when the instrument output is in the form of a change in resistance, capacitance or inductance. In such cases, the first requirement is to apply a signal conversion element to convert the measurement data into a more suitable form (which is usually a varying voltage). Following any necessary signal conversion, it is frequently necessary to use quite long cable lengths to carry the data between the point of measurement and the signal recorder. During such data transmission, the quality of the signal can degrade, and action has to be taken to minimise or compensate for this degradation using signal-processing techniques.

Finally, following any necessary signal processing, various kinds of recorder can be used to make permanent records of the transmitted measurement signal. Chapter 5 provides fuller discussion of these processes of signal conversion, transmission, processing and recording.

Documentation

As with every other aspect of an EMS, all procedures implemented for process monitoring, parameter measurement and instrument calibration must be fully documented. There is also a requirement expressed in ISO 14001 that all measurements made in support of the EMS should be recorded and kept in the form of written EMS performance information. These records must be preserved, stored carefully and kept for a length of time specified in the general EMS documentation. All such documented records must be traceable to the operation or activity associated with the particular environmental parameter measured, so that any deviation in pollution levels above the usual level can be readily traced to the source causing it. This enables remedial action to be taken without delay so that environmental damage is minimised.

Summary of measurement and calibration procedures

The list of necessary measurement and calibration procedures can be summarised as follows:

(1) Calibrate all measuring instruments under specified environmental conditions so that their measuring accuracy is known over the whole measurement range. Calibration must be performed by standard instruments whose characteristics are traceable to national reference standards.
(2) Establish the variation in instrument characteristics under environmental conditions different from the calibration conditions.
(3) Establish measurement procedures that minimise the effect of environmental condition changes on the measuring instruments.
(4) Calculate the bounds of measurement error of an instrument under normal operating conditions.
(5) By practical experimentation, determine the rate of change of instrument characteristics over a period of time.

(6) Hence, determine the frequency at which instruments should be recalibrated, and quantify the maximum possible measurement error when the instrument has drifted farthest from its specification immediately before calibration.
(7) Combine all instrument measurement error levels into a figure that expresses the cumulative error level of the whole measurement process (where a measurement system is composed of the outputs of more than one instrument).
(8) Document all measurement and calibration procedures.

2.2.3 Other engineering contributions to ISO 14001 environmental management systems

Apart from their duties in ensuring that measurement and calibration procedures are adequate, engineers have a number of other responsibilities in ensuring the successful operation of an EMS. Firstly, a primary responsibility is to analyse measurements made to identify any changes that may have an environmental impact. Secondly, they must design equipment that will minimise environmental damage caused by the company's manufacturing and other operations. Thirdly, they have a responsibility to formulate emergency procedures that should be followed to minimise the environmental impact of any malfunctions indicated by the measurements made. Fourthly, they have important contributions to make in designing products so that their environmental impact is minimised.

Analysis of measurements and detection of malfunctions

Measurements have to be analysed to identify and assess the importance of any deviation of measured parameters away from their normal levels. Where malfunctions in manufacturing systems and other activities of a company may lead to environmental damage, procedures must be instituted to identify such malfunctions promptly. Once identified, appropriate procedures must be instigated so that prompt remedial action is taken to minimise the environmental impact of the malfunctions.

Unfortunately, looking for changes in parameter values that indicate a malfunction becomes difficult if process parameter values vary naturally within a given range, due to random effects that are expected but unquantifiable. Such allowable variations have to be distinguished from other variations that are indicative of a process fault that may lead to environmental damage. Statistical process control, as covered in more detail in Chapter 12, can be a very useful aid in making this distinction. If significant changes have occurred that may affect the environment, appropriate documented procedures must be followed to find and correct the source of the change.

Finally, it is important to emphasise that all EMS performance data collected, together with any remedial action taken, must be recorded and kept with the EMS documentation. The length of time for which records need to be kept should also be documented.

Equipment design

Many problems that have an environmental impact can be solved at relatively low cost without the installation of special equipment. Risk analysis and reliability calculations (see Chapter 11) are often useful tools in predicting the likelihood and magnitude of faults and how much, if any, special equipment should be installed to meet EMS targets. If found to be necessary, special equipment may perform various functions like removal of pollutants to avoid air and water contamination in the environment, and noise and vibration reduction. The detailed design and installation of such equipment is obviously the responsibility of appropriate engineering personnel. In executing this task, proper management is essential to ensure that the implemented system operates in an efficient manner and performs the functions required of it. In particular, the special equipment itself must be subjected to reliability calculations to ensure that the risk of failure and consequential environmental damage is acceptably low. Finally, all such special equipment is an essential part of the EMS, and its description and mode of operation must be properly documented in the EMS manual.

Provision of emergency procedures

It is essential to establish emergency procedures that must be followed to minimise the environmental impact if accidents happen or other abnormal situations and faults arise unexpectedly. All such procedures should be fully documented in the EMS manual. However, the adage 'prevention is better than cure' is very relevant, and this means that the EMS should include procedures that try to foresee what incidents might arise and put systems in place that prevent such incidents from happening as far as possible. However, emergency procedures are still necessary in case such preventative measures break down or an unforeseen situation arises. However, if the preventative measures have been thorough, such emergency procedures will only have to be followed very rarely.

The first stage in emergency prevention is to examine all operations and activities in a company and assess the potential for accidents and emergencies occurring. This is considered in greater detail in Chapter 11. Hazardous incidents are particularly likely during abnormal operation of process plant during startup and shutdown operations. Chemical spills, explosions, fires and fractured pipes that allow the release of harmful liquids or gases are common hazards. For every such potential incident, the chain of events that would precede it should be analysed, so that a potential accident or emergency can be recognised in advance and appropriate action taken to prevent the incident actually happening. The second stage in emergency prevention is to ensure that all personnel involved know the symptoms of a potential emergency and take the prescribed action to prevent an emergency situation actually occurring. Company management must ensure that any necessary training associated with this has taken place.

If all emergencies were anticipated and appropriate preventative action was always taken, emergency situations would never arise, and so there would not be a need for emergency response procedures. Unfortunately, preventative actions are not always successful, either because of human or equipment failures, and so some potential emergency situations that are recognised as such do go on to become full emergencies. In addition, the initial assessment of operations to identify situations that may lead to accidents or emergencies will never be able to identify all potential incidents. It is inevitable that some incident will arise that no one had thought of. This can subsequently be added to the list of potential incidents for which proper planning can be made. However, such incidents cause problems the first time that they occur. Hence, whatever preventative actions are taken, some incidents and emergencies will still arise from time to time for which a proper response is required to prevent environmental damage.

As for the actual emergency-response system, little can be said of a general nature except that it must be designed to be as reliable as possible, since any failure would lead to environmental damage. Therefore, risk analysis and reliability assessments must be applied to the emergency-response systems designed. Appropriate procedures for this are discussed in Chapter 11. In addition, ISO 14001 requires that all emergency-response procedures be fully documented in the EMS manual.

Because every situation is different, the procedures to be followed in response to an emergency must be determined on a case-by-case basis. Usually, the response procedure involves shutting down the manufacturing operation or servicing function until the fault is rectified. However, this often has to be done in a controlled way to minimise environmental damage and prevent it temporarily becoming worse. If the response system involves calling personnel like firefighters to assist, the means of communicating with such personnel must be displayed prominently. If the response system involves the use of special equipment, the location of this and method of usage must be known to all relevant personnel.

Whatever form the emergency response system takes, ISO 14001 requires that it be tested periodically wherever it is practical to do so, and that any associated equipment (e.g. fire extinguishers) is properly maintained. Also, all relevant personnel must be fully conversant with the emergency-response procedures. In addition, if an emergency arises and the response system is implemented, this should be used as an opportunity to review the effectiveness of the response system and determine any improvements that should be made in it.

Product design

The way that products are designed can have a significant effect on the environment in respect of the necessary manufacturing processes for the product, the effect of the product during usage and the potential environmental effect when it comes to the end of its useful life. Thus, it is an engineer's task to properly consider the environmental impact at the design stage of a new product and minimise this impact as far as possible by including features like reduced energy consumption, increased longevity and recyclability when it comes to the end of its useful life. Other desirable product-design features include minimising packaging and minimising the use of toxic materials that might damage the environment. Products should also be designed such that the manufacturing process for them is as environmentally friendly as possible and minimises the use of energy, water and raw materials. Whilst these kind of changes can often only be made at the design stage of new products, it is sometimes possible to modify an existing product to facilitate the achievement and maintenance of environmental targets.

However, whilst environmentally friendly design is desirable, it must always be remembered that the design specifications of a product are heavily constrained by customer requirements. No degree of conformance to a particular environmental performance is going to satisfy a customer if the basic design of the product means that it is unable to do everything that the customer wants. If a transport company orders a fleet of lorries with a 40-tonne load capacity, it is unlikely to be satisfied if vehicles with a capacity of 30 tonnes are supplied, whatever level of reliability and fuel efficiency is claimed for them. However, whilst accepting that product specifications must meet customer demands, various design modifications are often possible that enable environmental targets to be met without affecting the product's ability to perform its required function. It is these possibilities that must be investigated at the design stage.

Modifications to operations to reduce environmental impact

It is impossible to give a comprehensive guide to all the modifications in procedures that can be made to reduce the environmental impact of manufacturing operations and other activities of companies. Every situation is different and must be considered on its own merits. Therefore, the starting point must be to write down all of the environmental impacts of a company's operations and then, considering each impact in turn, to write down appropriate ways of reducing that environmental impact. Brainstorming by a small group of people who are familiar with the operations involved is often a useful tool in this process. The following list of procedure modifications that have been adopted by various companies to reduce the environmental impact of their operations may also provide inspiration about procedures that can be adopted in other situations:

- Manufacturing products either partially or wholly using recycled materials.
- Servicing equipment at customer sites rather than transporting them back to the equipment supplier, if this reduces fuel used in vehicles.
- Reducing fuel used in transportation of goods – using rail or water instead of road, using sea instead of air, transporting full-loads rather than part-loads.
- Reducing energy use – especially avoiding waste in unnecessary heating and lighting in the workplace.
- Reusing as much water as possible – this can significantly reduce water consumption.
- Using ion-exchange technology (de-ionisers) rather than chemicals to clean water used in industrial processes – this reduces the amount of pollutants discharged in waste water.
- Using reverse osmosis technology to clean water used in industrial processes (even more environmentally friendly than de-ionisers).
- Servicing equipment regularly at prescribed intervals to avoid pollution through faults – faults can cause environmental pollution in various ways such as leaks of hydraulic fluids and various chemicals. Faults can also lead to reduced operating efficiency and increased energy consumption.
- Designing products to make them as environmentally friendly as possible – minimising the use of toxic materials in them, minimising their energy consumption and minimising packaging used.
- Reducing the quantity of chemicals used – for example, reducing the amount of chemicals needed for cleaning by first cleaning components with rags.

- Choosing alternative chemicals if possible that have less harmful environmental effects – there is often particular scope for finding alternatives in the chemicals that are used for cleaning and degreasing purposes. Some companies have used particular chemicals for many years, and may be unaware of less harmful alternatives that have evolved during this time.
- Using alternatives to freon for equipment-cooling purposes that have less harmful effects on the environment (particularly depletion of the Earth's ozone layer).
- Terminating any process that uses cyanide-based products – cyanide is so dangerous that alternatives must be found, irrespective of the cost implications.
- Using lead-free solder in jointing operations.
- Replacing solvent-based paints and other surface-coating materials with low-solvent, water-based materials – this significantly reduces air pollution by volatile organic compounds.
- Avoiding coatings such as varnishes that are applied to products for cosmetic reasons, if they are unnecessary technically to protect the product.
- Storing chemicals carefully – this may involve incorporating a double wall in the design of tanks holding liquid chemicals to avoid leakage that may cause pollution in watercourses. Also, lids should always be kept on containers of chemicals in powder form to prevent winds picking them up and causing airborne pollution. Lids are also necessary for containers holding materials that evaporate.
- Reducing the amount of waste created, and ensuring that any waste produced is disposed of in an environmentally safe way – this is considered in more detail later, in Chapter 10.
- Avoiding use of sprays as far as possible, since sprays always cause waste through unavoidable dispersion (for example of paint).
- Using automated techniques, for example robot paint spraying, rather than manual methods, if this will optimise the use of raw materials and minimise waste.
- In batch-processing manufacturing operations, maximising the size of batches, since waste is often generated in producing out-of-tolerance products whilst production parameters are being tuned at the start of production runs.
- When producing paper documents such as user manuals, printing on both sides of paper as far as possible. Also, considering whether an electronic copy would suffice instead of a paper copy.

2.3 Environmental Management System Costing

An essential factor in designing an EMS is the ability to make an accurate assessment of the costs involved in developing and operating the system. Whilst good environmental performance is necessary to maintain customer satisfaction and the goodwill of the general public, product price is also very important. It is no use making an environmentally friendly product in an environmentally friendly way if its price is too high. Therefore, the environmental targets have to be set with regard to the cost in achieving the targets. For this to be possible, the costs of designing and implementing all aspects of the EMS must be accurately known. These costs can be divided into the four areas associated with: (1) management, training and day-to-day

Figure 2.5 *Summary of environmental management system costs and benefits.*

operation of the EMS; (2) design and modifications aimed at preventing malfunctions in manufacturing operations that might have an environmental impact; (3) design and installation of equipment to reduce harmful emissions to the environment; and (4) inspection and appraisal costs for the EMS. Figure 2.5 summarises these costs.

Before going on to discuss environmental management costs in more detail, two general points need to be established. Firstly, it must be recognised that environmental management costs will change over a period of time as environmental targets change and new technology is developed to assist in meeting the targets. Secondly, the accounting system used to assess environmental management costs must be appropriate. Some accounting practices can grossly distort the sizes of both absolute and relative costs by including full overheads in direct labour charges rather than just marginal costs. Such unjustified overheads often make an EMS appear uneconomic, resulting in nonimplementation and a net loss to the company in terms of its reputation, and possibly resulting in fines for transgression of environmental legislation.

As well as using sensible accounting procedures, it is also important that the cost-estimates used in the quantification of EMS costs are as accurate as possible. Such credibility of cost-estimates is extremely important because, without credibility, the whole argument for implementing the EMS may be seriously weakened. Extreme care in cost estimates is vital because the exposure of a weakness in any single part of the estimate can undermine confidence in the whole. For this reason, costs should be produced by, or at least endorsed by, a company's accounts department, as this puts the necessary 'stamp of approval' on the estimates. In view of their crucial importance, independent corroboration of cost estimates, perhaps by consulting engineers, should also be sought.

Some EMS costs are well defined, such as the costs of training relevant personnel and the costs of employing additional persons to fill posts in environmental monitoring and management. Unfortunately, the quantification of other EMS cost factors is difficult, because they are generated for other reasons besides the pursuance of meeting environmental targets. For example, in estimating the day-to-day operational costs, technicians who already have a natural function in a company may expend additional time and effort in measuring parameters relevant to the EMS. Quantifying this extra time requires apportionment of an employee's time between environmental protection activities and other functions. This is particularly prone to inaccuracy, especially when the employee concerned is made responsible for defining this apportionment via time sheets, as the data are then reliant upon his/her personal, subjective judgement about which type of work is related to environmental protection and which is not.

The technical cost of redesigning and modifying existing manufacturing plant to optimise the environmental performance of a product can be quantified via an in-depth study by experienced engineering personnel of the technical points involved. However, if a new plant is being designed rather than an existing one being modified, the incremental cost in design modifications to improve environmental performance of the plant is difficult to quantify.

The cost of designing and implementing equipment to reduce harmful emissions from a plant can be estimated accurately, but this requires considerable skill and experience on the part of the engineer making the estimate. Inspection and appraisal costs can also usually be estimated accurately by skilled personnel, particularly if there is access to cost information for similar plants operated by other companies.

Thus, EMS costs can be divided into those that can be accurately estimated and those that cannot. However, with the caveat that bad estimates are counterproductive, attempts should always be made to estimate all costs if at all possible, as failure to present a full picture is an irritation to those mandated to take a decision about the viability of a proposed EMS. Even if certain costs cannot be estimated at all, it is still useful to mention them, because it keeps them in view and encourages future attempts to measure them.

Assessment of the financial benefits accruing from implementing environmental management procedures is even more difficult to quantify. The six major elements contributing to financial savings are: reduced energy consumption, reduced material usage because of the minimisation of waste, reduced insurance costs, avoidance of fines for transgression of environmental legislation, increased sales generated by a reputation for good environmental performance, and a general improvement in operating efficiency resulting from the more effective general management system that usually results when management is reorganised to satisfy the requirements of ISO 14001. Only the first two of these elements are readily quantifiable. The value of waste reduction can be calculated from measurements of the waste generated before and after the implementation of the EMS. Costs of insuring the company against claims arising from environmental damage are often reduced substantially once the company has a proven EMS in operation. The third cost element depends on the nature of the environmental damage and the policy of legislation-enforcers, whilst the final two elements are highly subjective and can therefore only be estimated. A simplistic way of measuring increased sales would be to monitor the sales figures over a period of time before and after the introduction of the EMS. However, this fails to take account of other significant factors, such as whether the total market for a product expands or contracts during the periods in which sales are considered, and what competitors in the marketplace are doing. An EMS can only serve to maintain or increase the share that a particular manufacturer has in a market. However, if the total market contracts, some decrease in sales may occur even with an EMS in operation. The important point, however, is that the decrease would have been even greater without the EMS.

2.4 Environmental Management System Audits

If an EMS is to be operated with the intention of assuring customers of a company and also the general public about the environmental performance of the company and its products, then the company has to win full confidence in the effectiveness of the environmental protection procedures in operation. Mention has already been made about the importance of allowing customers to have full access to EMS documentation, and also of allowing them to visit the company to see the systems in operation. However, whilst such access to documentation and the company is necessary, it is not sufficient. In addition, audits of the EMS being operated must be carried out on a regular basis and recorded in the EMS documentation to demonstrate that the system continues to be effective and that no departures have been made from the procedures laid down. Appropriately qualified internal company personnel can carry out most of these reviews. However, periodically, an external audit must be carried out by an independent third party if the EMS is to be certified to give it full credibility. Within the UK, the main providers of such an auditing service are auditors working on behalf of the British Standards Institute; similar bodies exist elsewhere around the world.

2.4.1 External audits

The requirements for external audits will be presented first, because these also give essential guidance about what a company should be doing when conducting an internal audit. External audits by a third party are necessary to confirm that the EMS implemented conforms to the requirements of ISO 14001, has been properly implemented, is properly maintained, is achieving the targets set, is reviewed at regular intervals and is accompanied by a full set of up-to-date documentation. Such audits must be repeated regularly, and they are mandatory before ISO 14001 certification can be achieved for an EMS. External audits must necessarily be thorough, although if the same audit procedures have been followed previously by internal reviews (as they should be) then there will be no difficulty in 'passing' the audit and gaining certification or recertification of the EMS operated. Such certification of EMS procedures is only required for those parts of its operations that a company wishes to claim accreditation for. The nature and small environmental impact of some operations may be such that it is not cost-effective to operate an EMS for them and, in such cases, it is perfectly permissible for a company to register and claim accreditation for environmental management procedures that just apply to part of its operations.

A major item of concern to external auditors will be to satisfy themselves that there is support at the highest level of the company concerned for the EMS operated. Therefore, the auditors will require brief discussions with the managing director of the company to confirm this before proceeding with the audit. Thereafter, the starting point of an environmental audit is to study the documented assessment made by the company of the environmental impact of its operations. The assessors will look to see whether an effective EMS has been set up that at least addresses the worst environmental impacts of the company's operations and also meets any relevant legislation. Once satisfied with the design of the EMS, the auditors will look for evidence that the system is being operated according to the procedures described in the system documentation, and that there has been no departure from these agreed procedures (other than agreed ones notified to the accrediting body). Documentary evidence from internal audits of the quality system will assist with this, and the auditors will expect to see such internal audit results. The auditors will particularly want to see evidence that the system is meeting the environmental targets set. To demonstrate this, records that confirm continual monitoring of system operation will have to be produced, and measurements made in a proper way – according to documented procedures using calibrated instruments traceable to reference standards – will be required.

The auditors will then examine other aspects, specified in ISO 14001, that are necessary to ensure that the EMS operates efficiently. They will examine the management structure supporting the EMS, the training records for company personnel having direct impact on the EMS, the communication channels provided, and the documentation that describes all aspects of the system. They will be especially keen to ensure that all personnel whose actions can have an effect on the environment are working in the correct manner. They will test this out by selecting people at random to ask them: what job they do, what training they have had, where the instructions are that tell them how to do the job, and what specific steps they take in support of the EMS being operated by the company. They will also particularly want to see evidence of EMS reviews and efforts to achieve continual improvements in environmental performance. Depending on the state of advancement of the EMS, this may include investigation into the efforts made by the company to improve the environmental performance of its suppliers and customers.

The auditors will have a checklist on which they mark down the conformance or otherwise of the EMS to each of the clauses in the ISO 14001 standard. Nonconformance will be graded in three categories: (1) those that are a major diversion from the standard and are likely to lead to environmental problems, (2) those that are less serious but might lead to environmental problems in some circumstances, and (3) minor faults that need correcting. Any faults in category (1) will certainly mean that the auditors do not certify the system. Whether faults in category (2) lead to certification or not depends on their number and severity, with the most favourable outcome being that provisional registration is given but re-auditing will be required sooner than usual. Category (3) faults will usually lead to certification, with a recommendation that the faults be corrected and an expectation that this will be done as soon as possible.

The outcome of the audit will therefore either be the issue of an accreditation certificate for the EMS operated, or, alternatively, a statement of the corrective action that needs to be taken before the system can be certified. Once an accreditation certificate is given, arrangements will be made to re-audit the system, perhaps on an annual basis. In between such full audits, unannounced surveillance visits might be made to monitor the continued effectiveness of the EMS procedures agreed.

The necessary frequency with which formal audits are carried out depends on the past history of the EMS and the company's reputation in environmental matters. A new EMS needs fairly frequent audits, but, once the system has operated satisfactorily for some time without major problems being identified, the time between audits can be extended.

2.4.2 Internal audits

Internal audits are necessary for two reasons. Firstly, as mentioned above, the external auditors will expect to see documentary evidence that internal audits of the EMS have been carried out at regular intervals. Secondly, it is a necessary step towards ensuring that the system 'passes' when subjected to the external audit.

The internal auditing procedure consists of systematically going through all the questions that the external auditors will ask. This seems simple enough, but a problem can arise in that the person responsible for EMS operation in a company may be too close to it and may fail to spot deficiencies. It is much better if two or more companies can get together and swap their EMS managers for the purpose of carrying out internal quality audits. Such audits by an outside but 'friendly' auditor can often highlight problems that the company's own EMS manager has failed to see through overfamiliarity with the system.

2.5 ISO 14001 Registration

Having covered the requirements set out in ISO 14001 in some detail in the preceding sections, it is appropriate now to give a brief summary of the main steps necessary to achieve registration of an EMS. The first step must always be to design and document the system to be implemented. In doing this, it is highly desirable to obtain the advice of consultants who have specialist knowledge of the industry in question unless the company already has personnel with substantial expertise in environmental management in that industry.

Once the system has been implemented, its operation must be carefully monitored by a series of internal audits until everything seems to be working satisfactorily. Once this stage is reached, the company can apply to their National Standards Organisation[†] for an external audit, which is a necessary prerequisite of registration. The accreditation body will appoint a registrar to audit the system, which will be carried out at a mutually agreed time. The audit will examine whether the design, documentation and implementation of the system are satisfactory and whether the environmental targets set are being met. Once any deficiencies that the registrar has identified have been put right, a certificate of registration will be issued and the company will be placed on a public register of companies who are operating to the appropriate ISO 14001 standards.

2.6 Publicity about Good Environmental Performance

Good environmental performance is necessary for maintaining and expanding the customer base for a product, as well as for satisfying the concerns of noncustomers in the general public. However, the benefits coming from the achievement of good environmental performance by the company and the production of environmentally friendly products are greatly diminished if customers and the general public are not made aware of this achievement. Communication about the EMS associated with a product is therefore essential. This is primarily the responsibility of the marketing department in a company, but other people also have important roles to play, such as in the provision of documentation which demonstrates that the EMS implemented is efficient and maintained satisfactorily. This documentation should show clearly that the system conforms to ISO 14001. Customers should be able to inspect this documentation at all reasonable times and, if necessary, visit the company to see the EMS procedures described in operation.

Reference

1. *ISO 14015, 2001, Environmental Management Systems – Environmental Assessment of Sites and Organisations*, International Organisation for Standards, Geneva.

[†] In the USA, the appropriate body is the American National Standards Institute Registrar Accreditation Board (ANSI/RAB), and in the UK it is the British Standards Institute. Other countries have similar bodies.

3

Measurement Systems in Environmental Management

Parameters associated with the operation of an EMS must be measured and recorded to the degree of accuracy specified in the EMS manual. As explained in the last chapter, the level of accuracy specified is set according to the requirements of the EMS. In some cases, high levels of accuracy will be required but, in other cases, the accuracy requirement will be quite modest. To achieve the specified level of measurement accuracy, the measurement system and the measuring instruments used within it must be carefully designed. The principal components in a measuring system are shown in Figure 3.1. The primary component is a sensor or transducer that captures the information about the magnitude of the variable measured. This is often followed by a variable conversion element that translates the output measurement into a more convenient form. After this, various signal-processing operations are applied that improve the quality of the measurement. The measurement then passes via a signal-transmission system to a data recorder. Before leaving this discussion on measuring system components, it should be mentioned that a particular measurement system will not necessarily contain all of the components identified in Figure 3.1. For example, variable conversion, signal processing or signal transmission may not be needed in particular cases. It should also be noted that many commercial instruments combine several measurement system elements within one casing. Furthermore, intelligent instruments contain additional sensors/transducers to measure and compensate for disturbances in the environmental conditions of measurement.

Several conditions must be satisfied to achieve the quality of measurements specified in the EMS. Firstly, suitable measuring instruments must be chosen that have static and dynamic characteristics that are appropriate to the needs of the measurement situation, as discussed in the first part of this chapter. Secondly, the conditions

ISO 14000 Environmental Management Standards: Engineering and Financial Aspects. Alan S. Morris.
 ISBN 0-470-85128-7

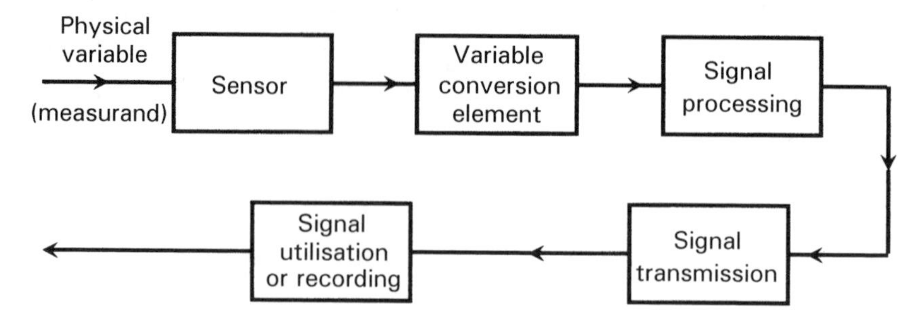

Figure 3.1 Principal components in a measurement system.

in which instruments will have to operate must be assessed and suitable instruments will have to be chosen that are as insensitive as possible to the operating environment. Thirdly, every measuring instrument should have a designated person responsible for it, who must ensure that the instrument is calibrated at the correct times by approved personnel, so that its measuring characteristics are guaranteed when it is used under specified environmental conditions (approved personnel means either staff within the company who have attended all the necessary courses relevant to the calibration duties, or subcontractors outside the company who are verified as being able to provide calibration services satisfactorily). Fourthly, having eliminated calibration errors, all other error sources in the measurement system must be identified and dealt with, as discussed in Chapter 4. Finally, the effect on accuracy must be considered of all other processes undergone by the measurements, including any variable conversion elements applied, signal processing, signal transmission and data recording, as discussed in Chapter 5.

3.1 Choosing Suitable Measuring Instruments

When choosing measuring instruments for a particular measurement situation, it is necessary to ensure that the instrument will satisfy the requirements specified by the EMS and, in particular, will not be adversely affected by the conditions in which it has to operate. The necessary background for this is an awareness of the nature of different kinds of instrument and knowledge of the various static and dynamic characteristics that govern the suitability of instruments in different applications.

3.1.1 Different types of instrument

A proper understanding of the fundamental nature of instruments is a necessary prerequisite for assessing the possible error levels in measurements, and ensuring that the performance of the instrument chosen is satisfactory. A convenient approach to this is to classify instruments into different types and to study the characteristics of each. These subclassifications are useful in broadly establishing attributes of particular instruments, such as accuracy, cost, and general applicability to different applications.

Deflection/null-type instruments

The pressure gauge in Figure 3.2(a) is also a good example of a deflection type of instrument, where the value of the quantity being measured is displayed in terms of the amount of movement of the pointer. In contrast, the dead-weight pressure gauge shown in Figure 3.2(b) is a null-type instrument. Here, weights are added on top of the piston until the piston reaches a datum level, known as the null point, where the downward force due to the weights is balanced by the upward force due to the fluid pressure. Pressure measurement is made in terms of the value of the weights needed to reach this null position.

The accuracy of these two instruments depends on different things. For the first one, it depends on the linearity and calibration of the spring, whilst for the second it relies on the calibration of the weights. As calibration of weights is much easier than careful choice and calibration of a linear-characteristic spring, this means that the second type of instrument will normally be more accurate. This is in accordance with the general rule that null-type instruments are more accurate than deflection types.

In terms of usage, the deflection-type instrument is clearly more convenient. It is far simpler to read the position of a pointer against a scale than to add and subtract weights until a null point is reached. Therefore, a deflection-type instrument is the one that would normally be used in the workplace. However, for calibration duties, the null-type instrument is preferable because of its superior accuracy. The extra effort required to use such an instrument is perfectly acceptable in this case because of the infrequent nature of calibration operations.

Active/passive instruments

Instruments are divided into active or passive ones according to whether the instrument output is entirely produced by the quantity being measured or whether the

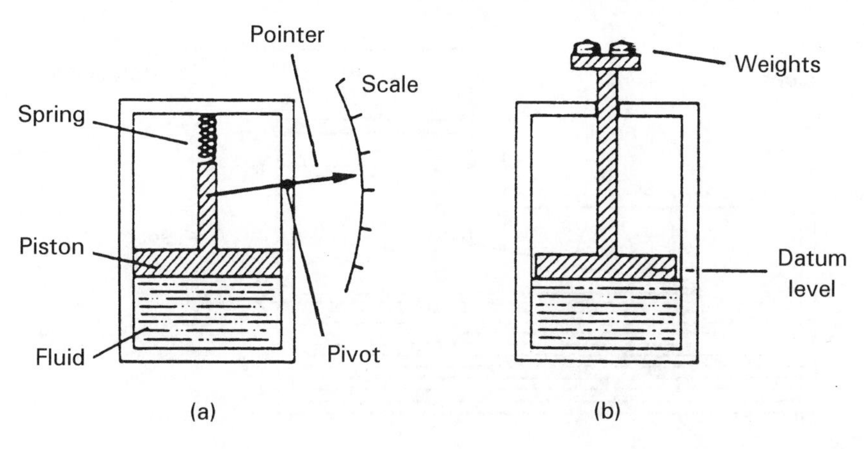

Figure 3.2 *Deflection/null types: (a) deflection-type pressure gauge; (b) dead-weight pressure gauge. From Morris (1997)* Measurement and Calibration Requirements, *© John Wiley & Sons, Ltd. Reproduced with permission.*

quantity being measured simply modulates the magnitude of some external power source. The pressure gauge shown in Figure 3.2(a) is an example of a passive instrument, because the energy expended in moving the pointer is derived entirely from the change in pressure measured: there is no other energy input to the system. A petrol tank level indicator, as sketched in Figure 3.3, is an example of an active instrument. The change in petrol level moves a potentiometer arm, and the output signal consists of a proportion of the external voltage source applied across the two ends of the potentiometer. The energy in the output signal comes from the external power source: the primary transducer float system is merely modulating the value of the voltage from this external power source. It should be noted that, whilst the external power source is usually in electrical form, in some cases it can be other forms of energy, such as pneumatic or hydraulic.

One very important difference between active and passive instruments is the level of measurement resolution obtained. With the simple pressure gauge shown, the amount of movement made by the pointer for a particular pressure change is defined by the nature of the instrument. Whilst it is possible to increase measurement resolution by making the pointer longer, such that the pointer tip moves through a longer arc, the scope for such improvement is clearly bounded by the practical limit on what is a convenient length for the pointer. However, in an active instrument, adjustment of the magnitude of the external energy input allows much greater control over measurement resolution. Incidentally, whilst the scope for improving measurement resolution is much greater, it is not infinite, because of limitations placed on the magnitude of the external energy input – in consideration of heating effects and for safety reasons.

In terms of cost, passive instruments are normally of a more simple construction than active ones and are therefore cheaper to manufacture. Therefore, choice between active and passive instruments for a particular application involves carefully balancing the measurement-resolution requirements against cost.

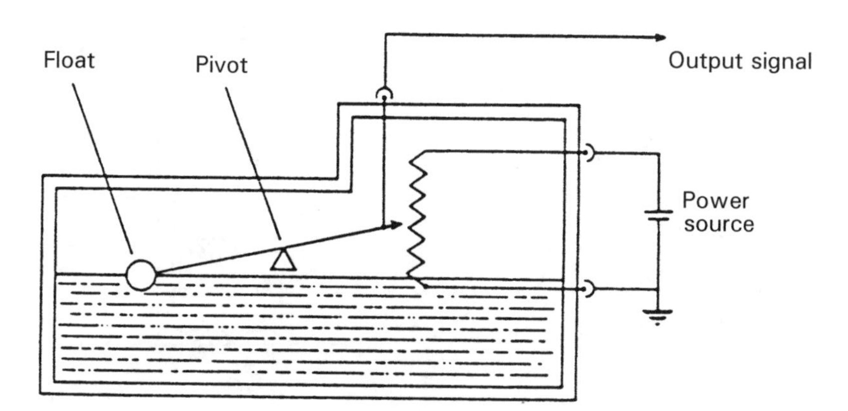

Figure 3.3 *Example of active instrument: petrol tank level indicator. From Morris (1997)* Measurement and Calibration Requirements, *© John Wiley & Sons, Ltd. Reproduced with permission.*

Analogue/digital instruments

An analogue instrument gives an output that varies continuously as the quantity being measured changes. The output can have an infinite number of values within the range that the instrument is designed to measure. The deflection-type of pressure gauge in Figure 3.2(a) is a good example of an analogue instrument. As the input value changes, the pointer moves with a smooth, continuous motion. Whilst the pointer can therefore be in an infinite number of positions within its range of movement, the number of different positions that the eye can discriminate between is strictly limited, this discrimination being dependent upon the size of the scale, and how finely it is divided.

A digital instrument, such as the rev-counter sketched in Figure 3.4, has an output that varies in discrete steps, and so can only have a finite number of values. The cam of the rev-counter is attached to the revolving body whose motion is being measured, and opens and closes a switch on each revolution. The switching operations are counted by an electronic counter. This system can only count whole revolutions and therefore cannot discriminate any motion that is less than a full revolution.

The distinction between analogue and digital instruments has become particularly important with the rapid growth in the application of computers in measurement and control systems. In such applications, an instrument whose output is in digital form is particularly advantageous, as it can be interfaced directly to the computer. In contrast, analogue instruments must be interfaced to the microcomputer by an analogue-to-digital (A/D) converter.

Intelligent/nonintelligent instruments

The term 'intelligent instrument' is used to describe a package that incorporates a digital processor as well as one or more of the measurement system components identified in Figure 3.1. Intelligent instruments are also sometimes referred to by other names, such as: *intelligent device*, *smart sensor* and *smart transmitter*. There is no formal definition for any of these alternative names, so that there is considerable overlap between the characteristics of particular devices and the name given to them. The processor within an intelligent instrument allows it to apply preprogrammed

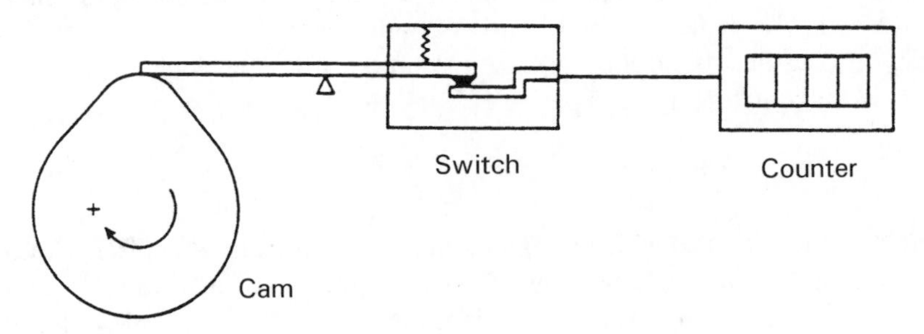

Figure 3.4 *Rev-counter. From Morris (1997)* Measurement and Calibration Requirements, *© John Wiley & Sons, Ltd. Reproduced with permission.*

signal-processing and data-manipulation algorithms to improve the quality of measurements, although the additional features inevitably make an intelligent instrument more expensive to buy than a comparable nonintelligent one. One important function of most intelligent instruments is to compensate measurements for systematic errors caused by environmental disturbances. To achieve this, they are provided with one or more secondary sensors to monitor the value of environmental disturbances, in addition to the primary sensor that measures the principal variable of interest. Although automatic compensation for environmental disturbances is a very important attribute of intelligent instruments, many versions of such devices also perform additional functions, such as:

- Providing switchable ranges (using several primary sensors within the instrument that each measure over a different range).
- Providing for remote adjustment and control of instrument parameters.
- Providing switchable output units (e.g. display in imperial or SI units).
- Linearisation of the output.
- Correction for the loading effect of measurement on the measured system.
- Providing signal damping with selectable time constants.
- Self-diagnosis of faults.

By contrast, nonintelligent instruments, as the name implies, do not have any form of computational power within them. Therefore, all mechanisms for carrying out any improvements to the quality of the measurements have to be achieved by devices external to the instrument.

3.1.2 Static instrument characteristics

The static characteristics of an instrument consist of a set of parameters that collectively describe the quality of the steady-state output measurement provided*. Some examples of static characteristics are: accuracy, sensitivity, linearity and the reaction to ambient temperature changes. All relevant static characteristics are given in the data sheet for a particular instrument, but it must be noted that values quoted in a data sheet only apply when the instrument is used under specified, standard calibration conditions. Due allowance must be made for variations in the characteristics when the instrument is used in other conditions. The important static characteristics to consider when choosing an instrument for a particular application are defined in the following paragraphs.

Accuracy

Accuracy is the extent to which a reading might be wrong, and is often quoted as a percentage of the full-scale reading of an instrument. If, for example, a pressure gauge of range 0–10 bar has a quoted inaccuracy of ±1.0% f.s. (±1% of full-scale

* 'Steady-state output measurement' means the non-changing output after any dynamic effects in the output reading have died out.

reading), then the maximum error to be expected in any reading is 0.1 bar. This means that when the instrument is reading 1.0 bar, the possible error is 10% of this value. For this reason, it is an important system design rule that instruments are chosen such that their range is appropriate to the spread of values being measured, in order that the best possible accuracy be maintained in instrument readings. Thus, if we were measuring pressures with expected values between 0 and 1 bar, we would not use an instrument with a range of 0–10 bar.

Tolerance

Tolerance is a term that is closely related to accuracy, and it defines the maximum error that is to be expected in some value. Whilst it is not, strictly speaking, a static characteristic of measuring instruments, it is mentioned here because the accuracy of some instruments is sometimes quoted as a tolerance figure. Tolerance, when used correctly, describes the maximum deviation of a manufactured component from some specified value. For example, if resistors have a quoted tolerance of 5%, one resistor chosen at random from a batch having a nominal value 1000 Ω might have an actual value anywhere between 950 Ω and 1050 Ω.

Precision/repeatability/reproducibility

Precision is a term that describes an instrument's degree of freedom from random errors. If a large number of readings are taken of the same quantity by a high-precision instrument, then the spread of readings will be very small. High precision does not imply anything about measurement accuracy. Hence, a high-precision instrument might actually have a low accuracy. Low-accuracy measurements from a high-precision instrument are usually caused by a bias in the measurements, which is removable by recalibration.

The terms 'repeatability' and 'reproducibility' mean approximately the same, but are applied in different contexts, as given below. *Repeatability* describes the closeness of output readings when the same input is applied repetitively over a short period of time, with the same measurement conditions, same instrument and observer, same location and same conditions of use maintained throughout. *Reproducibility* describes the closeness of output readings for the same input when there are changes in the method of measurement, observer, measuring instrument, location, conditions of use and time of measurement. Thus, both terms describe the spread of output readings for the same input. This spread is referred to as repeatability if the measurement conditions are constant, and as reproducibility if the measurement conditions vary.

The degree of repeatability or reproducibility in measurements is an alternative way of expressing precision. Figure 3.5 explains precision more clearly. This shows the results of testing three industrial robots that were programmed to place components at a particular point on a table. The target point was at the centre of the concentric circles shown, and the black dots represent the points where each robot actually deposited components at each attempt. Both the accuracy and precision of Robot 1 is shown to be low in this trial. Robot 2 consistently puts the component

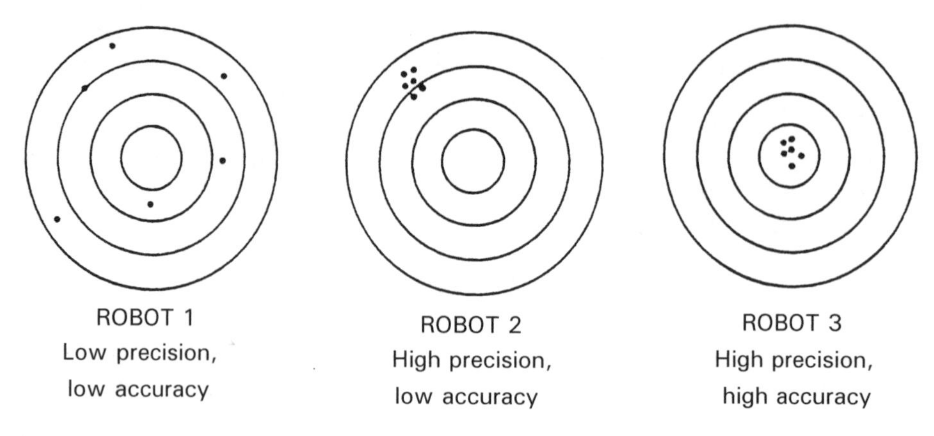

Figure 3.5 *Explanation of precision.*

down at approximately the same place, but this is the wrong point. Therefore, it has high precision but low accuracy. Finally, Robot 3 has both high precision and high accuracy, because it consistently places the component at the correct target position.

Range or span

The range or span of an instrument defines the minimum and maximum values of a quantity that the instrument is designed to measure. Instruments must not be used to measure values whose magnitude is outside the specified measurement range, since this could result in large measurement errors.

Linearity

It is normally desirable that the output reading of an instrument is linearly proportional to the quantity being measured. The Xs marked on Figure 3.6(a) show a plot of the typical output readings of an instrument when a sequence of input quantities are applied to it. Normal procedure is to draw a good-fit straight line through the Xs, as shown in this figure. (Whilst this can often be done with reasonable accuracy by eye, it is always preferable to apply a mathematical least-squares line-fitting technique.) The nonlinearity is then defined as the maximum deviation of any of the output readings marked X from this straight line. Nonlinearity is usually expressed as a percentage of the full-scale reading.

Sensitivity of measurement

The sensitivity of measurement is a measure of the change in instrument output that occurs when the quantity being measured changes by a given amount. Sensitivity is thus the ratio:

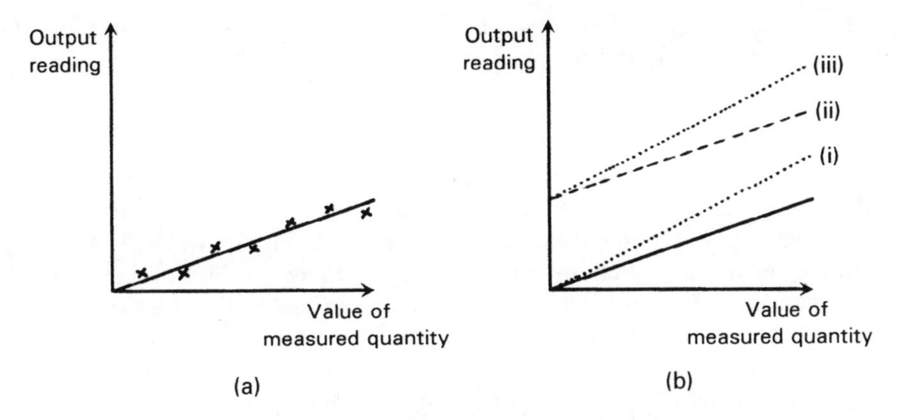

Figure 3.6 Instrument sensitivity: (a) standard instrument output characteristic; (b) effect on characteristic of drift: (i) sensitivity drift, (ii) zero drift (bias), (iii) sensitivity drift plus zero drift.

$$\frac{\text{Scale deflection}}{\text{Value of measured quantity}}$$

The sensitivity of measurement is therefore the slope of the straight line drawn on Figure 3.6(a). For example, if a pressure of 2 bars produces a deflection of 10 degrees in a pressure transducer, the sensitivity of the instrument is 5 degrees/bar (assuming that the relationship between pressure and the instrument reading is a straight-line one).

Sensitivity to disturbance

All calibrations and specifications of an instrument are only valid under controlled conditions of temperature, pressure, etc. These standard ambient conditions are usually defined in the instrument specification. As variations occur in the ambient temperature, etc., certain static instrument characteristics change, and the sensitivity to disturbance is a measure of the magnitude of this change. Such environmental changes affect instruments in two main ways, known as sensitivity drift and zero drift (bias).

Sensitivity drift (scale factor drift)

Sensitivity drift or scale factor drift defines the amount by which an instrument's sensitivity of measurement varies as ambient conditions change. Many components within an instrument are affected by environmental fluctuations, such as temperature changes: for instance, the modulus of elasticity of a spring is temperature-dependent. Line (i) on Figure 3.6(b) shows the typical effect of sensitivity drift on the output characteristic of an instrument. For the pressure gauge shown in Figure 3.2(a), in which the output characteristic is expressed in units of angular degrees/bar, the sensitivity drift would be expressed in units of the form (angular degree/bar)/°C if the spring was affected by temperature change.

Zero drift or bias

Zero drift, also known as bias, describes the effect where the zero reading of an instrument is modified by a change in ambient conditions. This causes a constant error over the full range of measurement of the instrument. Bathroom scales are a common example of instruments that are prone to bias. If there is a bias of 1 kg, then the reading would be 1 kg with no one standing on the scales. If someone of known weight 70 kg were to get on the scales, then the reading would be 71 kg, and if someone of known weight 100 kg were to get on the scales, the reading would be 101 kg. Instruments prone to zero drift normally have a means of adjustment (a thumbwheel in the case of bathroom scales) that allows the drift to be removed, so that measurements made with the instrument are unaffected. Typical units by which zero drift is measured are volts/°C, in the case of a voltmeter affected by ambient temperature changes. A typical change in the output characteristic of a pressure gauge subject to zero drift is shown by line (ii) in Figure 3.6(a). If the instrument suffers both zero drift and sensitivity drift at the same time, then the typical modification of the output characteristic is shown by line (iii) in Figure 3.6(b).

Resolution

When an instrument is showing a particular output reading, there is a lower limit on the magnitude of the change in the input measured quantity that produces an observable change in the instrument output. Resolution is sometimes specified as an absolute value and sometimes as a percentage of full-scale deflection. One of the major factors influencing the resolution of an instrument is how finely its output scale is divided into subdivisions. For example, a car speedometer has subdivisions of typically 20 km/h. This means that, when the pointer is between the scale markings, we cannot estimate speed more accurately than to the nearest 5 km/h. This figure of 5 km/h thus represents the resolution of the instrument.

3.1.3 Dynamic instrument characteristics

The static characteristics of a measuring instrument are concerned only with the steady-state reading that the instrument settles down to, such as the accuracy of the reading, etc. The dynamic characteristics describe the behaviour between the time that a measured quantity changes value and the time when the instrument output attains a steady value in response. As with static characteristics, any values for dynamic characteristics quoted in instrument data sheets only apply when the instrument is used under specified environmental conditions. Outside these calibration conditions, some variation in the dynamic parameters can be expected.

Various types of dynamic characteristics can be classified, known as zero-order, first-order and second-order characteristics. Fortunately, the practical effects of dynamic characteristics in the output of an instrument can be understood without resorting to formal mathematical analysis.

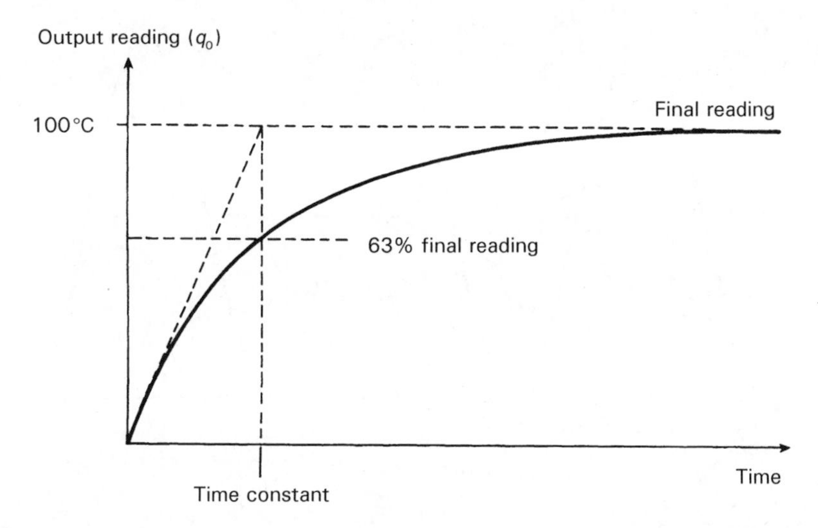

Figure 3.7 *First-order characteristic. From Morris (1997)* Measurement and Calibration Requirements, *© John Wiley & Sons, Ltd. Reproduced with permission.*

In a *zero-order instrument*, the dynamic characteristics are negligible, and the instrument output reaches its final reading almost instantaneously following a step change in the measured quantity applied at its input. A potentiometer, which measures motion, is a good example of such an instrument, where the output voltage changes approximately instantaneously as the slider is displaced along the potentiometer track.

In a *first-order instrument*, the output quantity q_o in response to a step change in the measured quantity q_i varies with time in the manner shown in Figure 3.7. The time constant τ of the step response is the time taken for the output quantity q_o to reach 63% of its final value. The liquid-in-glass thermometer is a good example of a first-order instrument. It is well known that, if a mercury thermometer at room temperature is plunged into boiling water, the mercury does not rise instantaneously to a level indicating 100 °C, but instead approaches a reading of 100 °C in the manner indicated by Figure 3.7. A large number of other instruments also belong to this first-order class. The main practical effect of first-order characteristics in an instrument is that the instrument must be allowed to settle to a steady reading before the output is read. Fortunately, the time constant of many first-order instruments is small relative to the dynamics of the process being measured, and so no serious problems are created.

It is convenient to describe the characteristics of *second-order instruments* in terms of three parameters: K (static sensitivity), ω (undamped natural frequency) and ε (damping ratio). The manner in which the output reading changes following a change in the measured quantity applied to its input depends on the value of these three parameters. The damping ratio parameter, ε, controls the shape of the output response, and the responses of a second-order instrument for various values of ε are shown in Figure 3.8. For case (A) where $\varepsilon = 0$, there is no damping, and the

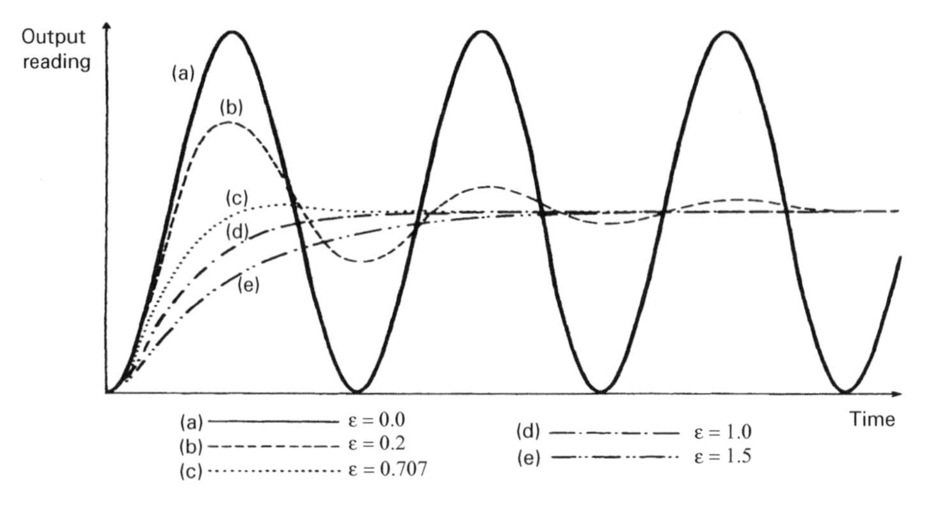

Figure 3.8 *Second-order characteristic.*

instrument output exhibits constant-amplitude oscillations when disturbed by any change in the physical quantity measured. For light damping of $\varepsilon = 0.2$, represented by case (B), the response to a step change in input is still oscillatory, but the oscillations gradually die down. Further increase in the value of ε reduces oscillations and overshoot still more, as shown by curves (C) and (D), and finally the response becomes very overdamped, as shown by curve (E) where the output reading creeps up slowly towards the correct reading. Clearly, the extreme response curves (A) and (E) are grossly unsuitable for any measuring instrument. If an instrument were to be only ever subjected to step inputs, then the design strategy would be to aim towards a damping ratio of 0.707, which gives the critically damped response (C). Unfortunately, most of the physical quantities that instruments are required to measure do not change in the mathematically convenient form of steps, but rather in the form of ramps of varying slopes. As the form of the input variable changes, so the best value for ε varies, and the choice of ε becomes a compromise between those values that are best for each type of input variable behaviour anticipated. Commercial second-order instruments, of which the accelerometer is a common example, are generally designed to have a damping ratio (ε) somewhere in the range of 0.6–0.8.

Thus, as for first-order instruments, it is necessary to allow second-order instruments to settle before the output reading is read, since there is a time lag between the measured quantity changing in value and the measuring instrument settling to a constant reading. This may limit the frequency at which the instrument output can be read, and can cause consequential difficulties in measuring rapidly changing variables.

3.1.4 Cost, durability and maintenance considerations in instrument choice

The static and dynamic characteristics discussed so far are those features that form the technical basis for a comparison between the relative merits of different instruments. However, in assessing the relative suitability of different instruments for a particular measurement situation, considerations of cost, durability and maintenance are also of great importance. Cost is very strongly correlated with the performance of an instrument, as measured by its characteristics. For example, increasing the accuracy or resolution of an instrument can only be done at a penalty of increasing its manufacturing cost. Therefore, instrument choice proceeds by specifying the minimum characteristics required by the measurement situation and then searching manufacturers' catalogues to find an instrument whose characteristics match those required. As far as accuracy is concerned, it is usual to specify maximum measurement uncertainty levels that are 10% of the tolerance levels of the parameter to be measured. To select an instrument whose accuracy and other characteristics are superior to the minimum levels required would only mean paying more than necessary for a level of performance that is greater than that needed.

As well as purchase cost, other important factors in the assessment exercise are the maintenance requirements and the instrument's durability. Maintenance requirements must be taken into account, as they also have cost implications. With regard to durability, it would not be sensible to spend £400 on a new instrument whose projected life was five years if an instrument of equivalent specification with a projected life of 10 years was available for £500. However, this consideration is not necessarily simple, as the projected life of instruments often depends on the conditions in which the instrument will have to operate.

As a general rule, a good assessment criterion is obtained if the total purchase cost and estimated maintenance costs of an instrument over its life are divided by the period of its expected life. The figure obtained is thus a cost per year. However, this rule becomes modified where instruments are being installed on a process whose life is expected to be limited, perhaps in the manufacture of a particular model of car. Then, the total costs can only be divided by the period of time for which an instrument is expected to be used, unless an alternative use for the instrument is envisaged at the end of this period.

To summarise therefore, instrument choice is a compromise between performance characteristics, ruggedness and durability, maintenance requirements, and purchase cost. To carry out such an evaluation properly, the instrument engineer must have a wide knowledge of the range of instruments available for measuring particular physical quantities, and he/she must also have a deep understanding of how instrument characteristics are affected by particular measurement situations and operating conditions.

3.2 Calibration of Measuring Instruments

Whatever instrument is chosen for a particular measurement application, its characteristics will change over a period of time and affect the relationship between the input and output. Changes in characteristics are brought about by factors such as mechanical wear, and the effects of dirt, dust, fumes and chemicals in the operating environment. To a great extent, the magnitude of the drift in characteristics depends on the amount of use that an instrument receives, and hence on the amount of wear and the length of time that it is subjected to the operating environment. However, some drift also occurs even in storage, as a result of ageing effects in components within the instrument. Thus, in order to maintain the accuracy of measurements made, all instruments should be calibrated at some predetermined frequency. It should also be emphasised that all elements in the measurement system, including the final signal recorder, must be included in the calibration exercise.

3.2.1 The calibration process

Calibration consists of comparing the output of the instrument being calibrated against the output of a standard instrument of known accuracy, when the same input (measured quantity) is applied to both instruments. During this calibration process, the instrument is tested over its whole range by repeating the comparison procedure for a range of inputs.

The instrument used as a standard for this procedure must be one that is kept solely for calibration duties. It must never be used for other purposes. Most particularly, it must not be regarded as a spare instrument that can be used for normal measurements if the instrument normally used for that purpose breaks down. Proper provision for instrument failures must be made by keeping a spare set of instruments. Standard calibration instruments must be kept totally separate.

To ensure that these conditions are met, the calibration function must be managed and executed in a professional manner. This will normally mean setting aside a particular place within the instrumentation department of a company where all calibration operations take place and where all instruments used for calibration are kept. As far as possible, this should take the form of a separate room, rather than a sectioned-off area in a room used for other purposes as well. This will enable better environmental control to be applied in the calibration area, and will also offer better protection against unauthorised handling or use of the calibration instruments.

Calibration instruments usually have a greater inherent accuracy (often at least ten times better) than the instruments that they are used to calibrate. Where instruments are only used for calibration purposes, this greater accuracy can often be achieved by specifying a type of instrument that would be unsuitable for normal measurements. For example, ruggedness is not required in calibration instruments, and freedom from this constraint opens up a much wider range of possible instruments. In practice, high-accuracy, null-type instruments are commonly used for calibration duties, because their requirement for a human operator is not a problem in these circumstances.

3.2.2 Standards laboratories

The calibration facilities provided within the instrumentation department of a company provide the first link in the calibration chain. An instrument used for calibration at this level is known as a working standard. As this working standard instrument is one that is kept by the instrumentation department for calibration duties, and for no other purpose, then it can be assumed that it will maintain its accuracy over a reasonable period of time, because use-related deterioration in accuracy is largely eliminated. However, over the longer term, even the characteristics of such a standard instrument will drift, mainly due to ageing effects in components within it. Therefore, over this longer term, a programme must be instituted for calibrating the working standard instrument against one of yet higher accuracy at appropriate intervals of time. The instrument used for calibrating working standard instruments is known as a secondary reference standard. This must obviously be a well-engineered instrument that gives high accuracy and is stabilised against drift in its performance over time. This implies that it will be an expensive instrument to buy. It also requires that the environmental conditions in which it is used are carefully controlled in respect of ambient temperature, humidity, etc.

Because of the expense involved in providing secondary reference standard instruments and the controlled environment that they need to operate in, the establishment of a company standards laboratory to provide such a calibration facility is economically viable only in the case of very large companies, where large numbers of instruments need to be calibrated across several factories. In the case of small- to medium-sized companies, the cost of buying and maintaining such equipment is not justified. Instead, they would normally use the services of one of the specialist companies that have developed a suitable standards laboratory for providing calibration at this level.

When the working standard instrument has been calibrated by an authorised standards laboratory, a calibration certificate will be issued[1]. This will contain at least the following information:

- the identification of the equipment calibrated;
- the calibration results obtained;
- the measurement uncertainty;
- any use limitations on the equipment calibrated;
- the date of calibration;
- the authority under which the certificate is issued.

3.2.3 Validation of standards laboratories

In the United Kingdom, the appropriate National Standards Organisation for validating standards laboratories is the National Physical Laboratory (in the United States of America, the equivalent body is the National Bureau of Standards). This has established a National Measurement Accreditation Service (NAMAS) that monitors both instrument calibration and mechanical testing laboratories. The formal structure for accrediting instrument calibration in standards laboratories is known as the British Calibration Service (BCS), and that for accrediting testing facilities is known as the National Testing Laboratory Accreditation Scheme (NATLAS).

Although each country has its own structure for the maintenance of standards, each of these different frameworks tends to be equivalent in its effect. To achieve confidence in the goods and services that move across national boundaries, international agreements have established the equivalence of the different accreditation schemes in existence.

A standards laboratory has to meet strict conditions[2] before it is approved. These conditions control laboratory management, environment, equipment and documentation. The person appointed as head of the laboratory must be suitably qualified, and independence of operation of the laboratory must be guaranteed. The management structure must be such that any pressure to rush or skip calibration procedures for production reasons can be resisted. As far as the laboratory environment is concerned, proper temperature and humidity control must be provided, and high standards of cleanliness and housekeeping must be maintained. All equipment used for calibration purposes must be maintained to reference standards, and supported by calibration certificates that establish this traceability. Finally, full documentation must be maintained. This should describe all calibration procedures, maintain an index system for recalibration of equipment, and include a full inventory of apparatus and traceability schedules. Having met these conditions, a standards laboratory becomes an accredited laboratory for providing calibration services and issuing calibration certificates. This accreditation is reviewed at approximately 12-monthly intervals to ensure that the laboratory is continuing to satisfy the conditions laid down for approval.

3.2.4 Primary reference standards

Primary reference standards describe the highest level of accuracy that is achievable in the measurement of any particular physical quantity. All items of equipment used in standards laboratories as secondary reference standards have to be calibrated themselves against primary reference standards at appropriate intervals of time. This procedure is acknowledged by the issue of a calibration certificate in the standard way. National standards organisations maintain suitable facilities for this calibration, although, in certain cases, such primary reference standards can be located outside national standards organisations. For example, the primary reference standard for dimension measurement is defined by the wavelength of the orange–red line of krypton light, and this can be realised in any laboratory equipped with an interferometer.

In certain cases (e.g. the measurement of viscosity), such primary reference standards are not available and reference standards for calibration are achieved by collaboration between several national standards organisations that perform measurements on identical samples under controlled conditions[3].

3.2.5 Traceability

What has emerged from the foregoing discussion is that calibration has a chain-like structure, in which every instrument in the chain is calibrated against a more accurate instrument immediately above it in the chain, as shown in Figure 3.9(a). All of the elements in the calibration chain must be known, so that the calibration of process instruments at the bottom of the chain is traceable to the fundamental measurement standards.

This knowledge of the full chain of instruments involved in the calibration procedure is known as *traceability*, and is specified as a mandatory requirement in satis-

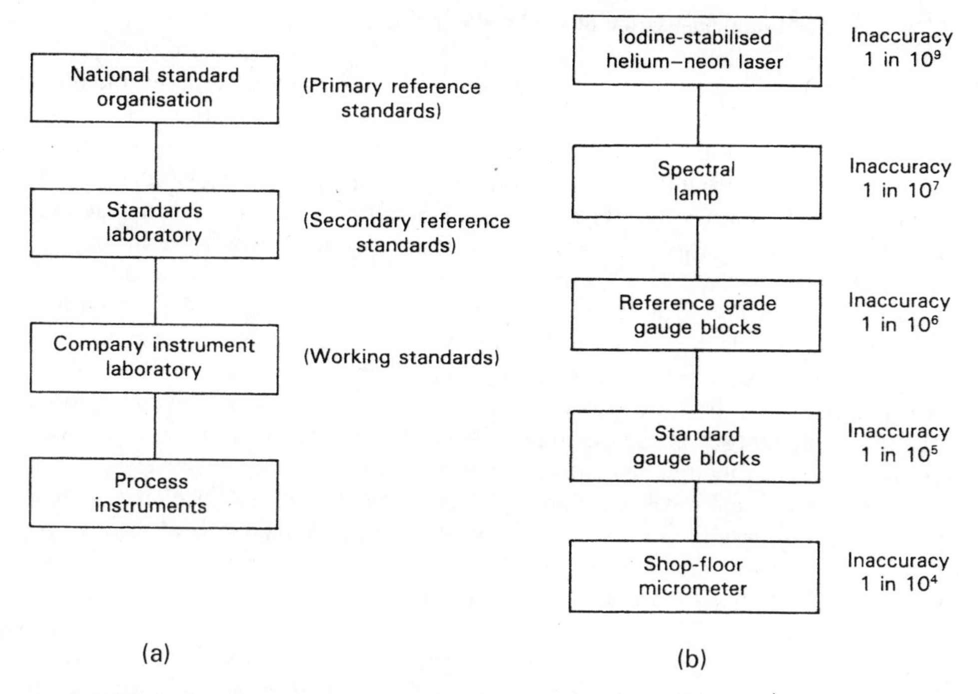

Figure 3.9 *Calibration chains: (a) typical structure; (b) calibration chain for micrometers. From Morris (1997)* Measurement and Calibration Requirements, *© John Wiley & Sons, Ltd. Reproduced with permission.*

fying standards such as ISO 9001 and ISO 14001. Documentation must exist which shows that process instruments are calibrated by standard instruments that are linked by a chain of increasing accuracy back to national reference standards. There must be clear evidence to show that there is no break in this chain.

To illustrate a typical calibration chain, consider the calibration of micrometers shown in Figure 3.9(b). A typical shop-floor micrometer has an uncertainty (inaccuracy) of less than 1 in 10^4. These would normally be calibrated in the instrumentation department laboratory of a company against laboratory-standard gauge blocks with a typical uncertainty of less than 1 in 10^5. A specialist calibration service company would provide facilities for calibrating these laboratory-standard gauge blocks against reference-grade gauge blocks with a typical uncertainty of less than 1 in 10^6. More accurate calibration equipment still is provided by national standards organisations. The National Physical Laboratory (UK) maintains two sets of standards for this type of calibration, a working standard and a primary standard. Spectral lamps are used to provide a working reference standard with an uncertainty of less than 1 in 10^7. The primary standard is provided by an iodine-stabilised helium–neon laser which has a specified uncertainty of less than 1 in 10^9. All of the links in this calibration chain must be shown in the measurement and calibration system documentation.

3.2.6 Practical implementation of calibration procedures

Having laid down these theoretical foundations of calibration procedures, the practical aspects of implementing these procedures must be considered. In practice, what is sensible, practical, achievable and affordable in any given situation may differ in substantial respects from the ideal. The most appropriate person to give advice about what standards of measurement accuracy and calibration are appropriate and acceptable in any given situation is a consultant who has a lot of experience in the particular industry involved in any situation.

As far as management of calibration procedures is concerned, it is important that the performance of all calibration operations is assigned as the clear responsibility of just one person. That person should have total control over the calibration function, and be able to limit access to the calibration laboratory to designated approved personnel only. Only by giving this appointed person total control over the calibration function, can the function be expected to operate efficiently and effectively. Lack of such rigid management will inevitably lead to unintentional neglect of the calibration system, and result in the use of equipment in an out-of-date state of calibration. Professional management is essential, so that the customer can be assured that an efficient calibration system is in operation, and that the accuracy of measurements is guaranteed.

As instrument calibration is an essential component in an EMS, the clause in ISO 14001 that requires all persons performing EMS-related functions to be adequately trained clearly extends to personnel who calibrate measuring instruments. Thus, the manager in charge of the calibration function must ensure that adequate training is provided and targeted at the particular needs of the calibration systems involved. People must understand what they need to know, and especially why they must have this information. Successful completion of training courses should be marked by the award of qualification certificates. These attest to the proficiency of personnel involved in calibration duties, and are a convenient way of demonstrating that the training requirement has been satisfied.

All instrument characteristics are affected to some extent by environmental conditions, and any parameters given in data sheets only apply for specified conditions. Therefore, as far as practicable, these same environmental conditions should be reproduced during calibration procedures. However, specification of the level of environmental control required should be considered carefully with due regard to the level of accuracy needed in the calibration procedure, as overspecification will lead to unnecessary expense. In practice, full air-conditioning is not normally required for calibration at this level, as it is very expensive, but sensible precautions should be taken to guard the area from extremes of heat or cold, and good standards of cleanliness should also be maintained.

For various reasons, it is not always possible to perform calibration operations in a controlled environment. For example, it may not be convenient or possible to remove instruments from process plant, and in such cases it is standard practice to calibrate them *in situ*. In these circumstances, appropriate corrections must be made for the deviation in the calibration environmental conditions away from those specified. However, this practice does not obviate the need to protect calibration instruments and to maintain them in constant conditions in a calibration laboratory at all times other than when they are involved in such calibration duties on plant.

The effect of environmental conditions that differ from those specified for calibration can be quantified by the following procedure: whilst keeping all other environmental parameters at some constant level, the environmental parameter under investigation is varied in steps over a range of values, causing the instrument output to vary in steps over a corresponding range of readings. This allows an input/output relationship to be drawn for that particular environmental parameter. Following this quantification of the effect of any environmental parameters that differ from the standard value specified for calibration, the instrument calibration exercise can proceed.

Determining the calibration frequency required

The characteristics of an instrument are only guaranteed immediately after calibration. Thereafter, as time progresses, the characteristics will change because of factors such as ageing effects, mechanical wear, long-term environmental changes, and the effects of dust, dirt, fumes and chemicals in the operating atmosphere. Fortunately, a certain amount of degradation in characteristics can be allowed before the instrument needs to be recalibrated. For example, if an instrument is required to measure a parameter to an accuracy of ±2% and its accuracy is ±1% following calibration, then its accuracy can degrade from ±1% to ±2% before recalibration is necessary.

Susceptibility to the factors that cause characteristics to change will vary according to the type of instrument involved and the frequency and conditions of usage. Often, an experienced engineer or the instrument manufacturer will be able to estimate the likely rate at which characteristics will change according to the conditions in which an instrument is used, and so the necessary calibration frequency can be calculated accordingly. However, it is prudent not to rely too much on such a priori predictions, because some significant effect on the instrument may have been overlooked. Thus, as long as the circumstances permit, it is preferable to start from basics in deriving the required calibration frequency, and not to use past-history information about the instrument and its operating environment. Applying this philosophy, the following frequency for checking the characteristics of new instruments might be appropriate (assuming 24 hours/day working):

week 1: once per day;
weeks 2–3: twice per week;
weeks 4–8: once per week;
months 3–6: twice per month;
months 7–12: once per month;
year 2: every three months;
thereafter: every six months.

The frequency of calibration checks should be reduced according to the above scheme, until a point is reached where deterioration in the instrument's accuracy is first detected. Comparison of the amount of performance degradation with the inaccuracy level that is permissible in the instrument will show whether the instrument should be calibrated immediately at this point or whether it can be safely left for a further period. If the above pattern of calibration checks were followed for an instrument, and the check at week 8 showed deterioration in accuracy that was close to the permissible limit, then this would determine that the calibration frequency for the instrument should be every eight weeks.

The above method of establishing the optimum calibration frequency is clearly an ideal which cannot always be achieved in practice, and indeed for some types of instrument this level of rigour is unnecessary. When used on many production processes, for instance, it would be unacceptable to interrupt production every hour to recheck instrument calibrations, unless a very good case could be made for why this was necessary. Also, the nature of some instruments, for example a mercury-in-glass thermometer, means that calibration checks will only ever be required infrequently. However, for instruments such as unprotected base-metal thermocouples, initial calibration checks after one hour of operation would not be at all inappropriate.

3.2.7 Procedure following calibration

When the instrument is calibrated against a standard instrument, its accuracy will be shown to be either inside or outside the required measurement accuracy limits. If the instrument is found to be inside the required measurement limits, the only course of action required is to record the calibration results in the instrument's record sheet and then put it back into use until the next scheduled time for calibration.

The options available if the instrument is found to be outside the required measurement limits depend on whether its characteristics can be adjusted and the extent to which this is possible. If the instrument has adjustment screws, these should be turned until the characteristics of the instrument are within the specified measurement limits. Following this, the adjustment screws must be sealed to prevent tampering during the instrument's subsequent use. In some cases, it is possible to redraw the output scale of the instrument. After such adjustments have been made, the instrument can be returned to its normal location for further use.

The second possible course of action if the instrument is outside measurement limits covers the case where no adjustment is possible or the range of possible adjustment is insufficient to bring the instrument back within measurement limits. In this event, the instrument must be withdrawn from use, and this withdrawal must be marked prominently on it to prevent it from being reused inadvertently. The options available then are to either send the instrument for repair, if this is feasible, or to scrap it.

3.2.8 Calibration procedure review

Whatever system and frequency of calibration are established, it is important to review these from time to time to ensure that the system remains effective and efficient. It may happen that a cheaper (but equally effective) method of calibration becomes available with the passage of time, and such an alternative system must clearly be adopted in the interests of cost-efficiency. However, the main item under scrutiny in this review is normally whether the calibration frequency is still appropriate. Records of the calibration history of the instrument will be the primary basis on which this review is made. It may happen that an instrument starts to go out of calibration more quickly after a period of time, either because of ageing factors within the instrument or because of changes in the operating environment. The conditions or mode of usage of the instrument may also be subject to change. As the environmental and usage conditions of an instrument may change beneficially as well as adversely, there is the possibility that the recommended calibration-interval may increase as well as decrease.

3.3 Documentation of Measurement and Calibration Systems

An essential element in the maintenance of measurement systems and the operation of calibration procedures is the provision of full documentation. The documentation must give a full description of the measurement requirements throughout the workplace, the instruments used, and the calibration system and procedures operated. Individual calibration records for each instrument must be included within this. This documentation is a necessary part of the EMS manual, although it may physically exist as a separate volume if this is more convenient. An overriding constraint on the style in which the documentation is presented is that it should be simple and easy to read. This is often greatly facilitated by a copious use of appendices.

The starting point in the documentation must be a statement of what measurement limits have been defined for each measurement system documented. It is customary to express the measurement limits as ± 2 standard deviations, i.e. within 95% confidence limits (see Chapter 4 for further explanation).

Following this, the instruments specified for each measurement situation must be listed. This list must be accompanied by full instructions about the proper use of the instruments concerned. These instructions will include details about any environmental control or other special precautions that must be taken to ensure that the instruments provide measurements of sufficient accuracy to meet the measurement limits defined. The proper training courses appropriate to personnel who will use the instruments must also be specified.

Having disposed of the question about the instruments that are used, the documentation must go on to cover the subject of calibration. A formal procedure for calibration must be defined, and the standard instruments used must be specified. This procedure must include instructions for the storage and handling of standard calibration instruments, and must specify the required environmental conditions under which calibration is to be performed. However, where a calibration procedure for a particular instrument uses standard practices that are documented elsewhere, it is sufficient to include reference to that standard practice in the documentation, rather than reproduce the whole procedure. Finally, whatever calibration system is established, the documentation must define a formal and regular review procedure that ensures its continued effectiveness. The results of each review must also be documented in a formal way.

An important part of calibration procedures is to maintain proper records of all calibrations carried out and the results obtained. A standard format for the recording of calibration results in record sheets should be defined in the documentation and, where appropriate, the documentation must also define the manner in which calibration results are to be recorded on the instruments themselves. A separate record, similar to that shown in Figure 3.10, must be kept for every measuring instrument, irrespective of whether it is in use or kept as a spare. Each record should include a description of the instrument, its serial number, the required calibration frequency and the person responsible for calibration, the date of each calibration and the calibration results in terms of the deviation from the required characteristics and the action taken to correct it.

The documentation must also specify procedures to be followed if an instrument is found to be outside the calibration limits. This may involve adjustment, redrawing its scale or withdrawing it, depending upon the nature of the discrepancy and the type of instrument involved. Withdrawn instruments will either be repaired or scrapped, but, until faults have been rectified, their status must be clearly marked on them to prevent them being accidentally put back into use.

Two other items must also be covered by the calibration document. The traceability of the calibration system back to national reference standards must be defined and supported by calibration certificates (see Section 3.2). Training procedures must also be documented, specifying the particular training courses to be attended by various personnel and what, if any, refresher courses are required.

<table>
<tr><td colspan="2">Type of instrument:</td><td>Company serial number:</td></tr>
<tr><td colspan="2">Manufacturer's part number:</td><td>Manufacturer's serial number:</td></tr>
<tr><td colspan="2">Measurement limit:</td><td>Date introduced:</td></tr>
<tr><td colspan="3">Location:</td></tr>
<tr><td colspan="3">Instructions for use:</td></tr>
<tr><td colspan="2">Calibration frequency:</td><td>Signature of person responsible for calibration:</td></tr>
<tr><td colspan="3">CALIBRATION RECORDS</td></tr>
<tr><td>Calibration date</td><td>Calibration results</td><td>Calibrated by</td></tr>
<tr><td></td><td></td><td></td></tr>
<tr><td></td><td></td><td></td></tr>
<tr><td></td><td></td><td></td></tr>
<tr><td></td><td></td><td></td></tr>
<tr><td></td><td></td><td></td></tr>
<tr><td></td><td></td><td></td></tr>
</table>

Figure 3.10 *Typical format for instrument record sheets. From Morris (1997)* Measurement and Calibration Requirements, *© John Wiley & Sons, Ltd. Reproduced with permission.*

All aspects of these documented calibration procedures will be given consideration as part of the periodic audit of the EMS. Whilst the basic responsibility for choosing a suitable interval between calibration checks rests with the engineers responsible for the instruments concerned, the auditor will require to see the results of tests which show that the calibration interval has been chosen correctly and that instruments are not going outside allowable measurement uncertainty limits between calibrations. Audits will check in particular for the existence of procedures that are instigated in response to instruments found to be out of calibration. Evidence that such procedures are effective in avoiding degradation in the environmental management function will also be required.

References

1. *NAMAS Document B 5103: Certificates of Calibration* (NAMAS Executive, National Physical Laboratory, Middlesex, UK), 1985.
2. *ISO/IEC GUIDE 25, 1990, General Requirements for the Calibration and Competence of Testing Laboratories* (International Organisation for Standards, Geneva).
3. *ISO 5725, 1986, Accuracy of Measurement Test Methods and Results* (International Organisation for Standards, Geneva) (also published by British Standards Institution as BS.ISO5725).

4

Measurement System Errors

Even if measurements are made carefully with instruments that are in a proper state of calibration and used in environmental conditions that are the same as those specified for calibration, measurement inaccuracy is still likely, because of errors that arise from a number of other sources. To maximise the quality of measurements, all errors must be reduced to the minimum possible level, and any remaining error must be quantified and compensated for. A prerequisite in this is a detailed analysis of the sources of error that exist. For this purpose, it is convenient to divide errors in measurement data into two groups, known as random errors and systematic errors.

Random errors are perturbations of the measurement either side of the true value, caused by random and unpredictable effects, such that positive errors and negative errors occur in approximately equal numbers for a series of measurements made of the same quantity. Such perturbations are mainly small, but large perturbations occur from time to time, again unpredictably. Random errors often arise when measurements are taken by human observation of an analogue meter, especially when this involves interpolation between scale points. Electrical noise can also be a source of random errors. To a large extent, random errors can be overcome by taking the same measurement a number of times and extracting a value by averaging or other statistical techniques, as discussed in Section 4.1.1. However, any quantification of the measurement value and statement of error bounds remains a statistical quantity. Because of the nature of random errors, and the fact that large perturbations in the measured quantity occur from time to time, the best that can be done is to express measurements in probabilistic terms: it may be possible to assign a 95% or even 99% confidence level that the measurement is a certain value within error bounds of say ±1%, but it is not possible to attach a 100% probability to measurement values that are subject to random errors.

Systematic errors describe errors in the output readings of a measurement system that are consistently on one side of the correct reading, i.e. either all the errors are

ISO 14000 Environmental Management Standards: Engineering and Financial Aspects. Alan S. Morris.
 ISBN 0-470-85128-7

positive or they are all negative. Two major sources of systematic errors are system disturbance during measurement and the effect of environmental inputs, as discussed in Sections 4.2.1 and 4.2.2. Other sources of systematic error include bent meter needles, the use of uncalibrated instruments, poor cabling practices and the thermal generation of electromotive forces (commonly known as thermal e.m.f's). The latter two sources are considered in Section 4.2.3. Even when systematic errors due to the above factors have been reduced or eliminated, some errors remain that are inherent in the manufacture of an instrument. These are quantified by the accuracy figure quoted in the specifications published in the instrument data sheet.

Finally, a word must be said about the distinction between random and systematic errors. Error sources in the measurement system must be examined carefully to determine what type of error is present, random or systematic, and to apply the appropriate treatment. In the case of manual data measurements, a human observer may make a different observation at each attempt, but it is often reasonable to assume that the errors are random, and that the mean of these readings is likely to be close to the correct value. However, this is only true as long as the human observer is not introducing a systematic parallax-induced error as well by persistently reading the position of a needle against the scale of an analogue meter from the same side rather than from directly above. In that case, correction would have to be made for this systematic error, or bias, in the measurements, before statistical techniques were applied to reduce the effect of random errors.

4.1 Random Errors

Random errors in measurements are caused by random, unpredictable variations in the measurement system, and they can only be quantified in probabilistic terms. In practice, because of their nature, random errors can largely be eliminated by calculating the mean or median of the measurements. The degree of confidence in the calculated mean/median values can be quantified by calculating the standard deviation or variance of the data, these being parameters which describe how the measurements are distributed about the mean/median value. All of these terms are explained more fully in Section 4.1.1.

Because of the unpredictability of random errors, any error bounds placed on measurements can only be quantified in probabilistic terms. Thus, if the possible error in a measurement subject to random errors is quoted as ±2% of the measured value, this only means that this is probably true, i.e. there is, say, a 95% probability that the error level does not exceed ±2%.

The distribution of measurement data about the mean value can be displayed graphically by frequency-distribution curves, as discussed in Section 4.1.2. Calculation of the area under the frequency-distribution curve gives the probability that the error will lie between any two chosen error levels.

4.1.1 Statistical analysis of data

In the analysis of measurements subject to random errors, various parameters can be extracted. Formal definitions of these and their means of calculation are given in the following sections. It should be noted that 'hand calculation' of these parameters is rarely necessary nowadays, as many standard computer packages are available for their calculation, and this facility is also provided by many personal calculators and also intelligent instruments.

Mean and median values

In any measurement situation subject to random errors, the normal technique is to take the same reading a number of times, ideally using different observers, and to extract the most likely value from the measurement data set. For a set of n measurements $x_1, x_2 \cdots x_n$, the most likely true value is the *mean*, given by:

$$x_{mean} = \frac{x_1 + x_2 + \cdots x_n}{n} \tag{4.1}$$

This is valid for all data sets where the measurement errors are distributed equally about the line of zero error, i.e. where the positive errors are balanced in quantity and magnitude by the negative errors. The average value of a set of measurements is also sometimes expressed by the *median* value, which is a close approximation to the mean value. The median is given by the middle value, when the measurements in the data set are written down in ascending order of magnitude.

Suppose that, in a particular measurement situation, a mass is measured by a beam-balance, and the set of readings in grams shown in Table 4.1 is obtained at a particular time by different observers. The mean value of this set of data is 81.18 g, calculated according to equation 4.1. The median value is 81.1 g, which is the middle value if the data values are written down in ascending order, starting at 80.5 g and ending at 81.8 g.

Standard deviation and variance

The probability that the mean or median value of a data set represents the true measurement value depends on how widely scattered the data values are. If the values of mass measurements in Table 4.1 had ranged from 79 g up to 83 g, confidence in the mean value would be much less. The spread of values about the mean is analysed by first calculating the deviation of each value from the mean value x_{mean}. For any general value x_i, the deviation d_i is given by: $d_i = x_i - x_{mean}$. The extent to which n measurement values are spread about the mean can now be expressed by the standard deviation σ, where σ is given by:

$$\sigma = \sqrt{\frac{d_1^2 + d_2^2 \cdots d_n^2}{n-1}} \tag{4.2}$$

Table 4.1 *Set of mass measurements subject to random errors*

81.6	81.1	81.4	80.9	81.1	80.5	81.3	80.8	81.2	81.8
81.1	81.5	81.0	81.3	81.1	80.8	81.3	81.6	81.1	

From Morris (1991) *Measurement and Calibration Requirements*, © John Wiley & Sons, Ltd. Reproduced with permission.

This spread can alternatively be expressed by the variance V, which is the square of the standard deviation, i.e. $V = \sigma^2$. Readers should note that some books give an alternative definition for σ and V, which is also used by some calculators for evaluating the standard deviation and variance functions. This alternative expression shown in equation (4.3) has (n) instead of ($n - 1$) in the denominator and is properly called the *root-mean-squared deviation*.

$$\text{deviation}_{\text{rms}} = \sqrt{\frac{d_1^2 + d_2^2 \cdots d_n^2}{n}} \tag{4.3}$$

This difference in definitions arises because the root-mean-squared deviation is defined for an infinite data set, but, in the case of measurements, data sets are not infinite. For a finite set of measurements (x_i) $i = 1,n$, the mean x_m will differ from the true mean μ of the infinite data set that the finite set is part of. If the true mean μ of a set of measurements was somehow known, then the deviations d_i could be calculated as the deviation of each data value from the true mean, and it would then be correct to calculate σ and V using (n) instead of ($n - 1$). However, in normal situations, using ($n - 1$) in the denominator of equation (4.2) produces a value that is statistically closer to the correct value.

Example 4.1

The following measurements were taken of the mass of dust particles collected in a hopper over a given time period (by weighing the hopper repeatedly and subtracting the mass of the hopper).

21.5 g, 22.1 g, 21.3 g, 21.7 g, 22.0 g, 22.2 g, 21.8 g, 21.4 g, 21.9 g, 22.1 g

Calculate the mean value, the deviations from the mean and the standard deviation.

Solution

Mean value = Σ(data values)/10 = 218/10 = 21.8 g.

Now draw a table of measurements and deviations:

Measurements:	21.5	22.1	21.3	21.7	22.0	22.2	21.8	21.4	21.9	22.1
Deviations from mean:	−0.3	+0.3	−0.5	−0.1	+0.2	+0.4	0.0	−0.4	+0.1	+0.3
(deviations)^2:	0.09	0.09	0.25	0.01	0.04	0.16	0.0	0.16	0.01	0.09

$\Sigma(\text{deviations})^2 = 0.90$

n = number of measurements = 10

$\Sigma(\text{deviations})^2/(n-1) = \Sigma(\text{deviations})^2/9 = 0.10$

$\sqrt{[\Sigma(\text{deviations})^2/9]} = 0.316$

Thus, standard deviation = 0.32 g (the nature of the measurements do not justify expressing the standard deviation to any accuracy greater than two decimal places).

4.1.2 Frequency distributions

A further and very powerful way of analysing the pattern in which measurements are spread around the mean value is to use graphical techniques. The simplest way of doing this is by means of a *histogram*, where bands of equal width across the range of measurement values are defined and the number of measurements within each band are counted. Figure 4.1 shows a histogram of measurements drawn from the set of mass data in Table 4.1 by choosing bands 0.3 g wide. There are, for instance, nine measurements in the range between 81.05 g and 81.35 g, and so the height of the histogram at this point is nine units (NB The scaling of the bands was deliberately chosen so that no measurements fell on the boundary between different bands and caused ambiguity about which band to put them in.) Such a histogram has the characteristic shape shown by truly random data, with symmetry about the mean value of the measurements. This symmetry is sufficient to demonstrate that the data only have random errors.

As the number of measurements increases, smaller bands can be defined for the histogram, which retains its basic shape, but then consists of a larger number of smaller steps on each side of the peak. In the limit, as the number of measurements approaches infinity, the histogram becomes a smooth curve known as the *frequency distribution curve* of the measurements, as shown in Figure 4.2. The ordinate of this curve is the frequency of occurrence of each measurement value, $F(x)$, the abscissa is the magnitude, x, and x_p is the most probable data value. If the errors are truly random, x_p is the mean value of the measurements.

Figures 4.1 and 4.2 can be converted to show the distribution of measurement errors by simply rescaling the abscissa in terms of the error values instead of the

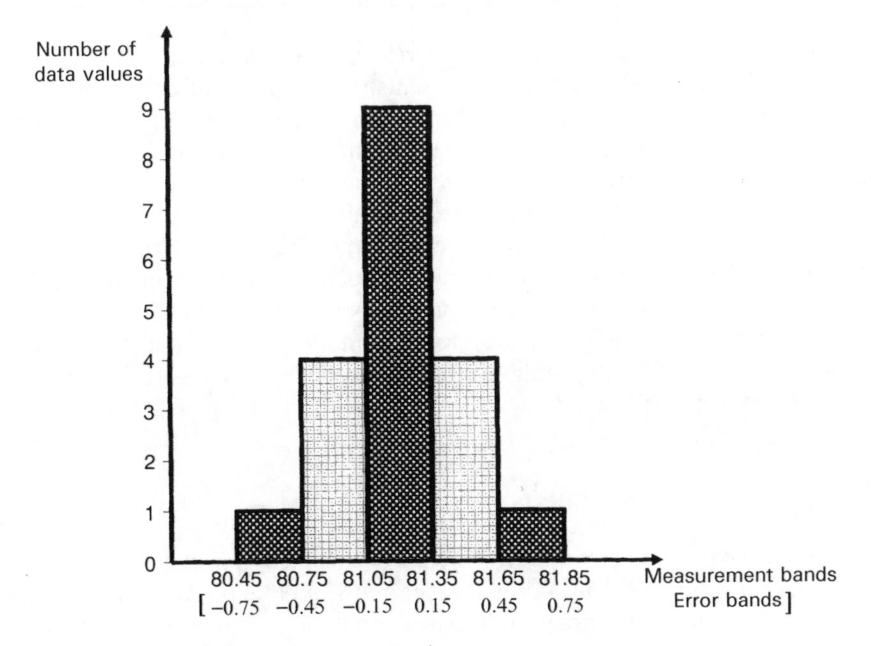

Figure 4.1 *Histogram of mass measurements and errors.*

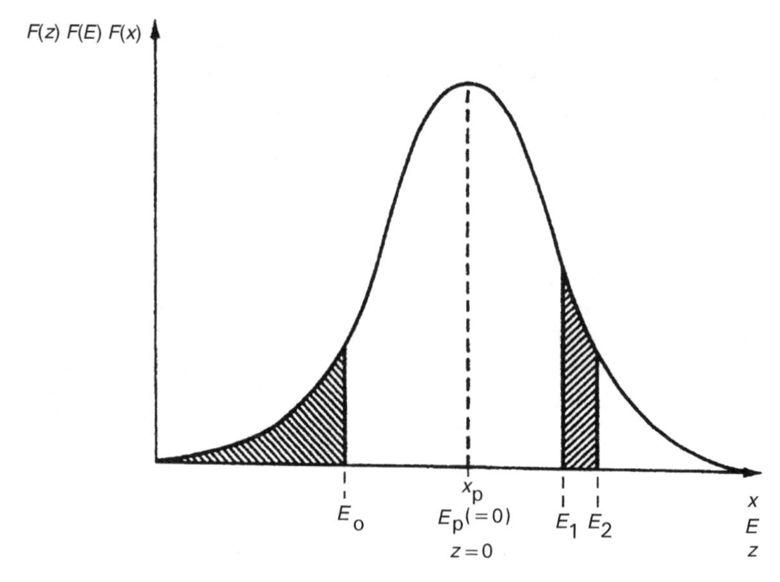

Figure 4.2 Frequency distribution curve of measurements and errors.

measurement values. To draw the histogram of errors, the mean of the measurement data values is calculated first, and then the error in each measurement in terms of its deviation from this mean value is calculated. Error bands of equal width are then defined and a histogram of errors drawn, according to the number of error values falling within each band. The histogram retains exactly the same shape, with symmetry about the line of zero error.

The frequency distribution curve of the errors is drawn in a similar way. The ordinate of the graph in Figure 4.2 then represents the frequency of occurrence of each error level, $F(E)$, and the abscissa is the error magnitude, E. The error magnitude E_p corresponding with the peak of the frequency distribution curve is the value of error which has the greatest probability. If the errors are entirely random in nature, then the value of E_p will equal zero. Any nonzero value of E_p indicates systematic errors in the data, in the form of a bias that is often removable by recalibration.

If the height of the frequency distribution of errors curve is normalised such that the area under it is unity, then the curve in this form is known as a *probability curve*, and the height $F(E)$ at any particular error magnitude E is known as the *probability density function* (p.d.f.). The condition that the area under the curve is unity can be expressed mathematically as:

$$\int_{-\infty}^{\infty} F(E)\mathrm{d}E = 1$$

The probability that the error in any one particular measurement lies between two levels E_1 and E_2 can be calculated by measuring the area under the curve contained between two vertical lines drawn through E_1 and E_2, as shown by the right-hand hatched area in Figure 4.2. This can be expressed mathematically as:

$$P(E_1 \leq E \leq E_2) = \int_{E_1}^{E_2} F(E)\mathrm{d}E \tag{4.4}$$

Expression (4.4) is often known as the *error function.*

Of particular importance for assessing the maximum error likely in any one measurement is the *cumulative distribution function* (c.d.f.). This is defined as the probability of observing a value less than or equal to E_0, and is expressed mathematically as:

$$P(E \leq E_0) = \int_{-\infty}^{E_0} F(E)\mathrm{d}E \tag{4.5}$$

Thus the c.d.f. is the area under the curve to the left of a vertical line drawn through E_0, as shown by the left-hand hatched area on Figure 4.2.

Three special types of frequency distribution known as the Gaussian, Binomial and Poisson distributions exist, and these are very important because most data sets show a fairly close fit to one of them. The distribution that is of relevance to data sets containing random measurement errors is the Gaussian type.

Gaussian distribution

A Gaussian curve (also known as normal distribution or bell-shaped distribution) is defined as a normalised frequency distribution where the frequency and magnitude of quantities are related by the expression:

$$F(x) = \frac{1}{\sigma\sqrt{2\pi}} \mathrm{e}^{\left[-(x-m)^2/2\sigma^2\right]} \tag{4.6}$$

where m is the mean value of the measurement set x, and the other quantities are as defined before. It is only applicable to data that have only random errors, i.e. where no systematic errors exist. Equation (4.5) is particularly useful for analysing a Gaussian set of measurements and predicting how many measurements lie within some particular defined range.

If the measurement errors E are calculated for all measurements such that $E = x - m$, then the curve of error frequency $F(E)$ plotted against error magnitude E is a Gaussian curve known as the error frequency distribution curve. The mathematical relationship between $F(E)$ and E can then be derived by modifying equation (4.5) to give:

$$F(E) = \frac{1}{\sigma\sqrt{2\pi}} \mathrm{e}^{-E^2/2\sigma^2} \tag{4.7}$$

Most measurement data sets, such as the values of mass in Table 4.1, fit to a Gaussian distribution curve, because, if errors are truly random, small deviations from the

Table 4.2 *Error function table*

Z	0.00	0.01	0.02	0.03	0.04	0.05	0.06	0.07	0.08	0.09
					$F(z)$					
0.0	0.5000	0.5040	0.5080	0.5120	0.5160	0.5199	0.5239	0.5279	0.5319	0.5359
0.1	0.5398	0.5438	0.5478	0.5517	0.5557	0.5596	0.5636	0.5675	0.5714	0.5753
0.2	0.5793	0.5832	0.5871	0.5910	0.5948	0.5987	0.6026	0.6064	0.6103	0.6141
0.3	0.6179	0.6217	0.6255	0.6293	0.6331	0.6368	0.6406	0.6443	0.6480	0.6517
0.4	0.6554	0.6591	0.6628	0.6664	0.6700	0.6736	0.6772	0.6808	0.6844	0.6879
0.5	0.6915	0.6950	0.6985	0.7019	0.7054	0.7088	0.7123	0.7157	0.7190	0.7224
0.6	0.7257	0.7291	0.7324	0.7357	0.7389	0.7422	0.7454	0.7486	0.7517	0.7549
0.7	0.7580	0.7611	0.7642	0.7673	0.7703	0.7734	0.7764	0.7793	0.7823	0.7852
0.8	0.7881	0.7910	0.7939	0.7967	0.7995	0.8023	0.8051	0.8078	0.8106	0.8133
0.9	0.8159	0.8186	0.8212	0.8238	0.8264	0.8289	0.8315	0.8340	0.8365	0.8389
1.0	0.8413	0.8438	0.8461	0.8485	0.8508	0.8531	0.8554	0.8577	0.8599	0.8621
1.1	0.8643	0.8665	0.8686	0.8708	0.8729	0.8749	0.8770	0.8790	0.8810	0.8830
1.2	0.8849	0.8869	0.8888	0.8906	0.8925	0.8943	0.8962	0.8980	0.8997	0.9015
1.3	0.9032	0.9049	0.9066	0.9082	0.9099	0.9115	0.9131	0.9147	0.9162	0.9177
1.4	0.9192	0.9207	0.9222	0.9236	0.9251	0.9265	0.9279	0.9292	0.9306	0.9319
1.5	0.9332	0.9345	0.9357	0.9370	0.9382	0.9394	0.9406	0.9418	0.9429	0.9441
1.6	0.9452	0.9463	0.9474	0.9484	0.9495	0.9505	0.9515	0.9525	0.9535	0.9545
1.7	0.9554	0.9564	0.9573	0.9582	0.9591	0.9599	0.9608	0.9616	0.9625	0.9633
1.8	0.9641	0.9648	0.9656	0.9664	0.9671	0.9678	0.9686	0.9693	0.9699	0.9706
1.9	0.9713	0.9719	0.9726	0.9732	0.9738	0.9744	0.9750	0.9756	0.9761	0.9767
2.0	0.9772	0.9778	0.9783	0.9788	0.9793	0.9798	0.9803	0.9808	0.9812	0.9817

2.1	0.9821	0.9826	0.9830	0.9834	0.9838	0.9842	0.9846	0.9850	0.9854	0.9857
2.2	0.9861	0.9864	0.9868	0.9871	0.9875	0.9878	0.9881	0.9884	0.9887	0.9890
2.3	0.9893	0.9896	0.9898	0.9901	0.9904	0.9906	0.9909	0.9911	0.9913	0.9916
2.4	0.9918	0.9920	0.9922	0.9924	0.9927	0.9929	0.9930	0.9932	0.9934	0.9936
2.5	0.9938	0.9940	0.9941	0.9943	0.9945	0.9946	0.9948	0.9949	0.9951	0.9952
2.6	0.9953	0.9955	0.9956	0.9957	0.9959	0.9960	0.9961	0.9962	0.9963	0.9964
2.7	0.9965	0.9966	0.9967	0.9968	0.9969	0.9970	0.9971	0.9972	0.9973	0.9974
2.8	0.9974	0.9975	0.9976	0.9977	0.9977	0.9978	0.9979	0.9979	0.9980	0.9981
2.9	0.9981	0.9982	0.9982	0.9983	0.9984	0.9984	0.9985	0.9985	0.9986	0.9986
3.0	0.9986	0.9987	0.9987	0.9988	0.9988	0.9989	0.9989	0.9989	0.9990	0.9990
3.1	0.9990	0.9991	0.9991	0.9991	0.9992	0.9992	0.9992	0.9992	0.9993	0.9993
3.2	0.9993	0.9993	0.9994	0.9994	0.9994	0.9994	0.9994	0.9995	0.9995	0.9995
3.3	0.9995	0.9995	0.9995	0.9996	0.9996	0.9996	0.9996	0.9996	0.9996	0.9996
3.4	0.9997	0.9997	0.9997	0.9997	0.9997	0.9997	0.9997	0.9997	0.9997	0.9998
3.5	0.9998	0.9998	0.9998	0.9998	0.9998	0.9998	0.9998	0.9998	0.9998	0.9998
3.6	0.9998	0.9998	0.9998	0.9999	0.9999	0.9999	0.9999	0.9999	0.9999	0.9999

From Morris (1997) *Measurement and Calibration Requirements*, © John Wiley & Sons, Ltd. Reproduced with permission.

mean value occur much more often than large deviations, i.e. the number of small errors is much larger than the number of large ones.

The Gaussian distribution curve is symmetrical about the line through the mean of the measurement values, which means that positive errors away from the mean value occur in equal quantities to negative errors in any data set containing measurements subject to random error. If the standard deviation is used as a unit of error, the curve can be used to determine what probability there is that the error in any particular measurement in a data set is greater than a certain value. By substituting expression (4.6) for $F(E)$ into the probability equation (4.3), the probability that the error lies in a band between error levels E_1 and E_2 can be expressed as:

$$P(E_1 \le E \le E_2) = \int_{E_1}^{E_2} \frac{1}{\sigma\sqrt{2\pi}} e^{\left(-E^2/2\sigma^2\right)} dE \tag{4.8}$$

Equation (4.7) can be simplified by making the substitution:

$$z = E/\sigma \tag{4.9}$$

(The quantity z is sometimes known as the *standard normal deviate.*) Then:

$$P(E_1 \le E \le E_2) = P(z_1 \le z \le z_2) = \int_{z_1}^{z_2} \frac{1}{\sigma\sqrt{2\pi}} e^{\left(-z^2/2\right)} dz = F(z) \tag{4.10}$$

Conversion to the parameter z merely applies a scaling operation, and the Gaussian distribution curve retains exactly the same shape but the axes of Figure 4.2 become $F(z)$ and z. However, even after carrying out this simplification, equation (4.9) still cannot be evaluated by the use of standard integrals, and numerical integration has to be used instead. To simplify the burden involved in this, values of $F(z)$, which is known as the *error function*, are normally read from precomputed tables of the integral $F(z)$ for various values of z. Table 4.2 shows such a tabulation of values and this is usually known as an *error function table.*

Error function tables

In Table 4.2, $F(z)$ represents the proportion of data values that are less than or equal to z and is equal to the area under the normalised probability curve to the left of z. Study of the table will show that $F(z) = 0.5$ for $z = 0$. This shows that, as expected, the number of data values ≤ 0 is 50% of the total. This must be so if the data only have random errors.

Use of error function tables

It will be observed that Table 4.2, in common with most published error function tables, only gives $F(z)$ for positive values of z. For negative values of z, the following relationship can be used because the frequency distribution curve is normalised:

$$F(-z) = 1 - F(z) \tag{4.11}$$

($F(-z)$ is the area under the curve to the left of $(-z)$, i.e. it represents the proportion of data values $\leq -z$.)

Example 4.2

How many measurements in a data set subject to random errors lie outside boundaries of $+\sigma$ and $-\sigma$, i.e. how many measurements have an error $<|\sigma|$?

Solution

The required number is represented by the sum of the two shaded areas in Figure 4.3. This can be expressed mathematically as:

$P(E < -\sigma \text{ or } E > +\sigma) = P(E < -\sigma) + P(E > +\sigma)$

For $E = -\sigma$, $z = -1.0$ (from equation 4.9), and using the error function table:

$P(E < -\sigma) = F(-1) = 1 - F(1) = 1 - 0.8413 = 0.1587$

Similarly, for $E = +\sigma$, $z = +1.0$, and the error function table gives:

$P(E > +\sigma) = 1 - P(E < +\sigma) = 1 - F(1) = 1 - 0.8413 = 0.1587$

(This last step is valid because the frequency distribution curve is normalised such that the total area under it is unity.)

Thus, $P(E < -\sigma) + P(E > +\sigma) = 0.1587 + 0.1587 = 0.3174 \approx 32\%$

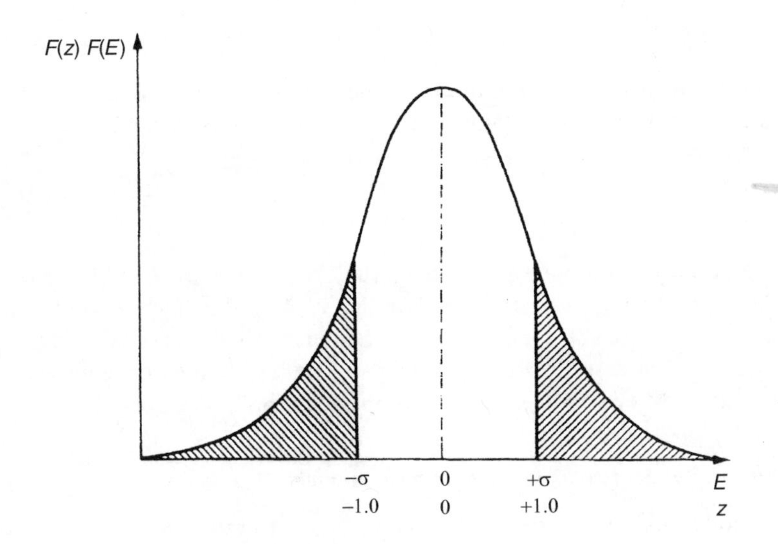

Figure 4.3 *Measurements outside ±σ boundaries.*

Therefore, 32% of the measurements lie outside the $\pm\sigma$ boundaries, i.e. 32% of the measurements have an error greater than $|\sigma|$. It follows that 68% of the measurements lie inside the boundaries of $\pm\sigma$.

The above analysis shows that, for Gaussian-distributed data values, 68% of the measurements have errors that lie within the bounds of $\pm\sigma$. Similar analysis shows that boundaries of $\pm 2\sigma$ contain 95.4% of data points, and extending the boundaries to $\pm 3\sigma$ encompasses 99.7% of data points. The probability of any data point lying outside particular error boundaries can therefore be expressed by the following table:

Deviation boundaries	Percentage of data points within boundary	Probability of any particular data point being outside boundary
$\pm\sigma$	68.0	32.0%
$\pm 2\sigma$	95.4	4.6%
$\pm 3\sigma$	99.7	0.3%

Standard error of the mean

The foregoing analysis is only strictly true for measurement sets containing infinite populations. It is not of course possible to obtain an infinite number of data values, and some error must therefore be expected in the calculated mean value of the practical, finite data set available. If several subsets are taken from an infinite data population, then, by the central limit theorem, the means of the subsets will form a Gaussian distribution about the mean of the infinite data set. The error in the mean of a finite data set is usually expressed as the standard error of the mean, α, which is calculated as:

$$\alpha = \sigma/\sqrt{n} \quad (4.12)$$

This tends towards zero as the number of measurements in the data set is expanded towards infinity. The value obtained from a set of n measurements, $x_1, x_2, \cdots x_n$ is then expressed as:

$$x = x_{\text{mean}} \pm \alpha \quad (4.13)$$

For the data set of mass measurements in Table 4.1, $n = 19$, $\sigma = 0.318$ and $\alpha = 0.073$. The mass can therefore be expressed as 81.18 ± 0.07 g (68% confidence limit). However, it is more usual to express measurements with 95% confidence limits ($\pm 2\sigma$ boundaries). In this case, $2\sigma = 0.636$, $2\alpha = 0.146$ and the value of the mass can therefore be expressed as 81.18 ± 0.15 g (95% confidence limits).

Estimation of random error in a single measurement

In many situations where measurements are subject to random errors, it is not practical to take repeated measurements and find the average value. Also, the averaging process becomes invalid if the measured quantity does not remain at a constant value, as is usually the case when process variables are being measured. Thus, if only one measurement can be made, some means of estimating the likely magnitude of error in it is required. The normal approach to this is to calculate the error within 95% confidence limits, i.e. to calculate the value of the deviation D such that 95% of the area under the probability curve lies within limits of $\pm D$. These limits correspond to a deviation of $\pm 1.96\sigma$. Thus, it is necessary to maintain the measured quantity at a constant value whilst a number of measurements are taken in order to create a reference measurement set from which σ can be calculated. Subsequently, the maximum likely deviation in a single measurement can be expressed as: deviation = $\pm 1.96\sigma$. However, this only expresses the likely maximum deviation of the measurement from the calculated mean of the reference measurement set, which is not the true value as observed earlier. Thus, the calculated value for the standard error of the mean has to be added to the likely maximum deviation value. Thus, the maximum likely error in a single measurement can be expressed as:

$$\text{Error} = \pm(1.96\sigma + \alpha) \qquad (4.14)$$

Example 4.3

Suppose that a standard mass is measured 30 times with the same instrument to create a reference data set, and that the calculated values of σ and α are $\sigma = 0.43$ and $\alpha = 0.08$. If the instrument is then used to measure an unknown mass and the reading is 105.6 kg, how should the mass value be expressed?

Solution

Using (4.14), $1.96\sigma + \alpha = 0.92$. The mass value should therefore be expressed as: 105.6 ± 0.9 kg.

Before leaving this matter, it must be emphasised that the maximum error specified for a measurement is only specified for the confidence limits defined. Thus, if the maximum error is specified as ±1% with 95% confidence limits, this means that there is still one chance in 20 that the error will exceed ±1%.

Rogue data points

In a set of measurements subject to random error, measurements with a very large error sometimes occur at random and unpredictable times, where the magnitude of the error is much larger than could reasonably be attributed to the expected random variations in measurement value. Sources of such abnormal error include sudden transient voltage surges on the mains power supply and incorrect recording of data (e.g. writing down 146.1 when the actual measured value was 164.1). It is accepted practice in such cases to discard these rogue measurements, and a threshold level of a $\pm 3\sigma$ deviation is often used to determine what should be discarded. It is extremely rare for measurement errors to exceed $\pm 3\sigma$ limits when only normal random effects are affecting the measured value.

4.2 Systematic Errors

Systematic errors in the output of many instruments are due to factors inherent in the manufacture of the instrument arising out of tolerances in the components of the instrument. They can also arise due to wear in instrument components over a period of time. In other cases, systematic errors are introduced by either the effect of environmental disturbances or through the disturbance of the measured system by the act of measurement. These various sources of systematic error, and ways in which the magnitude of the errors can be reduced, are discussed below.

4.2.1 System disturbance due to measurement

Disturbance of the measured system by the act of measurement is one source of systematic error. If we were to start with a beaker of hot water and wished to measure its temperature with a mercury-in-glass thermometer, then we should take the thermometer, which would be initially at room temperature, and plunge it into the water. In so doing, we would be introducing the relatively cold mass of the thermometer into the hot water and a heat transfer would take place between the water and the thermometer. This heat transfer would lower the temperature of the water. Whilst in this case the reduction in temperature would be so small as to be undetectable by the limited measurement resolution of such a thermometer, the effect is finite and clearly establishes the principle that, in nearly all measurement situations, the process of measurement disturbs the system and alters the values of the physical quantities being measured.

The magnitude of the disturbance varies from one measurement system to the next and is affected particularly by the type of instrument used for measurement. Ways of minimising disturbance of measured systems are an important consideration in instrument design. A prerequisite for this, however, is a full understanding of the mechanisms of system disturbance.

Measurements in electric circuits are particularly prone to errors induced through the loading effect on the circuit when instruments are applied to make voltage and current measurements. For most electrical networks, circuit analysis methods such as Thevenin's theorem are needed to analyse such loading effects. However, for the simple circuit shown in Figure 4.4, the analysis is fairly easy.

In this circuit, the voltage across resistor R_2 is to be measured by a voltmeter with resistance R_m. Here, R_m acts as a shunt resistance across R_2, decreasing the resistance between points A and B and so disturbing the circuit. The voltage E_m measured by the meter is therefore not the value of the voltage E_o that existed prior to measurement. The extent of the disturbance can be assessed by calculating the open-circuit voltage E_o and comparing it with E_m.

Starting with the unloaded circuit in Figure 4.4, the current I is given by Ohm's law as:

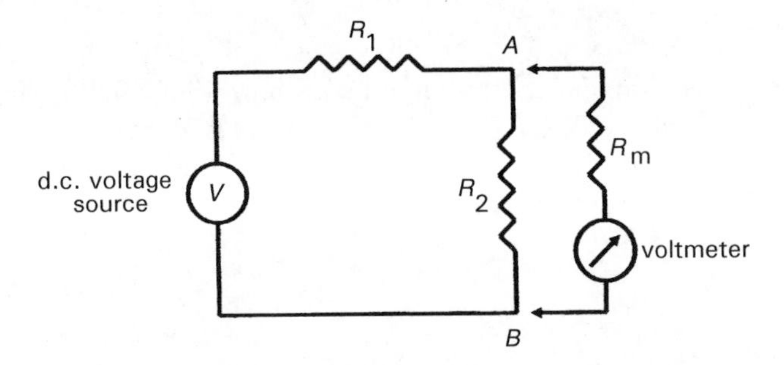

Figure 4.4 *Disturbance of an electrical circuit by loading.*

$$I = V/(R_1 + R_2)$$

Again, using Ohm's law, voltage across AB is then given by:

$$E_o = IR_2 = VR_2/(R_1 + R_2) \tag{4.15}$$

With the voltmeter added to the circuit, there are now two resistances in parallel across AB, R_2 and R_m, and the expression for the resistance across AB can be written as:

$$R_{AB} = R_2 R_m/(R_2 + R_m)$$

Then, replacing R_2 by R_{AB} in equation (4.15) above, the voltage E_m measured by the meter is given by:

$$E_m = \frac{VR_{AB}}{R_1 + R_{AB}} = \frac{VR_2 R_m}{(R_2 + R_m)} \times \frac{1}{R_1 + (R_2 R_m/R_2 + R_m)} = \frac{VR_2 R_m}{R_1 R_2 + R_m(R_1 + R_2)} \tag{4.16}$$

Thus, from equations (4.15) and (4.16):

$$\frac{E_m}{E_o} = \frac{VR_2 R_m}{[R_1 R_2 + R_m(R_1 + R_2)]} \times \frac{(R_1 + R_2)}{VR_2} = \frac{R_m(R_1 + R_2)}{R_1 R_2 + R_m(R_1 + R_2)} \tag{4.17}$$

If R_m is very large compared with R_1 and R_2, then $R_m (R_1 + R_2) >> R_1 R_2$ and then the denominator of equation (4.17) approaches $R_m (R_1 + R_2)$ and E_m/E_o approaches unity, and thus E_m approaches E_o. It is thus obvious that as R_m gets larger, the ratio E_m/E_o gets closer to unity, showing that the design strategy should be to make R_m as large as possible in order to minimise the disturbance of the measured system (note that we did not calculate the value of E_o, since this was not required in quantifying the effect of R_m).

Example 4.4

Suppose that the components of the circuit shown in Figure 4.4 have the following values: $R_1 = 500\,\Omega$; $R_2 = 500\,\Omega$. The voltage across AB is measured by a voltmeter whose internal resistance is $4750\,\Omega$. What is the measurement error caused by the resistance of the measuring instrument?

Solution

Proceeding by substituting the given component values into equation (4.17), we obtain: $\frac{E_m}{E_o} = \frac{4750 \times 1000}{(25 \times 10^4 + 4750 \times 1000)} = 0.95$. Thus the error in the measured value is 5%.

4.2.2 Environmental inputs to measurement systems

Environmental inputs are systematic measurement errors induced by changes in various environmental parameters, such as the ambient temperature, pressure and humidity. In some texts, environmental errors are given the alternative name *modifying inputs*, because they modify the output of a measurement system even if the real input (the measured quantity of interest) remains constant. These environmental factors do not affect every instrument to the same extent, and, indeed, some instruments are affected very little.

The magnitude of environmental inputs is quantified by the sensitivity drift and zero drift parameters defined previously in Chapter 3, both of which are generally included in the published specifications for an instrument. Without proper analysis, it is impossible to establish how much of an instrument's output is due to the real input, and how much is due to one or more environmental inputs. This is illustrated by the following example. Suppose that we have a small closed box weighing 0.1 kg when empty, which we think contains either a rat or a mouse. If we put the box on to bathroom scales and observe a reading of 1.0 kg, this does not immediately tell us what is in the box, because the reading may be due to one of three things:

(1) a 0.9 kg rat in the box (real input);
(2) an empty box with a 0.9 kg bias on the scales due to a temperature change (environmental input);
(3) A 0.4 kg mouse in the box together with a 0.5 kg bias (real plus environmental inputs).

Thus, the magnitude of any environmental input must be measured before the value of the measured quantity, which is the real input, can be determined from the output reading of an instrument.

In any general measurement situation, it is very difficult to avoid environmental inputs, because it is either impractical or impossible to control the environmental conditions surrounding the measurement system. System designers are therefore charged with the task of either reducing the susceptibility of measuring instruments to environmental inputs, or, alternatively, quantifying the effect of environmental inputs and correcting for them in the instrument output reading. The techniques used to deal with environmental inputs and minimise their effect on the final output measurement follow a number of routes, as discussed below.

Careful instrument design

Careful instrument design is the most useful weapon in the battle against environmental inputs, by reducing the sensitivity of an instrument to environmental inputs to as low a level as possible. In the design of strain gauges for instance, the element should be constructed from a material whose resistance has a very low temperature coefficient (i.e. the variation of the resistance with temperature is very small). However, for many instruments, it is not possible to reduce their sensitivity to environmental inputs to a satisfactory level by simple design adjustments, and other procedures have to be followed.

Method of opposing inputs

The method of opposing inputs compensates for the effect of an environmental input in a measurement system by introducing an equal and opposite environmental input that cancels it out. One example of how this technique is applied is the voltmeter shown in Figure 4.5. This consists of a coil suspended in a fixed magnetic field produced by a permanent magnet. When an unknown voltage is applied to the coil, the magnetic field due to the current interacts with the fixed field and causes the coil (and a pointer attached to the coil) to turn. If the coil resistance is sensitive to temperature, then any environmental input to the system in the form of a temperature change will alter the value of the coil current for a given applied voltage, and so alter the pointer output reading. Compensation for this is made by introducing a compensat-

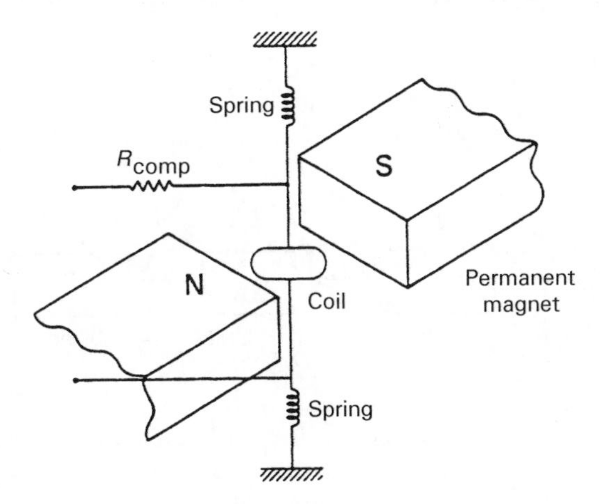

Figure 4.5 *Analogue voltmeter. From Morris (1997)* Measurement and Calibration Requirements, © *John Wiley & Sons, Ltd. Reproduced with permission.*

ing resistance R_{comp} into the circuit, where R_{comp} has a temperature coefficient which is equal in magnitude but opposite in sign to that of the coil.

High-gain feedback

The benefit of adding high-gain feedback to measurement systems is illustrated by considering the case of the voltage-measuring instrument whose block diagram is shown in Figure 4.6. In this system, the unknown voltage E_i is applied to a coil of torque constant K_c, and the torque induced turns a pointer against the restraining action of a spring with spring constant K_s. The effect of environmental inputs on the torque and spring constants is represented by variables D_c and D_s.

In the absence of environmental inputs, the displacement of the pointer X_o is given by:

$$X_o = K_c K_s E_i$$

In the presence of environmental inputs, both K_c and K_s change, and the relationship between X_o and E_i can be affected greatly. It therefore becomes difficult or impossible to calculate E_i from the measured value of X_o.

Now consider what happens if the system is converted into a high-gain, closed-loop one, as shown in Figure 4.7, by adding an amplifier of gain constant K_a and a feedback device with gain constant K_f. Assume also that the effect of environmental inputs on the values of K_a and K_f are represented by D_a and D_f. The feedback device

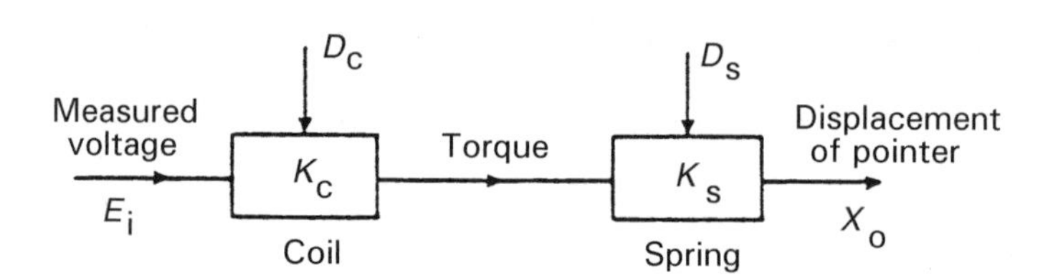

Figure 4.6 Block diagram of voltage-measuring instrument.

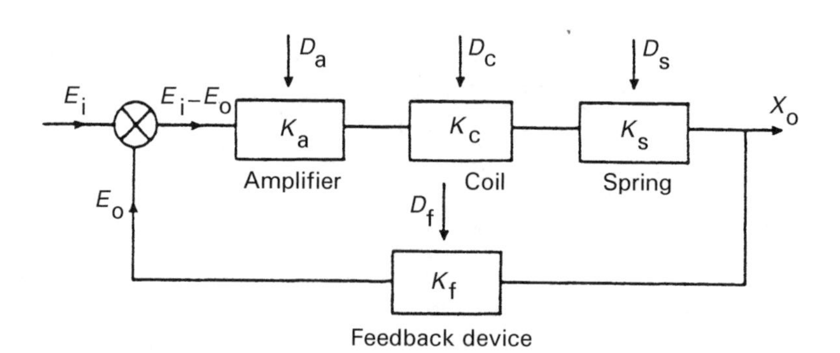

Figure 4.7 Conversion of system to high-gain, closed-loop form.

feeds back a voltage E_o proportional to the pointer displacement X_o. This is compared with the unknown voltage E_i by a comparator, and the error is amplified.

Writing down the equations of the system, we have:

$$E_o = K_f X_o \; ; \; X_o = (E_i - E_o) K_a K_c K_s = (E_i - K_f X_o) K_a K_c K_s$$

Thus:

$$E_i K_a K_c K_s = (1 + K_f K_a K_c K_s) X_o$$

that is,

$$X_o = \frac{K_a K_c K_s}{1 + K_f K_a K_c K_s} E_i \tag{4.18}$$

Because K_a is very large (it is a high-gain amplifier), the product $K_f \times K_a \times K_c \times K_s$ is very much greater than unity, and equation (4.18) can be approximated to:

$$X_o = E_i / K_f$$

This is a highly important result, because we have reduced the relationship between X_o and E_i to one that involves only K_f. The sensitivity of the gain constants K_a, K_c and K_s to the environmental inputs D_a, D_c and D_s has thereby been rendered irrelevant, and we only have to be concerned with one environmental input D_f. Conveniently, it is usually an easy matter to design a feedback device that is insensitive to environmental inputs: this is much easier than trying to make a coil or spring insensitive. Thus, high-gain feedback techniques are often a very effective way of reducing a measurement system's sensitivity to environmental inputs. However, one potential problem that must be mentioned is that there is a possibility that high-gain feedback will cause instability in the system. Therefore, any application of this method must include careful stability analysis of the system.

Signal filtering

One frequent problem in measurement systems is corruption of the output reading by periodic noise, often at a frequency of 50 Hz caused by pick-up through the close proximity of the measurement system to apparatus or current-carrying cables operating on a mains supply. Periodic noise corruption at higher frequencies is also often introduced by mechanical oscillation or vibration within some component of a measurement system. The amplitude of all such noise components can be substantially attenuated by the inclusion of filtering of an appropriate form in the system, as discussed at greater length in Chapter 6. Band-stop filters can be especially useful where the corruption is of one particular known frequency, or, more generally, low-pass filters are employed to attenuate all noise in the frequency range of 50 Hz and above.

Measurement systems with a low-level output, such as a bridge circuit measuring a strain-gauge resistance, are particularly prone to noise, and Figure 4.8(a) shows the typical corruption of a bridge output by 50 Hz pick-up. The beneficial effect of

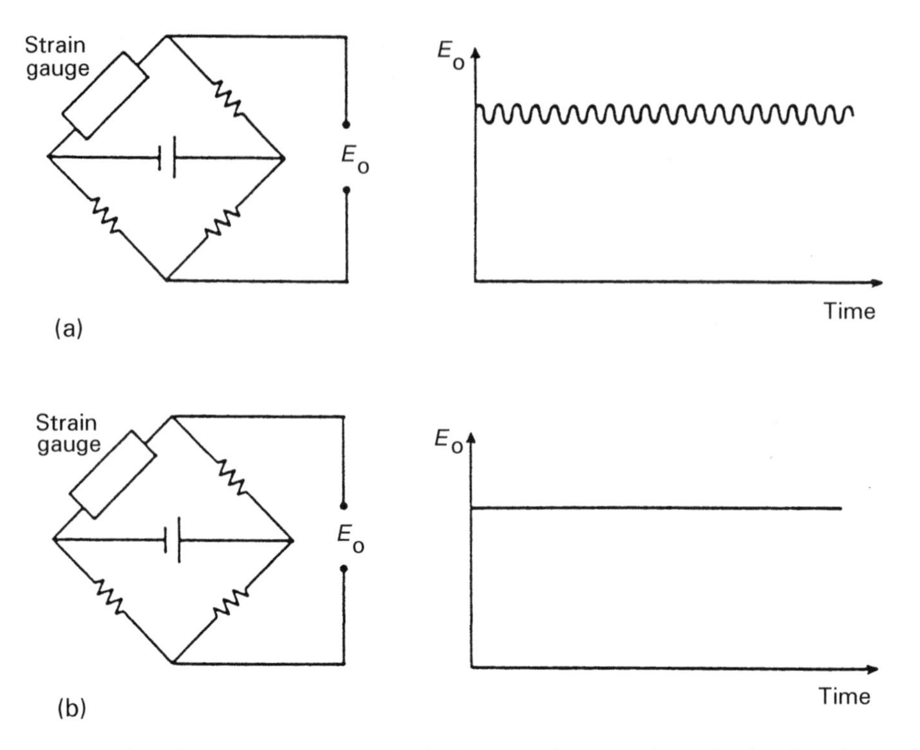

Figure 4.8 *Effect of noise: (a) corruption of a bridge output by noise; (b) result of applying low-pass filter. From Morris (1997)* Measurement and Calibration Requirements, *© John Wiley & Sons, Ltd. Reproduced with permission.*

putting a simple passive RC low-pass filter across the output is shown in Figure 4.8(b).

Compensation of output reading

If other techniques do not reduce sufficiently the errors due to environmental inputs, it may be necessary to calculate the error and compensate the output readings accordingly. This is a tedious procedure if done by hand, and it can only be applied if the following preconditions are satisfied.

(1) The physical mechanism by which a measurement sensor is affected by ambient condition changes must be fully understood, and all physical quantities that affect the sensor output must be identified.
(2) The effect of each ambient variable on the output characteristic of the measurement sensor must be quantified.
(3) Suitable secondary sensors for monitoring the value of all relevant ambient variables must be installed.

Condition (1) means that the thermal expansion/contraction of all elements within a measuring device must be considered, in order to evaluate how it will respond to ambient temperature changes. Similarly, the response, if any, to changes in ambient pressure, humidity and gravitational force must be examined. Quantification of the effect of each ambient variable on the characteristics of the measurement sensor is then necessary, as stated in condition (2). The analytical quantification of the ambient condition changes from the purely theoretical consideration of the construction of a measuring device is usually extremely complex, and so is normally avoided. Instead, the effect is quantified empirically by laboratory tests, in which the output characteristic is observed as the ambient environmental conditions are changed in a controlled manner. However, where such compensation of output readings is necessary, current practice is normally to use an intelligent instrument (see Section 4.3), since these calculate and apply the necessary compensation automatically.

4.2.3 Other sources of systematic error

Wear in instrument components

Systematic errors can frequently develop over a period of time, because of wear in instrument components. Recalibration often provides a full solution to this problem.

Connecting leads

In connecting together the components of a measurement system, a common source of error is the failure to take proper account of the resistance of connecting leads, or pipes in the case of pneumatically or hydraulically actuated measurement systems. In typical applications of a resistance thermometer, for instance, it is common to find the thermometer separated from other parts of the measurement system by perhaps 30 metres. The resistance of such a length of 7/0.0076 copper wire is $2.5\,\Omega$ and there is a further complication that such wire has a temperature coefficient of $1\,\mathrm{m}\Omega/^{\circ}\mathrm{C}$.

Therefore, careful consideration needs to be given to the choice of connecting leads. Not only should they be of adequate cross-section so that their resistance is minimised, but they should be adequately screened if they are thought likely to be subject to electrical or magnetic fields that could otherwise cause induced noise. Also, the routing of cables needs careful planning to minimise exposure to electrical or magnetic fields.

Thermal e.m.f.'s (thermally generated electromotive forces)

Whenever metals of two different types are connected together, a thermal e.m.f. is generated, which varies according to the temperature of the joint. This is known as the *thermoelectric effect* and is the physical principle on which temperature-measuring thermocouples operate (see Chapter 13). Such thermal e.m.f.'s are only a few millivolts in magnitude, and so the effect is only significant when the typical voltage output signals of a measurement system are of a similar low magnitude.

One such situation is where one e.m.f. measuring instrument is used to monitor the output of several thermocouples measuring the temperatures at different points in a process control system. This requires a means of automatically switching the output of each thermocouple to the measuring instrument in turn. Nickel–iron reed-relays with copper connecting leads are commonly used to provide this switching function. This introduces a thermocouple effect of magnitude 40 μV/°C between the reed-relay and the copper connecting leads. There is no problem if both ends of the reed-relay are at the same temperature because then the thermal e.m.f.'s will be equal and opposite and so cancel out. However, there are several recorded instances where, because of lack of awareness of the problem, poor design has resulted in the two ends of a reed-relay being at different temperatures and causing a net thermal e.m.f. The serious error that this introduces is clear. For a temperature difference between the two ends of only 2 °C, the thermal e.m.f. is 80 μV, which is very large compared with a typical thermocouple output level of 400 μV.

Another example of the difficulties that thermal e.m.f.'s can create becomes apparent in considering the following problem which was reported in a current-measuring system. This system had been designed such that the current in a particular part of a circuit was calculated by applying it to an accurately calibrated wire-wound resistor of value 100 Ω and measuring the voltage drop across the resistor. In calibration of the system, a known current of 20 μA was applied to the resistor and a voltage of 2.20 mV was measured by an accurate high-impedance instrument. Simple application of Ohm's law reveals that such a voltage reading indicates a current value of 22 μA. What then was the explanation for this discrepancy? The answer once again is a thermal e.m.f. Because the designer was not aware of thermal e.m.f.'s, the circuit had been constructed such that one side of the standard resistor was close to a power transistor, creating a difference in temperature between the two ends of the resistor of 2 °C. The thermal e.m.f. associated with this was sufficient to account for the 10% measurement error found.

4.3 Error Reduction Using Intelligent Instruments

An intelligent instrument is distinguished from a dumb (nonintelligent) instrument by the inclusion of a microcomputer, and by the addition of one or more extra transducers at its input. These additions inevitably add significantly to the instrument's cost, but large reductions in measurement errors can often be achieved by using them.

The inputs to an intelligent instrument are data from a *primary transducer* and additional data from one or more *secondary transducers*. The primary transducer measures the magnitude of the main quantity of interest, whilst the secondary transducers measure the magnitude of environmental parameters. For instance, in an intelligent mass-measuring instrument, the primary transducer is usually a load cell, and additional secondary transducers are provided to measure environmental inputs such as the ambient temperature and atmospheric pressure.

The microcomputer performs preprogrammed signal-processing functions and data manipulation algorithms on the data from the primary transducer (the measured quantity of interest), using data read from the secondary transducer(s), and outputs the processed measurement from the primary transducer for presentation at the instrument output. The effect of this computerisation of the signal-processing function is an improvement in the quality of the instrument output measurements and a general simplification of the signal-processing task. Some examples of the signal processing which a microprocessor within an intelligent instrument can readily perform include correction of the instrument output for bias caused by environmental variations (e.g. temperature changes), and conversion to produce a linear output from a transducer whose characteristic is fundamentally nonlinear. A fuller discussion about the techniques of digital signal processing can be found in Chapter 5. However, as far as the user is concerned, an intelligent instrument behaves as a black box, and no knowledge of its internal mode of operation is required in normal measurement situations.

One example of the benefit that intelligence can bring to instruments is in volume flow rate measurement, where the flow rate is inferred by measuring the differential pressure across an orifice plate placed in a fluid-carrying pipe (see Chapter 13 for more details). The flow rate is proportional to the square root of the difference in pressure across the orifice plate. For a given flow rate, this relationship is affected both by the temperature and by the mean pressure in the pipe, and changes in the ambient value of either of these cause measurement errors. A typical intelligent flow rate measuring instrument contains three transducers: a primary one measuring the pressure difference across an orifice plate and secondary ones measuring absolute pressure and temperature. The instrument is programmed to correct the output of the primary differential pressure transducer according to the values measured by the secondary transducers, using appropriate physical laws that quantify the effect of ambient temperature and pressure changes on the fundamental relationship between flow and differential pressure. The instrument is also normally programmed to convert the square root relationship between flow and signal output into a direct one, making the output much easier to interpret. Typical inaccuracy levels of such intelligent flow measuring instruments are ±0.1%, compared with ±0.5% for their non-intelligent equivalents.

4.3.1 Reduction of random errors

If a measurement system is subject to random errors, then intelligent instruments can be programmed to take a succession of measurements of a quantity within a short space of time, and perform simple averaging or other statistical techniques on the readings before displaying an output measurement. This is valid for reducing any form of random error, including those due to human observation deficiencies, electrical noise or other random fluctuations.

As well as displaying an average value obtained from a number of measurements, intelligent instruments are often able to display other statistical parameters about the measurements taken, such as the standard deviation, variance and standard error of the mean. All of these quantities could of course be calculated manually, but the great advantage of using intelligent instruments is their much higher processing speed and the avoidance of the arithmetic errors that are liable to occur if humans perform these functions.

4.3.2 Reduction of systematic errors

The inclusion of intelligence in instruments can bring about a gross reduction in the magnitude of systematic errors. For instance, in the case of electrical circuits that are disturbed by the loading effect of the measuring instrument, an intelligent instrument can readily correct for measurement errors by applying equations such as (4.17) with the resistance of the measuring instrument inserted. Intelligent instruments are particularly effective in improving the accuracy of measurements subject to environmental inputs. To do this, they implement the procedures for compensation of the output reading that were suggested earlier in Section 4.2.2 as manual procedures. Once the ambient variables affecting a measurement sensor have been identified and their effect quantified, an intelligent instrument can be designed which includes secondary transducers to monitor the value of the ambient variables.

4.4 Total Measurement System Errors

A measurement system often consists of several separate components, each of which is subject to systematic and/or random errors. Mechanisms have now been presented for quantifying the errors arising from each of these sources, and therefore the total error at the output of each measurement system component can be calculated. What remains to be investigated is how the errors associated with each measurement system component combine together, so that a total error calculation can be made for the complete measurement system.

All four mathematical operations of addition, subtraction, multiplication and division may be performed on measurements derived from different instruments/transducers in a measurement system. Formulae to predict the most likely maximum error in each case are given below. Each formula is a statistical estimate of composite error, which is valid providing that the magnitude of each measurement component error is relatively small (e.g. 1 to 2%).

Error in added measurement components

Let S be the sum of the outputs y and z of two separate measurement system components. If the maximum errors in y and z are $\pm ay$ and $\pm bz$ respectively (with $a << 1$ and $b << 1$), the most probable maximum error in S can be expressed by a quantity e, where e is calculated in terms of the *absolute* errors as:[2]

$$e = \sqrt{(ay)^2 + (bz)^2} \tag{4.19}$$

Thus S can be expressed as:

$$S = (y + z) \pm e,$$

or in the alternative form:

$$S = (y + z)(1 \pm f) \tag{4.20}$$

where $f = e/(y + z)$

Example 4.5

A circuit requirement for a resistance of 550 Ω is satisfied by connecting together two resistors of nominal values 220 Ω and 330 Ω in series. If each resistor has a tolerance of ±2%, the error in the sum calculated according to equations (4.19) and (4.20) is given by:

$$e = \sqrt{(0.02 \times 220)^2 + (0.02 \times 330)^2} = 7.93; \quad f = 7.93/50 = 0.0144$$

Thus the total resistance S can be expressed as: $S = 550\,\Omega \pm 7.93\,\Omega$

or $S = 550(1 \pm 0.0144)\Omega$ i.e. $S = 550\,\Omega \pm 1.4\%$

Error in subtracted measurement components

Let S be the difference $(y - z)$ between the outputs y and z of two separate measurement system components. If the maximum errors in y and z are $\pm ay$ and $\pm bz$ respectively (with $a \ll 1$ and $b \ll 1$), the most probable maximum error in S can also be expressed by the quantity e in equation (4.19), where e is calculated in terms of the *absolute* errors. Thus, the difference S can be expressed as:

$$S = (y - z) \pm e \quad \text{or} \quad S = (y - z)(1 \pm f) \quad \text{where } f = e/(y - z) \tag{4.21}$$

Example 4.6

A fluid flow rate is calculated from the difference in pressure measured on both sides of an orifice plate. If the pressure measurements are 10.0 bar and 9.5 bar and the errors in the pressure measuring instruments are specified as ±0.1%, the values for e and f calculated using (5.21) are:

$$e = \sqrt{(0.001 \times 10)^2 + (0.001 \times 9.5)^2} = 0.0138; \quad f = 0.0138/0.5 = 0.0276$$

This example illustrates very poignantly the relatively large error that can arise when calculations are made based on the difference between two measurements.

Error in multiplied measurement components

Let P be the product of the outputs y and z of two separate measurement system components. If the maximum errors are $\pm ay$ in y and $\pm bz$ in z (with $a \ll 1$ and $b \ll 1$), the most probable maximum error e in P can be written as:

$$e = \sqrt{a^2 + b^2} \tag{4.22}$$

Note that in the case of multiplicative errors, e is calculated in terms of the *fractional* errors in y and z (as opposed to the absolute error values used in calculating additive and subtractive errors). Thus, the product P can be expressed as:

$$P = yz \pm e \quad \text{or} \quad P = yz(1 \pm f) \quad \text{where } f = e/yz \tag{4.23}$$

Example 4.7

If the power in a circuit is calculated from measurements of voltage and current in which the calculated maximum errors are respectively: ±1% and ±2%, then the maximum likely error in the calculated power value, calculated using (4.22) is $\pm\sqrt{0.01^2 + 0.02^2} = \pm 0.022$ or $\pm 2.2\%$.

Error in divided measurement components

Let Q be the result of dividing the output measurement y of one system component by the output measurement z of another system component. If the maximum errors are $\pm ay$ in y and $\pm bz$ in z (with $a \ll 1$ and $b \ll 1$), the most probable maximum error e in Q, calculated in terms of the fractional errors in y and z, is also given by (4.22). Thus, Q can be expressed as:

$$Q = y/z \pm e \quad \text{or} \quad Q = y/z(1 \pm f) \quad \text{where } f = ez/y \tag{4.24}$$

Example 4.8

If the density of a substance is calculated from measurements of its mass and volume, where the respective errors are ±2% and ±3%, then the maximum likely error in the density value using (4.22) is $\pm\sqrt{0.02^2 + 0.003^2} = \pm 0.036$ or $\pm 3.6\%$.

References

1. Chatfield, C., 1983, *Statistics for Technology* (Chapman and Hall, London).
2. ANSI/ASME Standards, 1985, *ASME Performance Test Codes, Supplement on Instruments and Apparatus, Part 1: Measurement Uncertainty* (American Society of Mechanical Engineers, New York).

5

Measurement Signal Conversion, Processing, Transmission and Recording

The two previous chapters have been concerned with discussing the proper choice, use and calibration of measuring sensors/instruments and then analysing and compensating for the error sources that may affect the measurement obtained. However, detecting the value of some EMS parameter by a sensor is only the first part of the measurement process. After this, the measurement data usually undergo a number of other processes. Firstly, the output from the measuring sensor may not be in a form that can be transmitted to a data recorder, and therefore it has to be converted into an alternative form. Secondly, further signal processing may be necessary to improve the quality of the measurement data in other ways. Thirdly, a suitable way of transmitting the measurements to a data recorder has to be chosen. Finally, a suitable data recorder must be selected.

5.1 Variable Conversion Elements

The most convenient form of output from a measurement sensor is a varying electrical voltage. Unfortunately, many sensors have other forms of output, such as variation in electrical resistance, inductance, capacitance and current, and variation in the phase or frequency of an a.c. signal. If the output is in one of these non-voltage forms, conversion to a varying voltage can be achieved by a variable conversion element, as discussed below.

ISO 14000 Environmental Management Standards: Engineering and Financial Aspects. Alan S. Morris.
 ISBN 0-470-85128-7

5.1.1 Resistance, inductance and capacitance changes

A large number of measuring devices have an output that has the form of either a resistance, inductance or capacitance change. The most common way of converting such changes into a voltage change is to use a bridge circuit. This can accurately detect very small changes in sensor output about a nominal value and comes in two forms, d.c. for converting resistance changes and a.c. for converting inductance and capacitance variations. A bridge circuit has the basic structure shown in Figure 5.1, with four 'arms' containing electrical components, an excitation voltage V_i applied across points AC and the output V_o measured by a voltmeter applied across points BD. The component marked as Z_u represents the impedance of the sensor whose output has to be converted. The other components Z_1, Z_2 and Z_3 are resistive, inductive or capacitive impedances according to the type of bridge, with Z_2 and Z_3 set equal and Z_1 chosen to be equal to the nominal value of Z_u. Both null and deflection types of bridge exist. Null-type bridges give better measurement accuracy, but are tedious to use. Hence, null types are normally chosen for calibration duties and deflection type for other purposes.

In a *null-type bridge*, Z_1 is a variable resistor, inductor or capacitor that is adjusted until the bridge is 'balanced' and the output voltage V_o is zero. At this balance point:

$$Z_u = \frac{Z_1 Z_3}{Z_2} \quad \text{and} \quad Z_u = Z_1 \quad \text{if} \quad Z_2 = Z_3 \tag{5.1}$$

In the alternative *deflection-type bridge*, Z_1 is a fixed-value component. Provided that the voltmeter measuring the output voltage has high impedance and draws

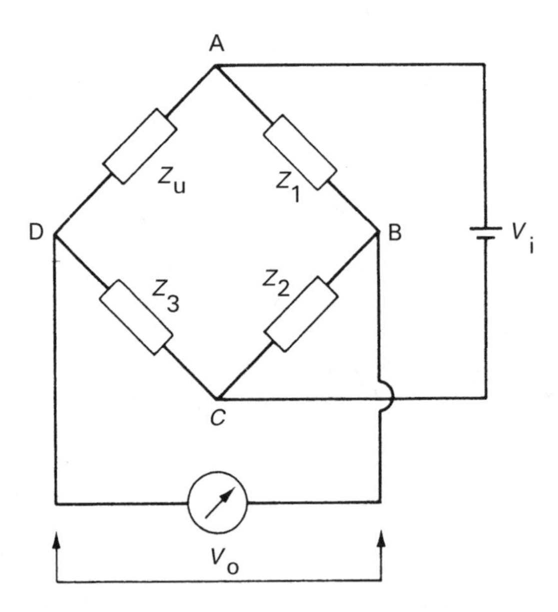

Figure 5.1 Structure of a bridge circuit.

negligible current from the system, the relationship between the output voltage and the value of the unknown component Z_u is:

$$V_o = V_i\left(\frac{Z_u}{Z_u + Z_3} - \frac{Z_1}{Z_1 + Z_2}\right) \tag{5.2}$$

This is a nonlinear relationship between V_o and Z_u but the linearity can be improved by making Z_2 and Z_3 substantially larger (e.g. 10 × larger) than Z_1 and the nominal value of Z_u. However, this improvement in linearity comes at the expense of decreasing the overall sensitivity of the measurement system (because, for a given change in Z_u, the change in V_o is reduced if Z_2 and Z_3 are increased).

D.c. bridge

In a d.c. bridge circuit, V_i is a d.c. voltage and Z_u is a sensor with varying resistance. Z_2 and Z_3 are fixed, equal resistances. Z_1 is either a variable resistor (usually a decade resistance box) or a fixed resistor according to whether the bridge is a null or deflection type. Replacing Z by R in equations (5.1) and (5.2), the equations for null and deflection type d.c. bridges are respectively:

$$R_u = R_1; \quad V_o = V_i\left(\frac{R_u}{R_u + R_3} - \frac{R_1}{R_1 + R_2}\right) \tag{5.3}$$

A.c. bridge

In an a.c. bridge circuit, V_i is an a.c. voltage and Z_u is a sensor with varying inductance or capacitance such that $Z_u = R_u + j\omega L_u$ or $Z_u = 1/j\omega C_u$, where L_u and C_u are the inductance or capacitance respectively of the sensor (NB it is impossible to fabricate an inductor that has no resistance, and so the impedance has both resistive R_u and inductive L_u terms). The Z_2 and Z_3 are normally fixed resistances. Substituting for Z in (5.1) and (5.2), the equations for null and deflection a.c. bridges measuring capacitance and inductance are therefore:

$$\begin{gathered} R_u + j\omega L_u = \frac{(R_1 + j\omega L_1)R_3}{R_2}; \quad \frac{1}{j\omega C_u} = \frac{(1/j\omega C_1)R_3}{R_2}; \\ V_o = V_i\left(\frac{R_u + j\omega L_u}{R_u + j\omega L_u + R_3} - \frac{R_1 + j\omega L_1}{R_1 + j\omega L_1 + R_2}\right); \\ V_o = V_i\left(\frac{1/j\omega C_u}{1/j\omega C_u + R_3} - \frac{1/j\omega C_1}{1/j\omega C_1 + R_2}\right) \end{gathered} \tag{5.4}$$

For deflection-type bridges, Z_1 is commonly just a fixed-value capacitor or a fixed-value inductor according to whether the sensor output is a varying capacitance or a varying inductance. For a null-type bridge measuring capacitance, Z_1 is usually a variable decade capacitance box. However, for a null-type bridge measuring inductance, a variable inductor is not commonly used since these are very expensive to manufacture. Instead, a special design of bridge such as a Maxwell bridge is used, that contains a variable capacitor instead of a variable inductor.

5.1.2 Electrical current changes

For measuring devices whose output is in the form of a varying current, the most common way of converting this change into a varying voltage is to place a resistor of known value in series with the current-carrying circuit and to measure the voltage drop across it. Any type of digital or analogue voltmeter can be used for the voltage measurement.

5.1.3 Frequency changes

Several measuring devices convert the measured quantity into a frequency change. Such frequency changes are commonly detected either by a digital counter-timer, a phase-locked loop or a Wien bridge.

The *digital counter-timer* is a very accurate and flexible electronic instrument that can measure frequency with uncertainty levels as low as 1 part in 10^8. It is capable of measuring all frequencies between d.c. and several gigahertz.

The *phase-locked loop* is a circuit consisting of a phase-sensitive detector, a voltage controlled oscillator (VCO), and amplifiers, connected in a closed loop system. The VCO becomes locked to the frequency of the input signal and gives a d.c. output voltage that is proportional to the input signal frequency.

The *Wien bridge* is a special form of a.c. bridge circuit that is used to measure frequencies in the audio range. The signal of unknown frequency f is applied as the excitation voltage V_i and, referring to Figure 5.1, the bridge components are: a resistor R_1 for Z_1, a resistor R_2 for Z_2, a variable resistor R_3 in parallel with a capacitor C_3 for Z_3, and a variable resistor R_4 in series with a capacitor C_4 in place of Z_u. At balance, the unknown frequency is calculated according to:

$$f = 1/(2\pi R_4 C_4)$$

The instrument is very accurate at audio frequencies, but errors due to losses in the capacitors and stray capacitance effects become significant at higher frequencies.

5.1.4 Phase changes

Several measuring devices have an output in the form of a phase change in an a.c. signal. These can be measured approximately by either an oscilloscope or a x–y plotter, but the only accurate means of measurement is to use an electronic counter-timer instrument. The phase-sensitive detector, although primarily a device for improving signal quality as discussed later in this chapter, can also be used to measure the phase difference between two signals of identical frequency.

5.2 Signal Processing

Whether a measurement sensor gives a varying-voltage output directly or whether a signal conversion element has to be used to convert the output to voltage form, some subsequent processing of the voltage signal is normally required to improve the quality of the measurement data. Such processing is usually necessary, whether or not a significant transmission distance is involved between the measurement sensor and the rest of the measurement system.

Signal processing is necessary whenever the electrical signals that carry measurement data are corrupted by induced noise. Induced noise arises both within the measurement circuit itself and also during the transmission of measurement signals to remote points. The aim when designing measurement systems is always to reduce such induced noise voltage levels as far as possible. However, it is not usually possible to eliminate all such noise, and signal processing has to be applied to deal with any noise that remains. Noise voltages can exist either in serial mode or common mode forms. Serial-mode noise voltages act in series with the output voltage from the measurement sensor, and often cause very significant errors in the output measurement signal. The extent of series-mode noise corruption is measured by a quantity known as the *signal-to-noise ratio*, which is defined as:

$$\text{Signal-to-noise ratio} = 20 \log_{10}(V_s / V_n)$$

where V_s is the mean voltage level of the signal and V_n is the mean voltage level of the noise. In the case of a.c. noise voltages, the root-mean-squared value is used as the mean.

Common-mode noise voltages are less serious, because they cause the potential of both sides of a signal circuit to be raised by the same level, and thus the magnitude of the output measurement signal is unchanged. However, they have to be considered carefully, since they can be converted into series-mode voltages in certain circumstances.

Noise can be generated from sources both external and internal to the measurement system. Induced noise from external sources arises in measurement systems for a number of reasons that include their proximity to mains-powered equipment and cables (causing noise at the mains frequency), proximity to fluorescent lighting circuits (causing noise at twice the mains frequency), proximity to equipment operating at audio and radio frequencies (causing noise at the corresponding frequency), switching of nearby d.c. and a.c. circuits, and corona discharge (both of the latter causing induced spikes and transients). Internal noise includes potentials due to thermoelectric effects and electrochemical action. The thermoelectric effect describes the creation of a thermoelectric potential (sometimes called a *thermal e.m.f.*) wherever metals of two different types are connected together. Electrochemical potentials arise due to electrochemical action, poorly soldered joints being a common source.

Signal processing is designed to compensate both for errors in the raw (unprocessed) data captured by the sensor and also for transmission-induced errors. The form that it takes depends on the nature of the raw output signals from the measurement sensors. Signal amplification, attenuation, linearisation, filtering and bias removal are all particular forms of signal processing that are applied according to the form of correction required in the raw signal. Mention will also be made in the following sections of certain other special-purpose devices and circuits that are used to manipulate signals. Signal-processing functions are normally carried out after transmission of the measurement signal from the sensor. However, amplification is often needed before transmission.

The implementation of signal-processing procedures can be carried out either by analogue techniques or by digital computation. Choice between these is largely determined by the degree of accuracy required in the signal processing procedure. Analogue signal processing involves the use of various electronic circuits, usually built around the operational amplifier, whereas digital signal processing uses software modules on a digital computer to condition the input measurement data. Digital signal processing is inherently more accurate than analogue techniques, but this advantage is greatly reduced in the case of measurements coming from analogue transducers, which have to be converted by an analogue to digital converter prior to digital processing, thereby introducing conversion errors. Digital processing is also slower than analogue processing. Therefore, it is common practice to use analogue techniques for all signal-processing tasks, except where the accuracy of this is insufficient. It should also be noted that, where measurements are made by an inherently inaccurate measuring device, the extra accuracy provided by digital signal processing is insignificant and therefore inappropriate.

For the purpose of explaining the procedures involved, this chapter concentrates on analogue signal processing, and concludes with a relatively brief discussion of equivalent digital signal-processing techniques. One particular reason for this method of treatment is that some prior analogue signal conditioning is often necessary even when the major part of the signal processing is carried out digitally.

5.2.1 Analogue signal processing

Signal amplification

Signal amplification is frequently needed, both for increasing the signal level of transducers such as thermocouples that have a low-magnitude output, and also for compensating for attenuation of signals during transmission from remote sensors. Amplification by analogue means is usually carried out by an operational amplifier. This is normally required to have high-input impedance so that its loading effect on the transducer output signal is minimised. In some circumstances, such as when amplifying the output signal from accelerometers and some optical detectors, the amplifier must also have a high-frequency response, to avoid distortion of the output reading.

The operational amplifier is an electronic device that has two input terminals and one output terminal, the two inputs being known as the inverting input and noninverting input respectively. When connected as shown in Figure 5.2(a), it provides amplification but also changes the polarity of the signal. The raw (unprocessed) signal V_i is connected to the inverting input through a resistor R_1 and the noninverting

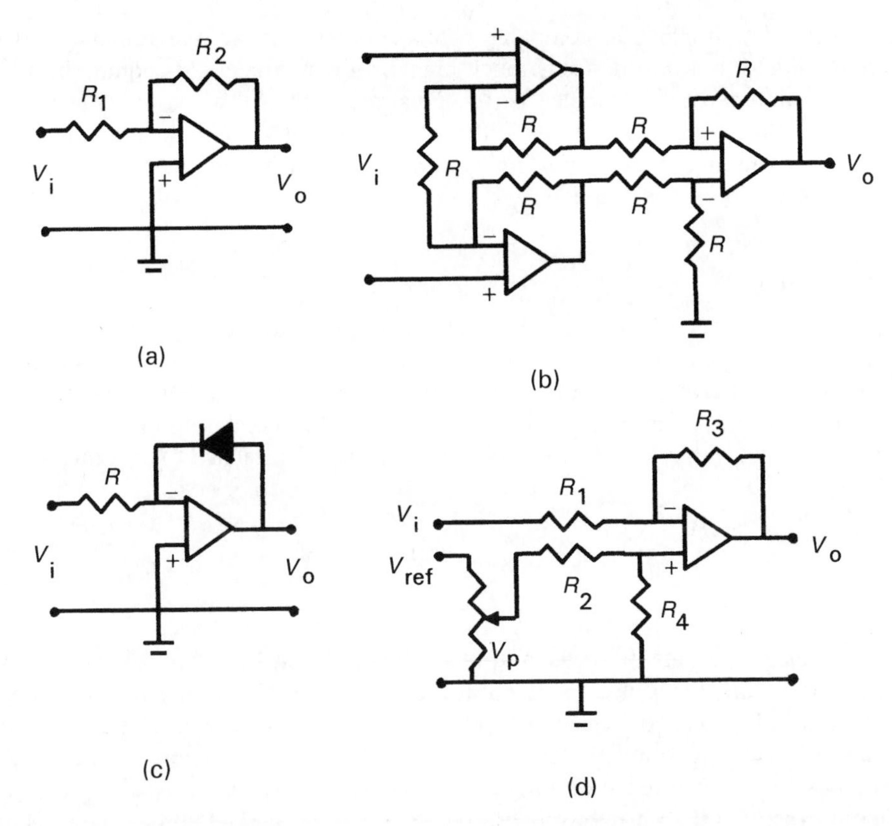

Figure 5.2 *Operational amplifier circuits for signal processing: (a) connected as an inverting amplifier; (b) instrumentation amplifier; (c) connected for signal linearisation; (d) differential amplification mode to remove bias.*

input is connected to ground (earth). A feedback path is provided from the output terminal through a resistor R_2 to the inverting input terminal. Assuming ideal operational amplifier characteristics, the processed signal V_o at the output terminal is then related to the voltage V_i at the input terminal by the expression:

$$V_o = -\frac{R_2 V_i}{R_1} \tag{5.5}$$

The amount of signal amplification is therefore defined by the relative values of R_1 and R_2. This ratio between R_1 and R_2 is known as the amplifier gain or closed-loop gain. If $R_1 = 1\,\text{M}\Omega$ and $R_2 = 10\,\text{M}\Omega$, an amplification factor of 10 is obtained (i.e. gain = 10). If necessary, the polarity change can be corrected by feeding the signal through a further amplifier configured for unity gain ($R_1 = R_2$).

If the output voltage level from a sensor is very small, a special circuit designed to amplify low-level signals known as an *instrumentation amplifier* is used. This has three standard operational amplifiers, as shown in Figure 5.2(b). The advantage of the instrumentation amplifier as compared with a standard operational amplifier is that its differential input impedance is much higher. In consequence, its common mode noise rejection capability is much better. This means that, if a twisted-wire pair is used to connect a transducer to the differential inputs of the amplifier, any induced noise will contaminate each wire equally and will be rejected by the common mode rejection capacity of the amplifier.

Signal attenuation

One method of attenuating signals by analogue means is to use a potentiometer connected in a voltage-dividing circuit, with the input voltage V_i applied across the full length of the resistive element and the output V_o being the voltage between the potentiometer slider and earth. For the slider positioned a distance X_w along the resistance element of total length X_t, the processed signal voltage V_o is related to the input signal voltage by:

$$V_o = \frac{X_w V_i}{X_t} \tag{5.6}$$

Unfortunately, the potentiometer output is often affected by the impedance of the device (or circuit) connected to its output terminals. To avoid this, an operational amplifier can be used, connected as shown in Figure 5.2(a) but with R_1 chosen to be greater than R_2. Equation 5.6 is still valid and therefore, if R_1 is chosen to be $10\,\text{M}\Omega$ and R_2 as $1\,\text{M}\Omega$, an attenuation factor of ten is achieved (gain = 0.1). However, this is more expensive than using a potentiometer, since the operational amplifier is an active device that needs a power supply.

Signal linearisation

Several types of transducer used in measuring instruments give an output that is a nonlinear function of the measured input quantity. Usually, such signals can be linearised by special operational amplifier configurations that have an equal and opposite nonlinear relationship between the amplifier input and output terminals. For example, light intensity transducers typically have an exponential relationship between the output signal and the input light intensity of the form:

$$V_o = K\mathrm{e}^{-\alpha Q} \tag{5.7}$$

where Q is the light intensity, V_o is the voltage level of the output signal, and K and α are constants. If a diode is placed in the feedback path between the input and output terminals of the amplifier as shown in Figure 5.2(c), the relationship between the amplifier output V_o and input V_1 is:

$$V_o = C\log_e(V_1) \tag{5.8}$$

If the output of the light intensity transducer with a characteristic given by equation (5.7) is conditioned by an amplifier with a characteristic given by equation (5.8), the voltage level of the processed signal is:

$$V_o = C\log_e(K) - \alpha CQ \tag{5.9}$$

Expression (5.9) shows that the output signal now varies linearly with light intensity Q, but with an offset of $C\log_e(K)$. This offset or bias would normally be removed by further signal conditioning, as described below.

Bias removal

Sometimes, either because of the nature of the measurement transducer itself, or as a result of other signal-conditioning operations such as linearisation as just described, a bias exists in the output signal. This can be expressed mathematically for a physical quantity x and measurement signal y as:

$$y = Kx + B \tag{5.10}$$

where B is the bias. Bias can be removed by an operational amplifier connected in differential amplification mode, as shown in Figure 5.2(d). Referring to this circuit, for $R_1 = R_2$ and $R_3 = R_4$, the output V_o is given by:

$$V_o = (R_3/R_1)(V_b - V_i) \tag{5.11}$$

where V_i is the unprocessed signal y equal to $(Kx + B)$ and V_b is the output voltage from a potentiometer supplied by a known reference voltage V_{ref}, that is set such that $V_b = B$. Now, substituting these values for V_i and V_b into equation (5.11), y can be written as:

$$y = K'x \tag{5.12}$$

where the new constant K' is related to K according to $K' = -K(R_3/R_1)$. It is clear that the bias has been successfully removed and equation (5.12) is now a linear relationship between the measurement signal y and the measured quantity x.

Signal filtering

Signal filtering consists of processing a signal to remove a certain band of frequencies within it. The band of frequencies removed can be either at the low-frequency end of the frequency spectrum, at the high-frequency end, at both ends, or in the middle of the spectrum. Filters to perform each of these operations are known respectively as low-pass, high-pass, band-pass and band-stop. All such filtering operations can be carried out by either analogue or digital methods.

The range of frequencies passed by a filter is known as the *pass band*, the range not passed is known as the *stop band*, and the boundary between the two ranges is known as the *cut-off frequency*. To illustrate this, consider a signal whose frequency spectrum is such that all frequency components in the frequency range from zero to infinity have equal magnitude. If this signal is applied to an ideal filter, then the outputs for a low-pass filter, high-pass filter, band-pass filter and band-stop filter respectively are shown in Figure 5.3. Note that for the latter two types, the bands are defined by a pair of frequencies, rather than by a single cut-off frequency.

Unfortunately, perfect filtering is impossible. Unwanted frequency components are not erased completely, and the filtered signal retains some components (of relatively low magnitude) in the unwanted frequency range. Also, there is a small amount of attenuation of frequencies within the pass-band, which increases as the cut-off frequency is approached. Figure 5.4 shows the typical output characteristics of a

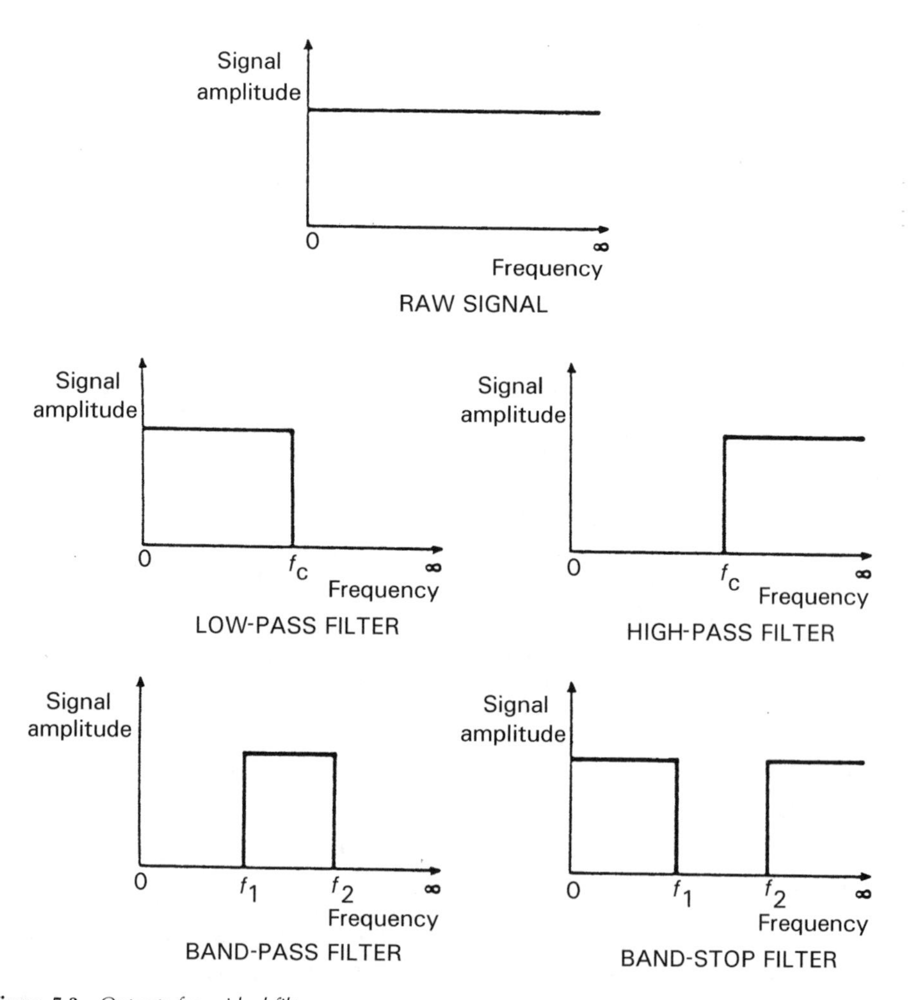

Figure 5.3 *Outputs from ideal filter.*

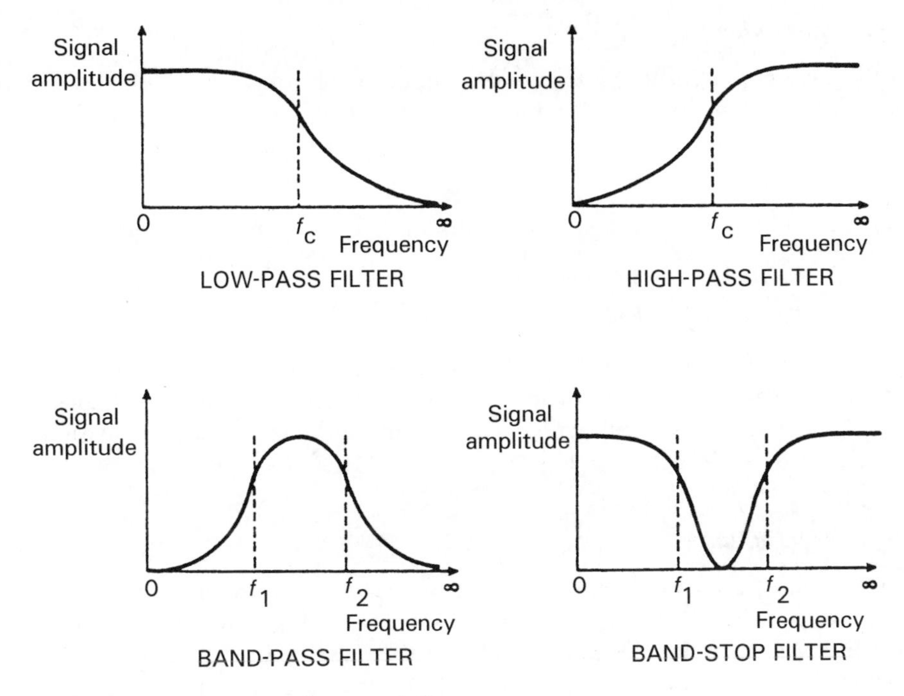

Figure 5.4 *Outputs from practical constant-k filters.*

practical constant-k* filter designed respectively for high-pass, low-pass, band-pass and band-stop filtering. Filter design is concerned with trying to obtain frequency rejection characteristics that are as close to the ideal as possible. However, improvement in characteristics is only achieved at the expense of greater complexity in the design. Therefore, the filter chosen for any given situation is a compromise between performance, complexity and cost.

In the majority of measurement situations, the physical quantity being measured has a value that is either constant or only changing slowly with time. In these circumstances, the most common types of signal corruption are high-frequency noise components, for which a low-pass filter is required. In a few cases, the measured signal itself has a high frequency, for instance when mechanical vibrations are being monitored, and a high-pass filter is needed to attenuate low-frequency noise components. Band-stop filters are used where a measurement signal is corrupted by noise at a particular frequency. Such noise is frequently due to mechanical vibrations or proximity of the measurement circuit to other electrical apparatus. Both passive and active implementations exist for analogue filters, but a detailed discussion on filter designs is not appropriate in this text. However, there are a large number of specialised texts that cover filter design.

* 'constant-k' is a term used to describe a common type of passive filter.

Signal integration

If there is a requirement to integrate the output signal from a transducer, this can be achieved with an operational amplifier connected in the configuration shown in Figure 5.5(a). This integrates the input signal V_i, such that the output signal V_o is given by:

$$V_o = -(1/RC)\int V_i \mathrm{d}t$$

Pre-amplifier (voltage follower)

The pre-amplifier, also known as a voltage follower, has very high input impedance and its main application is to reduce the load on the measured system. It consists of a unity gain amplifier circuit with a short circuit in the feedback path, as shown in Figure 5.5(b), such that: $V_o = V_i$.

Voltage comparator

The output of a voltage comparator switches between positive and negative values, according to whether the difference between the two input signals to it is positive or negative. An operational amplifier connected as shown in Figure 5.5(c) gives an output that switches between positive and negative saturation levels, according to whether ($V_1 - V_2$) is greater than or less than zero. Alternatively, the voltage of a

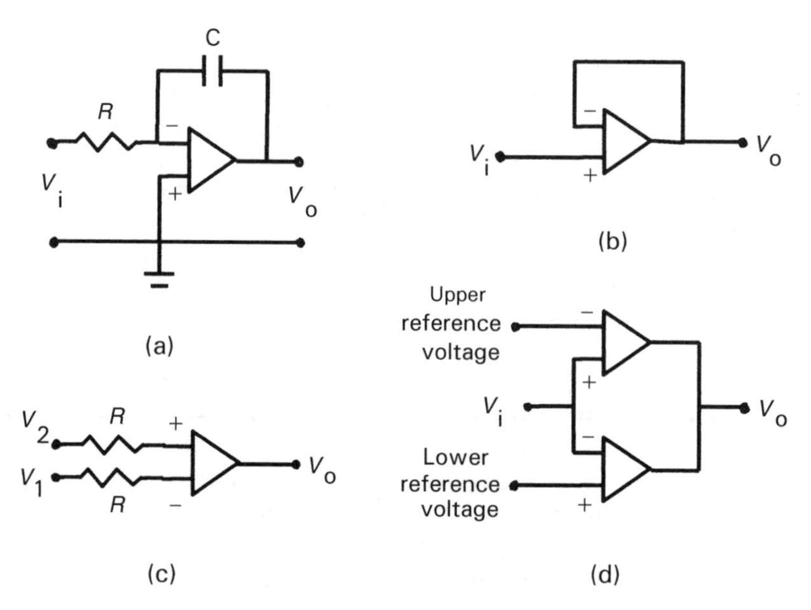

Figure 5.5 *Further signal-processing circuits using operational amplifiers: (a) connected for signal integration; (b) connected as a pre-amplifier (voltage follower); (c) comparison of two voltage signals; (d) comparison of signal against reference value.*

single input signal can be compared against positive and negative reference levels with the circuit shown in Figure 5.5(d).

Signal addition, subtraction and multiplication

Summation: Two or more input signals can be summed by an operational amplifier connected in signal-inversion mode, as shown in Figure 5.6(a). For input signal voltages V_1, V_2 and V_3, the output voltage V_o is given by:

$$V_o = -(V_1 + V_2 + V_3)$$

Subtraction: Signal subtraction is required in applications like volume flow rate measurement using an obstruction device (see Chapter 13), in which the pressure either side of the obstruction is measured by a pressure sensor having a voltage output. The difference between such voltages is typically small and is usually corrupted by common-mode noise voltages. The differential amplifier circuit shown in Figure 5.6(b) is commonly used to subtract the two signals. This provides adequate signal amplification and also attenuates the common-mode noise. The output voltage V_o is given by:

$$V_o = (R_3/R_1)(V_B - V_A)$$

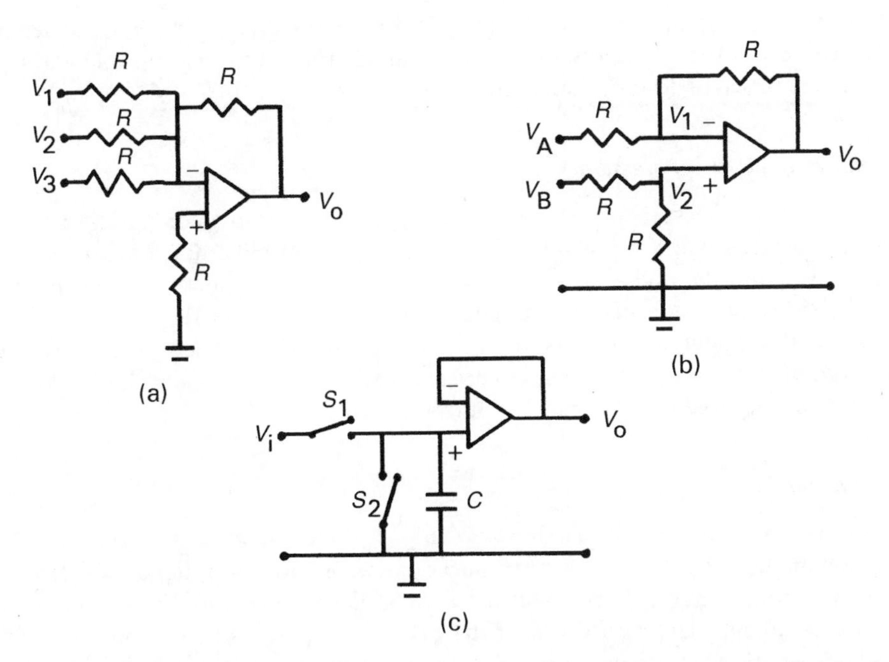

Figure 5.6 *Further signal-processing circuits using operational amplifiers: (a) connected for signal addition; (b) connected as differential amplifier for signal subtraction; (c) connected as 'sample and hold circuit'.*

Multiplication: Great care must be taken when choosing a signal multiplier because, whilst many circuits exist for multiplying two analogue signals together, most of them are two-quadrant types that only work for signals of a single polarity, i.e. both positive or both negative. Such schemes are unsuitable for general analogue signal processing, where the signals to be multiplied may be of changing polarity. For analogue signal processing, a four-quadrant multiplier is required such as a Hall-effect multiplier or a translinear multiplier.

Phase-sensitive detector

One function of a phase-sensitive detector is to measure the phase difference between two signals that have the same frequency. For two input signals of amplitude V_1 and V_2 and frequency f, the output is given by $V_1 V_2 \cos\phi$, where ϕ is the phase difference between the signals. In many cases, the phase difference is adjusted to zero ($\cos\phi = 1$) so that the output is a maximum.

A phase-sensitive detector can also be used as a cross-correlator to enhance the quality of measurement signals that have a poor signal-to-noise ratio. For this role, the detector requires: firstly, a clean reference voltage at the same frequency as the measurement signal, and secondly, phase-control circuits to make the phases of the reference and measurement signals coincide. Commercial instruments known as *lock-in amplifiers* are available that combine a phase-sensitive detector with other components. Phase-sensitive detectors are known by several alternative names, two examples of which are synchronous demodulator and synchronous detector. They can also exist physically in a number of alternative forms that include both transformer-based and fully electronic circuits.

Analogue-to-digital conversion

If the analogue measurement signals are going to be input to a computer, there is a fundamental mismatch between the analogue form of output data from sensors and transducers and the digital form of data required by a computer. This problem is solved by the provision of an analogue-to-digital converter in the computer-input interface. If the computer then computes a control signal that needs to be output to an actuator which requires a signal in analogue form, a *digital-to-analogue converter* will also be required in the computer output interface.

Sample and hold circuit

A sample and hold circuit, as shown in Figure 5.6(c), is needed at the interface between an analogue instrument/transducer and an analogue-to-digital converter. It holds the input signal at a constant level whilst the analogue-to-digital conversion process is taking place and prevents the conversion errors which would probably result if variations in the measured signal were allowed to pass through to the converter. The input signal is applied to the circuit for a very short time duration with switch S_1 closed and S_2 open, after which S_1 is opened and the signal level is then held until, when the next sample is required, the circuit is reset by closing S_2.

5.2.2 Digital signal processing

Digital techniques achieve much greater levels of accuracy in signal processing than equivalent analogue methods. However, the time taken to process a signal digitally is longer than that required to carry out the same operation by analogue techniques, and the equipment required is more expensive. Whilst digital signal-processing elements in a measurement system can exist as separate units, it is more usual to find them as an integral part of an intelligent instrument. However, their construction and mode of operation are the same irrespective of whether they exist physically as separate boxes or within an intelligent instrument.

Because most measuring sensors and instruments have an output signal in analogue form, analogue-to-digital conversion is required at the interface with a digital computer. This involves using a sample and hold circuit to periodically sample the analogue signal and hold it at a constant level whilst it is converted to a digital value. The conversion takes a certain finite time, during which the analogue measurement signal is usually changing in value. The next sample of the analogue signal cannot be taken until the conversion of the last sample to digital form is completed. Therefore, the representation of a continuous analogue signal within a digital computer is a discrete sequence of samples whose pattern only approximately follows the shape of the original signal. The process of conversion between a continuous analogue sine wave signal at a frequency of approximately 0.75 Hz, and a discrete digital equivalent is illustrated in Figure 5.7. With the rate of sampling shown of approximately 11 samples per second, reconstruction of the samples matches the original analogue signal very well. However, if the rate of sampling is decreased, the fit between the reconstructed samples and the original signal becomes less accurate. If the rate of sampling is very much less than the frequency of the raw analogue signal, such as 1 sample per second, only the samples marked 'X' in Figure 5.7 are obtained. Fitting a line through these Xs incorrectly estimates the signal frequency as approximately

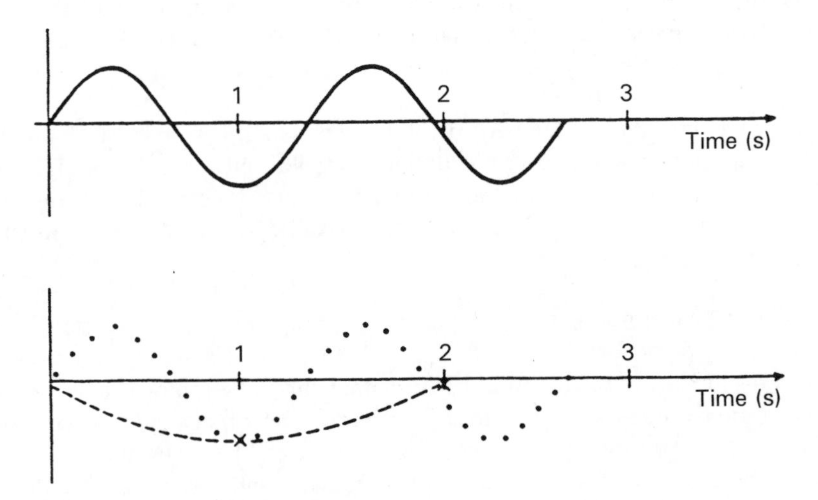

Figure 5.7 Conversion of continuous analougue signal to discrete sampled signal. From Morris (1997) Measurement and Calibration Requirements, © John Wiley & Sons, Ltd. Reproduced with permission.

0.25 cycles per second. This phenomenon, whereby the process of sampling transmutes a high-frequency signal into a lower frequency one, is known as *aliasing*. To avoid aliasing, the sampling rate must be at least twice the highest frequency in the analogue signal sampled. In practice, sampling rates of between five and 10 times the highest frequency signal are normally chosen, so that the discrete sampled signal is a close approximation to the original analogue signal in amplitude as well as frequency. However, problems can still arise if the analogue signal is corrupted by high-frequency noise, since aliasing can transmute noise components into the same frequency range as the measurement signal, thus giving erroneous results. In such circumstances, it is necessary to apply analogue signal filtering before the digitisation procedure.

One further factor that affects accuracy during analogue-to-digital conversion is quantisation. *Quantisation* describes the procedure whereby the continuous analogue signal is converted into a number of discrete levels. At any particular value of the analogue signal, the digital representation is either the discrete level immediately above this value or the discrete level immediately below this value. If the difference between two successive discrete levels is represented by the parameter Q, then the maximum error in each digital sample of the raw analogue signal is $\pm Q/2$. This error is known as the quantisation error and is clearly proportional to the resolution of the analogue-to-digital converter, i.e. to the number of bits used to represent the samples in digital form.

Once a satisfactory digital representation has been obtained, the procedures of signal amplification, signal attenuation and bias removal become trivial. For signal amplification and attenuation, all samples have to be multiplied or divided by a fixed constant. Bias removal involves simply adding or subtracting a fixed constant from each sample of the signal.

Signal linearisation requires a priori knowledge of the type of nonlinearity involved, in the form of a mathematical equation that expresses the relationship between the output measurements from an instrument and the value of the physical quantity being measured. This equation is used to calculate the value of the measured physical quantity corresponding to each discrete sample of the measurement signal.

Digital signal processing can perform all low-pass, high-pass, band-pass and band-stop filtering functions. However, the design of digital filters requires a high level of theoretical knowledge, including the use of z-transform theory, which is outside the scope of this book. However, there are many specialised texts that cover this subject.

5.3 Signal Transmission

Signal transmission from the point of measurement to the point where the data is processed and recorded is required for two main reasons. Firstly, there are cost savings in having signal processing and recording equipment in one location rather than having various items of such equipment scattered around several sites. Secondly, measurement sensors and instruments often have to operate in hostile conditions involving high temperatures, flames, fumes, smoke, steam and large volumes of dust.

Such environments are unsuitable for the components that are required to process and record the measurement data, and so the data must be transmitted to somewhere where the environment is less harsh.

It is inevitable that, during transmission from the measurement point to another location, there will be some degradation in the quality of the signals. Hence, the main aim in designing signal transmission systems is to minimise this degradation. The most serious difficulty associated with long-distance transmission of data in the form of a voltage signal is attenuation of the signal due to the resistance of the signal wires. This causes particular difficulties in the case of sensors that have a low-magnitude output, and it obviously becomes worse as the transmission distance increases. One way to compensate for attenuation is to amplify the signal prior to transmission. However, unless transmission distances are very large, the problem can be avoided altogether if the conductors used to carry the signals are of adequate cross-section. Nevertheless, if a reasonable signal-to-noise ratio is to be obtained for low-output sensors, signal amplification before transmission still remains necessary.

Contamination of the measurement signal by noise during transmission is a further problem. Typical noise sources include radiated electromagnetic fields from electrical machinery and power cables, induced fields through wiring loops, and voltage spikes on the mains power supply. Fortunately, the magnitude of this noise contamination can be greatly reduced if the wires carrying the signals are adequately protected inside an earthed, braided, metal shield that is itself isolated from the signal wires. This provides a high degree of noise protection, especially against capacitive-induced noise due to the proximity of signal wires to high-current power conductors. Noise can also be minimised by careful design and location of the signal wires, routing them as far away as possible from noise sources and twisting the signal wires together along their entire length (known as a 'twisted pair') to reduce inductive coupling.

Many other techniques of improving the quality of signal transmission exist. Most of these consist of converting the measurement signal prior to transmission from its usual varying-voltage form into some other form that is less susceptible to contamination during transmission. It is then usually converted back to a varying voltage after transmission. A number of different nonvoltage forms are suitable for signal transmission, as presented below.

5.3.1 Current loop transmission

The signal-attenuation effect of conductor resistances can be minimised if varying voltage signals are transmitted as varying current signals. This requires a voltage-to-current converter of the form shown in Figure 5.8(a), which is commonly known as a 4–20 mA current loop interface. Two voltage-controlled current sources are used, one (I_1) providing a constant 4 mA output that is used as the power supply current, and the other (I_2) providing a variable 0–16 mA output that is proportional to the input voltage level. The net output current therefore varies between 4 and 20 mA. This is a very commonly used means of connecting remote instruments to a central control room. After transmission, conversion back from current to voltage form is required. This can be achieved with an operational amplifier connected as shown

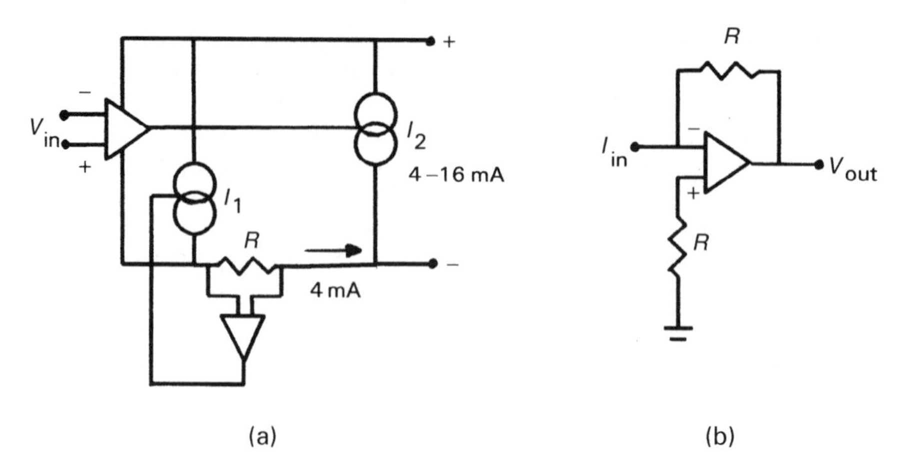

Figure 5.8 Current loop transmission: (a) voltage-to-current converter (current loop interface); (b) current-to-voltage converter.

in Figure 5.8(b). The output voltage V_{out} is simply related to the input current I by: $V_{out} = IR$.

5.3.2 Transmission using an a.c. carrier

Another solution to the problem of noise corruption in a low-level, d.c. voltage measurement signal is to transfer the signal on to an a.c. carrier system before transmission and extract it from the carrier at the end of the transmission line. Both amplitude modulation (AM) and frequency modulation (FM) can be used for this.

The AM consists of translating the voltage signal into variations in the amplitude of a carrier sine wave. An a.c. bridge circuit is commonly used for this, as part of the system for converting the outputs of sensors that have a varying resistance (R), capacitance (C) or inductance (L) form of output. Referring to equation (5.4), if the input voltage V_i is a sinusoidal voltage of frequency ω, the output voltage V_o is at the same frequency ω and has a varying amplitude according to the value (R_u, C_u or L_u) of the sensor. Thus, the input voltage acts as an a.c. carrier at a frequency ω that is typically several kHz. After shifting the d.c. signal on to the high-frequency a.c. carrier, a high-pass filter can be applied to the AM signal. This successfully rejects noise in the form of low-frequency drift voltages and mains interference. At the end of the transmission line, demodulation is carried out to extract the measurement signal from the carrier.

FM involves translating the analogue voltage signal variations into frequency variations on a high-frequency carrier signal. This achieves even better noise rejection than AM. Figure 5.9(a) shows a voltage-to-frequency conversion circuit, in which the analogue voltage signal input is integrated and applied to the input of a comparator that is preset to a certain threshold voltage level. When this threshold level is reached, the comparator generates an output pulse that resets the integrator and is also applied

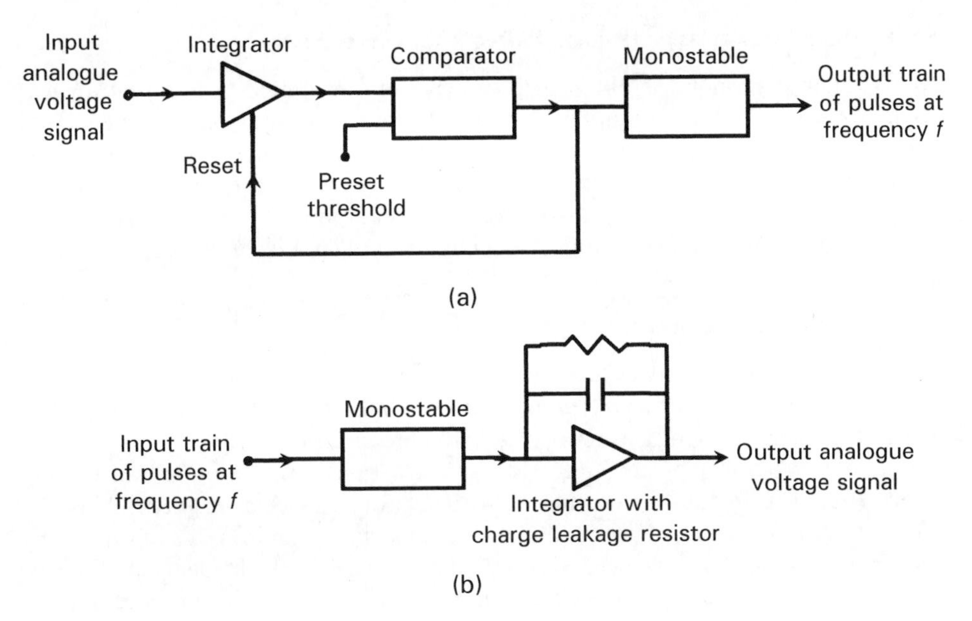

Figure 5.9 A.c. carrier transmission: (a) voltage-to-frequency conversion circuit; (b) frequency-to-voltage converter.

to a monostable. This causes the frequency f of the output pulse train to be proportional to the amplitude of the input analogue voltage.

At the end of the transmission line, the FM signal is usually converted back to an analogue voltage by a frequency-to-voltage converter such as that shown in Figure 5.9(b). In this, the input pulse train is applied to an integrator that charges up for a specified time. The charge on the integrator decays through a leakage resistor, and a balance voltage is established between the input charge on the integrator and the decaying charge at the output. This output balance voltage is proportional to the input pulse train at frequency f.

5.3.3 Pneumatic transmission

In recent years, pneumatic transmission tends to have been replaced by other alternatives in most new implementations of instrumentation systems, although many examples can still be found in operation in the process industries. Pneumatic transmission consists of transmitting an analogue voltage measurement signal as a varying pneumatic pressure level that is usually in the range of 3–15 p.s.i. (Imperial units are still commonly used in process industries, although the equivalent range in SI units is 207–1034 mbar, which is often rounded to 200–1000 mbar in metric systems.) A few systems also use alternative ranges of 3–27 p.s.i. or 6–48 p.s.i. Pneumatic transmission has the advantage of being intrinsically safe, and it provides similar levels of noise immunity to current loop transmission. However, one disadvantage of using air as the transmission medium is that transmission speed is much less than with electrical or optical transmission.

5.3.4 Digital transmission (voltage-to-frequency conversion)

High immunity to noise is obtained if a signal is transmitted in a digital format. Digital transmission involves applying the analogue voltage measurement signal to a voltage-to-frequency converter circuit. This converts the voltage variations into corresponding frequency variations that can be readily transmitted in digital format. After transmission, a frequency-to-voltage converter is applied to reconvert the signal back into a varying analogue voltage if an analogue signal recorder is to be used.

5.3.5 Fibre-optic transmission

Fibre-optic signal transmission involves transforming electrical signals into a modulated light wave that is transmitted along a fibre optic cable. Then, at the receiving end of the cable, the light is transformed back into electrical form.

Light has a number of advantages over electricity as a medium for transmitting information: it is immune to corruption by neighbouring electromagnetic fields, the attenuation over a given transmitted distance is much less, and it is also intrinsically safe. However, there is an associated cost penalty because of the higher cost of a fibre optic system compared with the cost of metal conductors. The primary reason for this penalty is the high cost of the terminating transducers that perform the signal conversion function at each end of the cable.

The light transmitting cable contains at least one, but more often a bundle, of glass or plastic fibres. This is terminated at each end by a transducer, as shown in Figure 5.10. At the input end, the transducer converts the signal from the electrical form, in which most signals originate, into light. At the output end, the transducer converts the transmitted light back into an electrical form suitable for use by data recording, manipulation and display systems. These two transducers are often known as the transmitter and receiver respectively.

Signals are normally transmitted along a fibre optic cable in digital format, although analogue transmission is sometimes used. If there is a requirement to transmit more than one signal, it is more economic to multiplex the signals on to a single cable rather than transmit the signals separately on multiple cables. *Time division multiplexing* involves switching the analogue signals in turn, in a synchronised sequential manner, into an analogue-to-digital converter that outputs on to the transmission line. At the other end of the transmission line, a digital-to-analogue converter transforms the digital signal back into analogue form, and it is then switched in turn on to separate analogue signal lines.

5.3.6 Optical wireless telemetry

Wireless telemetry is particularly relevant to environmental measurement applications, where measuring devices are often in remote locations. It allows signal trans-

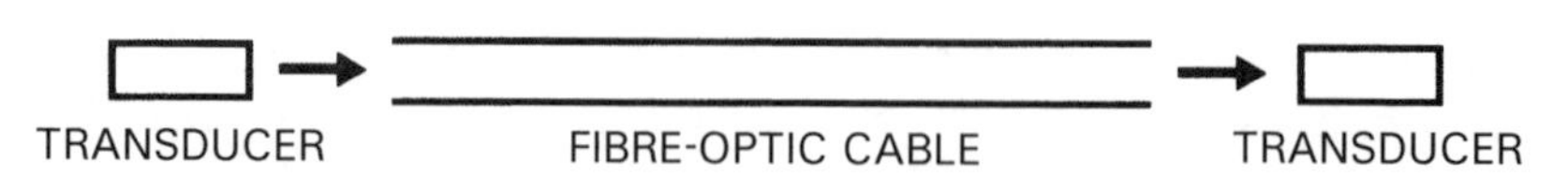

Figure 5.10 Fibre-optic signal transmission.

mission to take place without laying down a physical link in the form of electrical or fibre-optic cable. This can be achieved using either radio or light waves to carry the transmitted signal across a plain air path between a transmitter and a receiver.

Optical wireless transmission was first developed in the early 1980s. It is immune to electromagnetic interference and consists of a light source (usually infrared) transmitting encoded data information across an open, unprotected air path to a light detector. Three distinct modes of optical telemetry are possible, known as point-to-point, directed and diffuse:

Point-to-point telemetry uses a narrow beam of light to carry the measurement data. A data speed of 5 Mbit/s is possible over a transmission distance of 1000 m but a higher speed of 20 Mbit/s is possible over smaller transmission distances up to 200 m.

Directed telemetry can only be used over relatively short distances up to about 70 m. It transmits a slightly divergent beam of light that is directed towards reflective surfaces, such as the walls and ceilings in a room. This produces a wide area of coverage and means that the transmitted signal can be received at a number of points. Over a transmission distance of 70 m, the maximum possible transmission speed is only 1 Mbit/s, but a speed of 10 Mbit/s is possible if the transmission distance is limited to 20 m.

Diffuse telemetry has an even more divergent beam than directed telemetry. This provides an increased area of coverage, but reduces the transmission speed and range. At the maximum range of 20 m, the maximum speed of transmission is 500 kbit/s, although this increases to 2 MBit/s at a reduced range of 10 m.

In practice, optical wireless telemetry is not commonly used, because data transmitted across an open, unprotected air path are susceptible to random interruption. Consequently, alternative forms of transmission are usually preferred.

5.3.7 Radio telemetry (radio wireless transmission)

Radio telemetry is also particularly relevant in environmental applications, and can be used for transmission distances up to 400 miles. A particular advantage of radio transmission is that it does not require a physical link, which is often difficult to install for electrical or fibre-optic transmission systems. In consequence of this, despite being capable of transmitting over distances up to 600 km, it is often used to transmit measurement data over quite short distances as well. The great advantage that radio telemetry has over optical wireless transmission through an air medium is that radio waves are attenuated much less by obstacles between the energy transmitter and receiver.

In radio telemetry, data are usually transmitted in a frequency modulated (FM) format. This actually involves two separate stages of frequency modulation, and the system is consequently known as an FM/FM system. In the first stage, the analogue voltage signal is converted by a voltage-to-frequency converter into a varying frequency around the centre frequency (typically 10 kHz) of a subcarrier. In the second stage, the subcarrier is translated into the radio frequency range as modulations on a typical carrier frequency of 217.5 MHz.* After transmission, a demodulator is

* Carrier frequencies are subject to national agreements and vary in different countries.

applied to return the signal back into a varying voltage form. A single radio telemetry system is able to transmit measurement data from multiple sensors by allocating a different subcarrier frequency for the data from each sensor. It is also suitable for use with sensors mounted on mobile units.

The inaccuracy of radio telemetry is typically ±1%. In theory, it is very reliable because, although the radio frequency waveband is relatively crowded, specific frequencies within it are allocated to specific usage under national agreements that are normally backed by legislation. Interference is avoided by licensing each frequency to only one user in a particular area, and limiting the transmission range through limits on the power level of transmitted signals, such that there is no interference to other licensed users of the same frequency in other areas. Unfortunately, interference can still occur in practice, due both to adverse atmospheric conditions extending the transmission range beyond that expected into adjoining areas, and also due to unauthorised transmissions by other parties at the wavelengths licensed to registered users. There is a legal solution to this latter problem, although some time may elapse before the offending transmission is stopped successfully.

5.4 Signal Recording

ISO 14001 sets great emphasis on documentation, and records of measurements made as part of operating an EMS are therefore very important. Hence, once measurement signals have been captured, processed and transmitted, they must be recorded and stored in some permanent form. The oldest form of data recorder available is the mechanical chart recorder. This provides a simple, cheap and reliable means of recording data. Despite the fact that the basic technology involved is very old, mechanical chart recorders continue to be popular, although recent models almost always incorporate a microprocessor to enhance the recording facilities provided. The main alternative for environmental applications are digital recorders, although these are often used in conjunction with a microprocessor-enhanced mechanical chart recorder to provide permanent data records. These options are discussed in greater detail below.

5.4.1 Mechanical chart recorders

The simplest form of mechanical chart recorder is the *galvanometric recorder*. This works on the same principle as a moving coil voltmeter, except that the pointer carries a pen and draws a trace on chart paper instead of moving against a graduated scale. The measurement signal is applied to the coil, and the angular deflection of this and its attached pointer is proportional to the magnitude of the signal applied. By using a motor running at constant speed to drive the chart paper, a time history of the measured signal is produced. Two basic designs exist, known as the strip recorder and the circular recorder. These differ in the manner in which the chart paper is driven. In the *strip recorder*, as sketched in Figure 5.11(a), the paper is driven out from a roll, and the time axis of the trace on the chart is therefore straight. In the alternative *circular recorder*, as sketched in figure 5.11(b), the recording chart is

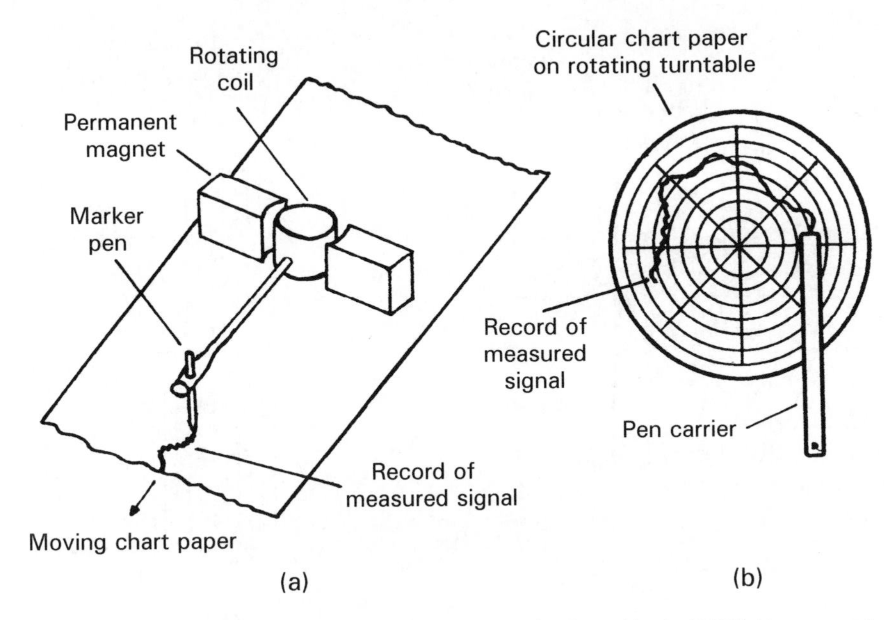

Figure 5.11 *Galvanometric chart recorders: (a) strip recorder. From Morris (1997)* Measurement and Calibration Requirements, *© John Wiley & Sons, Ltd. Reproduced with permission; (b) circular recorder.*

circular and is carried on a rotating platform, thus making the time axis of the trace circular. In both cases, it is normal to have several separate channels in the recorder so that different voltage signals can be recorded simultaneously. Strip recorders with up to 12 channels are commonly available, as well as circular recorders with up to four channels.

The accuracy with which data values can be read from chart recordings is limited by the inherent design of the recorder. Inspection of Figure 5.12(a) shows that the pen displacement y across the chart is given by $y = R\sin\theta$. This sine relationship between the input signal and the displacement y is nonlinear, and results in an error of 0.7% for deflections of ±10°. However, in practice, the error is greater than this, because, with the pen moving in an arc, it is difficult to relate the magnitude of deflection with the time axis accurately. One way of overcoming this for strip recorders is to print a grid on the chart paper in the form of circular arcs, as illustrated in Figure 5.12(b). Unfortunately, measurement errors still occur in reading this type of chart, because interpolation for points drawn between the curved grid lines is difficult. Therefore, the typical uncertainty in measurements is likely to be about ±2% and the best measurement resolution is normally 1% of full-scale reading. Design limitations also mean that the maximum frequency allowable for input signals is only 30 Hz, although this is not usually a problem for environmental measurement applications. Some improvement in accuracy over a standard galvanometric recorder is obtained in the *potentiometric recorder*, where a typical inaccuracy of ±0.1% of full scale (f.s.) and a measurement resolution of 0.2% f.s. is achievable. Such instruments employ a servo system, where the pen is driven by a servomotor and a potentiometer on the pen feeds back a signal proportional to pen position. This position signal is

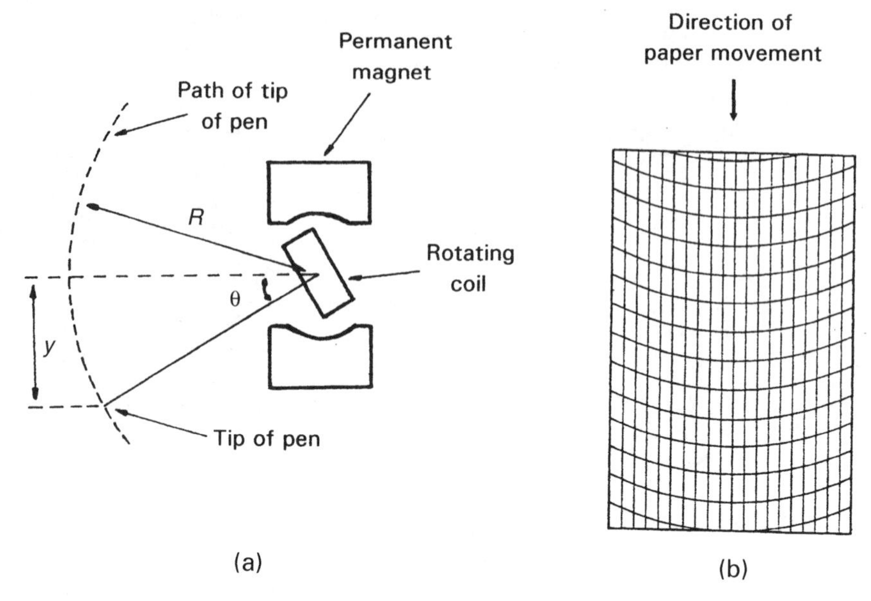

Figure 5.12 *Galvanometric recorder analysis: (a)* y *versus theta relationship; (b) curvilinear chart paper to improve reading accuracy. From Morris (1997)* Measurement and Calibration Requirements, © *John Wiley & Sons, Ltd. Reproduced with permission.*

compared with the measured signal, and the difference is applied as an error signal that drives the motor. However, the instrument has a slow response time in the range of 0.2–2.0 seconds as a consequence of this electromechanical balancing mechanism, which means that the recorder can only record signals that are constant or only varying slowly.

As mentioned previously, whilst both strip and circular chart recorders continue to be widely available, almost all current models incorporate a microprocessor. The microprocessor provides a number of enhanced features, including the ability to select recording range and chart speed. It also allows specification of alarm modes and levels to provide an automatic response if measured variables go outside acceptable limits. Additional information such as names, times and dates of variables recorded can also be printed on charts.

However, even with a microprocessor to enhance performance, pen-based chart recorders still have limitations. Firstly, they are liable to suffer from blockages in the ink supply to the pen. Secondly, the nonlinearity in pen movement, and the fact that the pen moves in an arc rather than a straight line, causes errors in reading measurement values from recordings on the chart. This has stimulated the development of chart recorders where the trace is produced entirely digitally, and where the only moving parts are in the chart-drive mechanism. Examples of these are the thermal recorder and the inkjet recorder.

The *thermal recorder* uses thermal matrix printing technology to print on heat-sensitive paper. Models capable of recording up to 16 channels of data are available. Its main disadvantage is the cost of the special chart paper needed.

The *inkjet recorder* is another variation that uses inkjet printing technology to record multiple channels of data. Unlike, the thermal recorder, only standard chart paper is required.

5.4.2 Digital recorders

For many years, the only way to record signals at frequencies higher than 80 kHz was to use a standard digital computer in a procedure known as *data logging*. As measurement signals are usually in analogue form, a prerequisite for recording is an analogue-to-digital (A–D) converter board to sample the analogue signals and convert them to digital form. Correct choice of the sampling interval is necessary to ensure that an accurate digital record of the signal is obtained without aliasing problems, as explained in Section 5.2.2. Some prior analogue signal conditioning may also be required in some circumstances.

More recently, purpose-designed digital recorders have become available for this purpose. These are usually multichannel, and are available from many suppliers. Typically, a 10-bit A–D converter is used, which gives a 0.1% measurement resolution. Alternatively, a 12-bit converter giving 0.025% resolution is sometimes used. Typical specifications for digital recorders are frequency response of 25 kHz, maximum sampling frequency of 200 MHz, and data storage up to 4000 data points per channel.

Many digital recorders display recorded measurements graphically on a computer monitor incorporated into the recorder, and are often known as *paperless recorders* in this form. However, the lack of a hard copy of the recorded data in permanent form does not usually satisfy EMS documentation requirements. In such cases, it is possible to use the digital recorder in conjunction with some form of chart recorder to provide the necessary permanent record. Alternatively, the recorded measurement data can be output in numerical form on alphanumeric digital printing devices like matrix, inkjet and laser printers. However, whilst numeric data provides greater accuracy, analogue records on a chart recorder are often preferable in environmental monitoring, since the graphical display of the time history of a variable highlights trends in monitored parameter values much more readily

5.4.3 Obsolete recording technologies

Whilst microprocessor-enhanced chart recorders and digital recorders are the only kinds of recorder that are now commonly manufactured, many companies have other recorders still in use that are based on older technologies. In case such devices are available to record environmental parameters, it is prudent to briefly discuss their main features.

Ultraviolet recorders

These work on very similar principles to a standard galvanometric chart recorder, except that a narrow mirror is mounted on the rotating coil instead of a pointer and pen system. This achieves a large reduction in system inertia and spring constants and greatly improves the frequency response, so that signals at frequencies up to

13 kHz can be recorded with a typical inaccuracy of ±2% f.s. The mirror reflects an ultraviolet light beam on to ultraviolet sensitive paper. Usually, several of these mirror-galvanometer systems are mounted in parallel to provide a multichannel recording capability. Whilst it is possible to obtain satisfactory permanent signal recordings by this method, special precautions are necessary to protect the ultraviolet-sensitive paper from light before use. Also, a fixing lacquer has to be sprayed on to the paper after recording, otherwise ambient light will eventually obliterate the traces. Such instruments must also be handled with extreme care, because the mirror galvanometers and their delicate mounting systems are easily damaged by relatively small shocks.

Magnetic tape recorders

These are designed for measurement data recording, and they use technology similar to present-day audio and video tapes, except that wider tape, typically 25 mm, is used. Such instruments can record signals up to 80 kHz in frequency. As the speed of the tape transport can be switched between several values, signals can be recorded at high speed and replayed at a lower speed. Such time-scaling of the recorded information allows a hard copy of the signal behaviour to be obtained from instruments such as ultraviolet and galvanometric recorders whose bandwidth is insufficient to allow direct signal recording. A 200 Hz signal cannot be recorded directly on a chart recorder, but if it is recorded on a magnetic tape recorder running at high speed and then replayed at a speed ten times lower, its frequency will be time-scaled to 20 Hz, which is then within the frequency range that a chart recorder can handle. Instrumentation tape recorders typically have between four and 10 channels, allowing many signals to be recorded simultaneously. Whilst the technology is now largely obsolete, some magnetic tape flight recorders are still produced for avionic applications.

6

Quantification and Effects of Air Pollution

6.1 Air Pollution Sources and Effects

The greatest sources of air pollution in the world are natural processes in the form of decaying vegetation, forest fires and volcanic activity. However, the global distribution of such pollution sources and natural dispersion effects mean that the concentration in any particular area is relatively small. Man-made pollution in the world, although much smaller in total extent than natural pollution, is concentrated in relatively small geographical areas and, within those areas, its effect is therefore much greater than the effect of natural pollution.

Air pollutants can be subdivided into the following classifications:

- Particulate matter
- Gaseous sulphur compounds
- Gaseous nitrogen compounds
- Carbon monoxide
- Carbon dioxide
- Organic compounds
- Halogen compounds
- Other inorganic compounds

Atmospheric pollution by these substances leads to many health problems, including bronchitis, emphysema and heart disease. Toxic air pollutants, such as many hydrocarbon compounds, cause particular health problems. Many pollutants also cause odours that emanate from gases that are not considered toxic in themselves but certainly constitute a nuisance that demands attention. For example, sulphur dioxide

ISO 14000 Environmental Management Standards: Engineering and Financial Aspects. Alan S. Morris.
 ISBN 0-470-85128-7

produces an unpleasant taste sensation at concentrations above 0.3 ppm in air. Above 3.0 ppm, the odour from the gas is regarded as being very irritating.

The largest sources of man-made air pollution are the by-products of combustion processes in manufacturing operations. Combustion consists of a process whereby hydrogen and carbon in a fuel combine with oxygen (usually derived from atmospheric air). If the combustion process is complete, then the only products are carbon dioxide and water vapour, accompanied by the emission of light and heat. However, combustion is incomplete in many situations. Incomplete combustion occurs when the air-to-fuel ratio is poor, or when the combustion temperature is either too high or too low. The polluting products that result from incomplete combustion include carbon monoxide, oxides of nitrogen and sulphur, unburned hydrocarbons and ash.

Pollutants can also be subdivided into primary and secondary ones. Primary pollutants are emitted directly from pollutant sources, whereas secondary ones are formed by chemical reactions between primary pollutants and natural components in the Earth's atmosphere. For example, the primary pollutants of sulphur dioxide and hydrogen sulphide combine with atmospheric oxygen and water vapour to produce secondary pollutants such as sulphur trioxide and sulphuric acid.

Various primary pollutants, such as organic compounds and nitrogen oxide, decompose under the action of sunlight and other atmospheric parameters to yield a particular class of secondary pollutants known as *photochemical oxidants*. These secondary pollutants include ozone, hydrogen peroxide and formic acid. Ozone has particularly serious effects, as it attacks rubber (including vehicle tyres), cellulose, nylon and acrylic materials, as well as damaging vegetation.

Chlorofluorocarbons (CFCs) are also major pollutants. These, together with carbon dioxide, are thought to be depleting the ozone layer in the Earth's atmosphere and allowing increased amounts of ultraviolet radiation to penetrate the atmosphere, leading to the phenomenon known as 'global warming'. Both carbon dioxide and water vapour also absorb some of the long-wave radiation emitted from the Earth's surface, thereby contributing further to global warming. However, particulate pollutants in the atmosphere block some of the solar radiation to the Earth, causing a cooling effect. There is in fact little hard scientific evidence about which of these separate heating and cooling effects is going to be greatest over the long term, i.e. whether pollution will lead to global warming or global cooling. We may even be lucky and find that the heating and cooling effects cancel out, meaning that the mean temperature of the Earth would neither rise nor fall.

The effects and longevity of particulate pollutants depends on their typical size. Particles less than 0.1 μm in diameter undergo random Brownian motion in air, and have a lifetime that can vary from a few seconds up to several months. Particles between 0.1 μm and 2 μm in diameter tend to be removed from air naturally by rainwater, although this might only mean that they become waterborne pollutants instead of airborne ones. Finally, particles greater than 20 μm in diameter are heavier than air and so settle under gravity.

6.1.1 Emission inventory

Preparation of an emission inventory is a key element in developing an air-quality management strategy. The emission inventory documents all polluting emissions from a given source. Emissions are quantified in terms of emission factors, which define the typical levels of various pollutants from a source, measured either as a mass concentration or as a volumetric concentration. See Reference 1 for tables of emission factors for various pollutant sources.

6.1.2 Limits of acceptable pollution

The acceptability or otherwise of particular pollutant emission levels depends on the legislation relevant to the pollutant, the dispersion characteristics of the pollutant, and the distance between the pollutant source and human population centres likely to be affected. Whilst legislation on maximum acceptable levels of various pollutants varies somewhat from country to country, there is a common aim of setting permissible pollutant concentrations at levels below that at which harm may come to humans, either directly or through the food chain. Pollutants that are not directly dangerous to humans but are harmful in other ways, such as by contributing to 'global warming', are also strictly controlled in most countries. Tables of typical values of limits for various pollutants can be found in Reference 2.

6.1.3 Dispersion of pollutants

General air motion transports pollutants downwind. However, a degree of turbulent air motion always exists that distributes a proportion of the pollutants in directions that are not the same as the general wind direction. There is also a gradient of decreasing concentration as the distance from the source increases.

The dispersion pattern of small particles is very similar to that of gases, as they are readily carried along in the wind. However, the concentration gradient of pollutants in the form of medium-sized and large particles is strongly affected by their aerodynamic characteristics of size, shape and mass. These parameters determine the extent to which particles are carried by the wind. Many particles tend to fall to the ground, but are then carried back up into the air by gusts of wind and turbulent air conditions. Variations in meteorological conditions also determine the extent to which pollutants are carried up into the atmosphere. Certain conditions will cause pollutants to remain close to the ground, whereas other conditions will cause them to be carried high into the atmosphere.

Prediction of wind velocity and direction is a complicated meteorological problem. Differential heating of air over the Earth's land and water surface, and also between polar and equatorial regions, causes pressure gradients. Air (described as a 'wind') flows from high pressure to low-pressure areas, but not in a straight path, since the Coriolis force due to the Earth's rotation also has an effect. The air flow is also disturbed by variations in the land terrain and by obstacles such as buildings and trees. The obstacles cause unstable conditions in the air and produce vertical components of velocity that carry pollutants upwards and reduce their dispersion along the Earth's surface. Turbulence effects in unstable airflow conditions also cause pollutant dispersion away from the mean horizontal wind direction.

Beyond this brief introduction to patterns of pollution dispersion, readers requiring more detailed information are directed to References 1 and 3.

6.2 Measurement of Air Quality: Particulate Matter Content

Pollutants in the form of particulate matter can be divided into fine particles (<2.5 μm) and coarse particles (>2.5 μm). The terms *fumes* and *mist* are also used to distinguish between the size of pollutant particles. Fumes covers particle sizes between 0.001 μm and 1 μm, and mists cover particle sizes between 0.1 μm and 10 μm. Thus the descriptions of fumes and mists overlap for a certain band of particle size. Fumes are alternatively known as smoke and describe the effect where pollutant particles are formed by condensation, sublimation or chemical reaction. Mists are composed of liquid particles formed by condensation.

Particulate content in air is usually measured in units of micrograms per cubic metre of air (μg/m^3). Measurement techniques depend on whether the particles are lighter than or heavier than air. Light particles suspended in air are usually measured by driving the air at a constant rate through a filter for a given period of time. The mass of particles collected, divided by the total volume of air that has flowed, allows the mass concentration of the particles to be determined.

In the case of heavy particles that settle under gravity, the earliest known techniques for measurement consisted of collecting the particles in a bucket of known open area for a given length of time. A later development of this technique uses two separate collection buckets, so that both dry deposition and deposition during rainfall can be measured. To achieve this, a canopy is employed that covers one or other bucket. Normally, the canopy covers the 'wet' bucket, but it is moved to cover the 'dry' bucket by a mechanism that is activated by a rainfall sensor.

6.2.1 Measurement at the point of emission

Particulate content in flue gases can be measured in several ways, as explained in the following paragraphs.

Charge transfer sensors

Charge transfer sensors (also known as *triboelectric probes* and *particle impingement sensors*) consist of an insulated probe positioned in the gas flue. As particles in the gas impinge on the probe, an electrostatic charge is generated that can be measured electronically. The output from the sensor gives the particle mass flowing per unit time. Thus, measurement of and correction for gas flow velocity is required if the flow velocity is not constant. This constraint also means that this type of sensor only works in laminar (i.e. not turbulent) gas flow conditions. There is also a lower limit on the particle size and mass that can be measured. In the case of small particles in a low-velocity flow, a boundary layer can develop around the probe that deflects the particles. This prevents them impinging on the probe and contributing to the electrostatic charge build-up.

Opacity measurement

The particle content in gaseous waste products can be continuously monitored at the point of emission from a chimney by various optical systems that measure the opacity of the gaseous waste. Opacity is defined as the percentage of light that is prevented from passing through the gas because of absorption or scattering by particles within the gas. Opacity increases as the concentration of particles increases, and so measurement of opacity allows the concentration to be calculated. The simplest method of quantifying opacity is visual estimation. This involves comparing the greyness of the plume at the outlet of a chimney against a standard chart known as a *Ringelmann chart*. This shows six standard levels of greyness from 0 (white) to 5 (black). Whilst this method is easy to apply, it is highly subjective and prone to errors, especially in adverse weather conditions.

Opacity can be measured more precisely by mounting a light source and a photodetector inside the chimney stack, in an arrangement that is often called a *transmissometer* system. The transmission of light between the source and detector is attenuated according to the density of particles in the flowing gases. However, good measurement accuracy is only achieved if the amount of attenuation is significant. Hence, this technique is not suitable for measuring low particle densities, particularly when the chimney or flue size is such that the distance between the light source and detector is only small.

The measurement system used can be either single-pass or double-pass, as shown in Figure 6.1 (a) and (b). In the single-pass system, the light source is mounted at one side of the chimney and the detector on the other side whereas, for the double-pass system, both source and detector are placed on the same side and a reflector is placed at the other side of the chimney. This latter arrangement increases the measurement sensitivity and is very useful when the concentration of pollutants, and hence the opacity, is relatively small. With either arrangement, accurate alignment of the optical components is necessary to ensure that the transmitted light is received properly at the receiver. Build-up of particles on the optical surfaces of the transmitter and receiver (plus on the mirror in a double-pass system) is also a potential problem with this type of device, since such particles have the same attenuating effect on light transmission as particles in the flowing gases. One way to overcome this is to recalibrate the system at times when the chimney is not in use, in order to compensate for particles on the optical system. An alternative solution is to direct a clean jet of air continuously over the optical system elements.

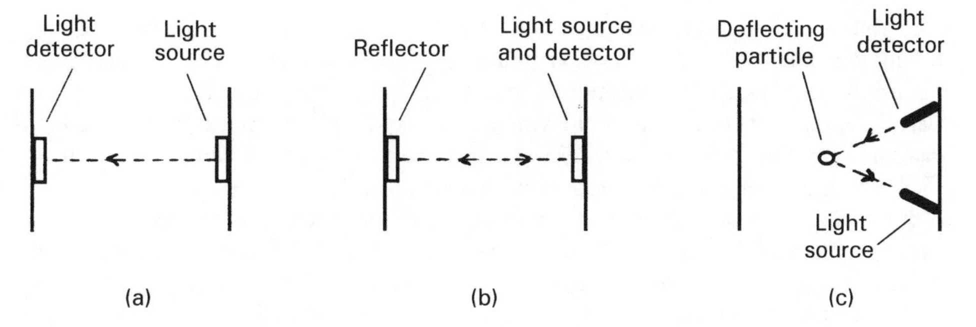

Figure 6.1 *(a) Single-pass opacity sensor; (b) double-pass opacity sensor; (c) backscatter sensor.*

Backscatter sensors

The backscatter sensor is an alternative optical technique that works best with low particle densities in small chimneys and flues. It thus complements the transmissometer, which works best with high particle densities in large chimneys/flues. In the backscatter sensor, light from a source is directed into the chimney at an angle of about 45° to the direction of gas flow and a detector measures the amount of light that is reflected from particles in the flowing gas, as shown in Figure 6.1(c). This technique is capable of measuring particle densities as small as a few $\mu g/m^3$.

6.3 Measurement of Air Quality: Concentration of Polluting Gaseous Products

Measurement of gaseous air pollutants is commonly expressed on a *parts-per-million (ppm)* volumetric basis and very small concentrations are sometimes expressed as *parts-per-billion (ppb)*. However, if the mass concentration is required, this can be obtained using the following conversion formula:

$$C_{\mathrm{m}} = (C_{\mathrm{v}} \times M \times 1000)/K$$

where C_{m} is the mass concentration in $\mu g/m^3$, C_{v} is the volumetric concentration in ppm, M is the molecular weight of the pollutant, and K is a constant that varies with atmospheric pressure and temperature. The value of K is 24.5 under standard conditions of 25 °C and normal atmospheric pressure (1.013 bar).

Example

The concentration of sulphur dioxide in an air sample is 0.5 ppm when measured at 25 °C and standard atmospheric pressure of 1.013 bar. Calculate the mass concentration given that the molecular weight of sulphur dioxide is 64.0 g/mole.

Applying the formula above, $C_{\mathrm{m}} = (0.5 \times 64.0 \times 1000)/24.5 = 1306\,\mu g/m^3$.

6.3.1 Sampling

The pollutant concentration in a gaseous product is normally calculated by analysing a sample of the product. However, if pollution concentration is to be calculated from a sample, it is essential that the sample accurately reproduces the sampled gas. Because pollutant concentration may be uneven across the cross-section of a flowing gas, the sampling process must involve a traverse across the whole cross-section, so that the sample collected is as representative as possible.

It is also very important that the sampling point for the sampling process is chosen carefully. Because bends and other obstructions in a gas-carrying pipe cause turbulance in the gas, the distance between the sampling point and such points of disturbance must be at least eight times the diameter of the pipe, to ensure that the sample collected is representative. Another important condition in sampling is that the equipment used to collect and analyse the gas sample must be inert with respect to the pollutants in the gas. This requirement can be difficult to satisfy in some circumstances, because many plastics and even stainless steel react chemically with some substances. Finally, the temperature at which sample collection and analysis equipment is held is also important. This must be above the dew point of condensible vapours in the sample, to avoid condensation, but below the temperature at which chemical reactions take place within the sampled gas.

More detailed discussion on sampling can be found in References 4 and 5.

6.3.2 General measurement techniques

Measurement techniques can be divided into 'wet chemistry', laboratory-based methods and on-line sensors. Laboratory techniques have been used for many years but they are slow in operation and unsuitable for the continuous monitoring of pollutants that is required in many situations. In consequence, laboratory techniques have been largely superseded by on-line sensors, and this chapter will therefore only cover the latter. Reference 5 does provide coverage of wet-chemistry laboratory techniques if these are required.

Many types of on-line sensor are available, some of which can be obtained in hand-held forms. All of the various types of sensor are summarised below. These have various merits, and choice between them follows the usual rule of comparing price with performance in any given application.

Electrochemical cells

Electrochemical cells produce an electric current that is a function of gas concentration. The rate of flow of pollutant through a semi-permeable membrane at the input to the cell is proportional to the pollutant concentration. The pollutant undergoes an electrochemical reaction with an oxidising electrode within the cell, producing electrons and hence an output current that is a function of the pollutant concentration. This technique can be used to measure hydrogen sulphide, sulphur dioxide, carbon monoxide, chlorine, ammonia, some hydrocarbons and all nitrogen oxides. Unfortunately, electrochemical cells have a relatively short life (one to two years) and cannot be used above 50 °C, but they have the advantage of having quite a low cost and being portable.

Electrical conductivity sensor

The electrical conductivity sensor is based on similar principles to the electrochemical cell. However, rather than having an output in the form of a varying current, it is the pollutant-induced conductivity change in the electrolyte that is measured. It is particularly suitable for quantifying ammonia, sulphur dioxide and carbon dioxide at concentrations between 1 ppm and 100 ppm.

Infrared/ultraviolet detectors

Several measurement techniques use infrared or ultraviolet light. These come in one of two forms, which are (1) sampling systems such as non-dispersive detectors that operate on a sample of gas extracted from a flue or chimney, and (2) across-duct systems that measure the whole gas flow through a flue or chimney. Techniques in both classes rely on selective absorption of light at specific wavelengths that are characteristic of particular gases. The peak absorption wavelengths for common pollutant gases are given in Table 6.1.

Nondispersive infrared detector

This type of detector, which is also known by several alternative names, including *infrared spectrometer*, *infrared photometer* and *nondispersive filter spectrometer*, relies on the selective absorption of infrared radiation at a particular wavelength that is characteristic of the gas detected. A chopped infrared source at the top of the detector passes into two central chambers, called the reference cell and the sample cell respectively, as shown in Figure 6.2. Filters are inserted between the infrared source

Table 6.1 *Peak absorption wavelengths for various gases*

Gas	Wavelength for peak absorption	
	Infrared (μm)	Ultraviolet (nm)
Ammonia (NH_3)	—	200
Carbon monoxide (CO)	4.60	—
Carbon dioxide (CO_2)	4.35	—
Chlorine (Cl)	—	280–380
Hydrogen sulphide (H_2S)	—	200–230
Methane (CH_4)	3.30	—
Nitric oxide (NO)	5.30	—
Nitrogen dioxide (NO_2)	3.70	380–420
Nitrous oxide (N_2O)	4.50	—
Sulphur dioxide (SO_2)	4.00	285

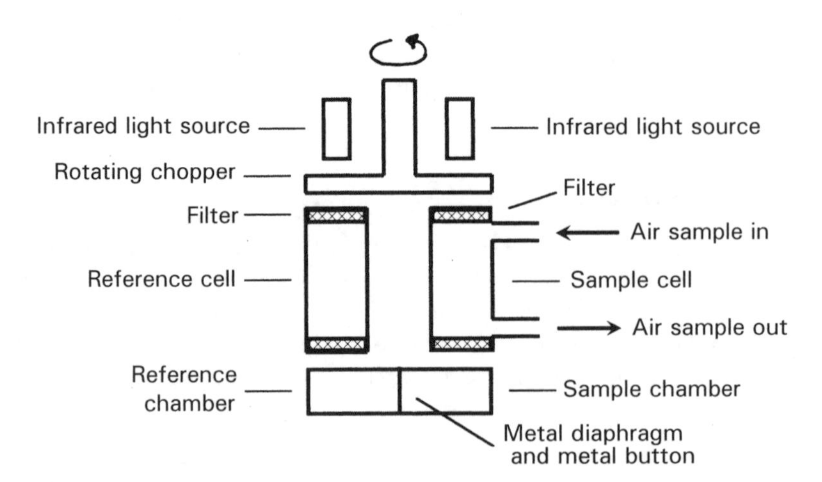

Figure 6.2 *Nondispersive infrared sensor.*

and the central chambers, to block all infrared wavelengths except those within the range that are absorbed by the gas being detected. The reference cell is filled with an inert gas (usually nitrogen), and the sample of air being measured is passed through the sample cell. The lower part of the detector consists of a reference chamber and a sample chamber that are separated by a metal diaphragm and a metal button. Both of these lower cells are filled with the pollutant gas that the instrument is set up to detect (for example, if the cell is set up to measure the concentration of sulphur dioxide in the gas sample, then the lower cells are filled with pure sulphur dioxide).

The central reference cell does not absorb any infrared radiation, and so all radiation passing through it enters the lower reference chamber, where it is absorbed, causing the temperature of the gas in it to rise. On the other side of the instrument, the pollutant gas in the air sample being measured in the sample cell absorbs some of the infrared radiation passing through, and so a lesser amount of radiation passes through for absorption in the lower sample chamber. In consequence, the temperature rise in the lower sample chamber is less than that in the reference chamber. Because both the lower chambers are of fixed volume, the differential temperature between the two chambers causes a differential pressure between them. This differential pressure causes the metal diaphragm to deflect, changing the capacitance between the diaphragm and the metal button, and allowing the concentration of gas in the air sample to be calculated from the capacitance change. Alternatively, the temperature difference between the two chambers can be measured directly by temperature sensors in the chambers.

A necessary part of a nondispersive detector is a sampler that extracts representative samples of the flowing gas, filters them to remove particulate matter, adjusts the temperature and pressure to standard values and then transmits the sample into the sample cell of the detector. One potential problem with the instrument is that accuracy is seriously impaired if the instrument is affected by vibration. However, recent developments have largely overcome the effect of any vibration, and allow low gas concentrations to be detected. For example, carbon monoxide concentrations as low as 20 ppm can be measured.

This technique can be used to measure sulphur dioxide, carbon monoxide, ammonia, all nitrogen oxides and some hydrocarbons. Unfortunately, instruments using this technique are quite expensive and have a relatively high power consumption. They are used particularly in conditions where other types of sensor are unsuitable.

Nondispersive ultraviolet detector

This has almost identical structure to the nondispersive infrared detector, except that an ultraviolet light source is used instead of an infrared one.

Fourier transform infrared (FTIR) detector

The FTIR detector allows many different gas components to be detected simultaneously. Its performance is generally better than that of nondispersive infrared detectors. This is a fairly new technique, with an expanding range of applications, but growth in use is limited by the relatively high cost and requirement for regular maintenance of this type of instrument.

Infrared/ultraviolet absorption detectors

Both infrared absorption detectors and ultraviolet absorption detectors use an energy source and receiver system mounted on opposite sides of a flue or chimney carrying pollutant gases. The energy source is either infrared or ultraviolet light, and the light wavelength used is specifically chosen to detect a particular gas. The degree of light absorption indicates the concentration of the gas detected. It is normally necessary to direct air jets across the optical components to prevent measurements being affected by dust in the system. As well as the list of gases shown in Table 6.1, ultraviolet absorption detectors are also used to measure some hydrocarbon pollutants.

Catalytic oxidisation sensor

This device, also known alternatively as a *calorimetric sensor* or *electro-catalytic sensor*, measures the heat evolved during catalytic oxidation of reducing gases. In one form of this type of sensor, a platinum coil is coated with a catalyst, and the resistance change due to the heat generated in the coil is measured when it is exposed to the pollutant. In an alternative form, the catalyst is deposited on the surface of a ceramic bead, with the temperature rise in the bead being measured by a thermocouple embedded in the bead.

Catalytic oxidation sensors are used to measure carbon monoxide, methane and various hydrocarbons. The devices are robust, cheap and simple to use, but have low sensitivity (the minimum gas concentration measureable is 500 ppm). Also, measurement becomes inaccurate at very high pollutant concentrations, because insufficient oxygen is available for combustion of all the pollutant. Thus, the measurement range is quite limited. A further problem is that the catalyst is susceptible to contamination by a number of substances (in particular any compound containing silicon, phosphorus, lead or sulphur). Catalysts that are resistant to contamination are now available, but at greater cost.

Catalytic gate FET (field effect transistor)

This device uses the catalytic dissociation of hydrogen at a palladium catalytic gate in a FET. The dissociation causes a voltage drop at the gate and increases the output current from the transistor for a given externally applied gate voltage. The extent to which the output current/gate voltage changes depends on the concentration of the gas detected. These are cheap devices that are used for measuring hydrogen sulphide, ammonia and some hydrocarbons, particularly at high concentrations.

Semiconductor sensors

In this sensor, the absorption of a pollutant gas by a heated semiconductor causes changes in the surface conductivity of the semiconductor. Hence, the concentration of the gas detected is determined by measuring the resistance change in the semiconductor. Usually, a metal oxide semiconductor such as tin oxide is used, but organic materials such as polypyrrol are used in some detectors. Detectors of this type are used to detect a wide range of both inorganic and organic pollutants at concentrations between 10 ppm and 1000 ppm. Whilst semiconductor sensors have the virtue of being relatively inexpensive, they have a nonlinear form of response and have poor accuracy.

Biosensors

Biosensors are a relatively new development that can measure both airborne and waterborne pollutants. A biosensor consists of a biological component such as an enzyme or cell attached to the surface of a transducing element within a probe. When biosensors are exposed to a pollutant, the output can take several forms, such as electron emission (electric current) or the emission of a gas like oxygen or ammonia within the probe. This output is detected by appropriate means. The internal structure of biosensors varies according to the method used to detect the output. Enzyme-based biosensors with a gaseous form of output have a structure of the form shown in Figure 6.3 and produce measurements quickly (around 60 s). The active element is a membrane contained within a jacket at the tip of a probe, with an enzyme attached to the membrane that is specific to the pollutant detected. As the pollutant diffuses through the enzyme, it is oxidised, producing an output gas (typically hydrogen peroxide) whose magnitude is proportional to the concentration of pollutant in the sample. The gas is detected by an electrode (often platinum), producing a measureable output current. More sophisticated variants have a three-membrane system with an outer membrane to protect the membrane containing the enzyme and an inner membrane to protect the electrode from fouling by the pollutant.

Biosensors provide an on-line monitoring capability at relatively low cost and require only a very small sample of polluted air or water for operation. They have a long life and can typically take hundreds of measurements before the membrane system jacket needs replacing. Different kinds of biosensor can be used to detect various metallic and organic pollutants. In general, cell-based biosensors measure pollutants across a broad spectrum rather than measuring specific pollutants. In

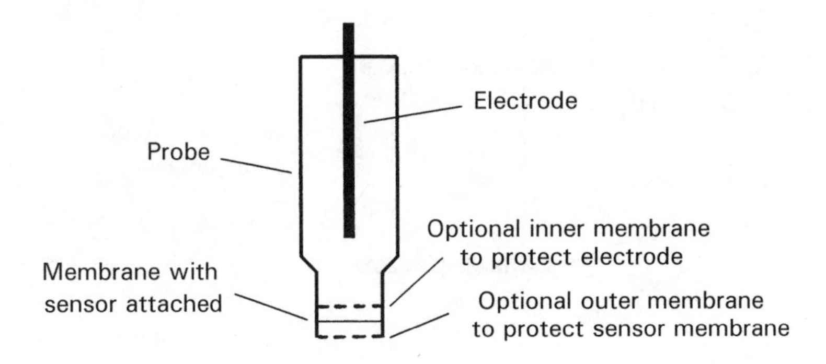

Figure 6.3 Biosensor.

contrast, enzyme-based biosensors are specifically tailored to detect a single pollutant. However, a basic enzyme-based sensor unit can usually detect a range of different pollutants by interchanging membrane jackets that are specific to particular pollutants. Greater detail can be found, if required, in References 6–7.

Photoionisation detector

The principle of the photoionisation detector is shown in Figure 6.4. The ionisation chamber has electrodes at the top and bottom, and pollutant molecules within it are ionised by photons from an ultraviolet (UV) discharge lamp, producing an output current between the electrodes. The UV lamp has to be carefully chosen, so that pollutant molecules are ionised but other molecules in air (e.g. oxygen, nitrogen, carbon monoxide) are not. A 11.7 eV lamp is generally suitable, since common pollutant molecules have an ionisation potential less than this and are ionised, whereas other components in air have an ionisation potential greater than 12 eV and are not ionised. Both organic and inorganic pollutants can be detected, although the primary use is for measuring hydrocarbons at concentration levels between 0.1 ppm and 1000 ppm. If several different pollutants are present, the detector can often be made selective to just one by using lamps of lower power (typically 9.5 eV or 10.2 eV), since the ionisation potentials of different pollutants straddle these values. Unfortunately, the photoionisation detector suffers from reliability problems, owing to the need to replace the discharge lamp at regular intervals, and hence its use is not as widespread as the measurement techniques mentioned earlier in this chapter.

Thermal conductivity sensor

This device, also known as a *katharometer*, compares the thermal conductivity of polluted air with that of pure air. Its main use is in measuring gases at medium to high concentrations (>0.1%).

Flame ionisation detector

The flame ionisation detector is used to detect hydrocarbon pollutants, and has the form shown in Figure 6.5. The air being analysed is mixed with hydrogen and burnt

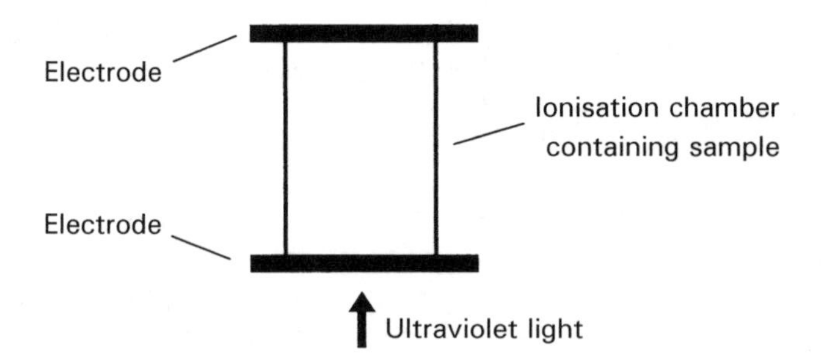

Figure 6.4 *Photoionisation detector.*

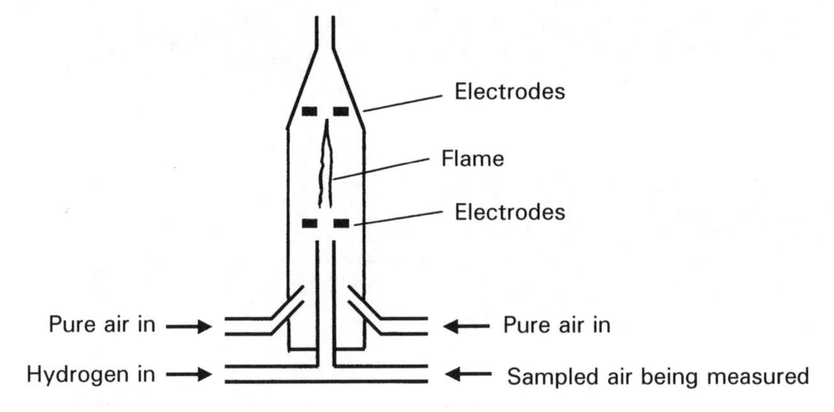

Figure 6.5 *Flame-ionisation detector.*

at the end of a metal capillary tube that has a supply of pure air flowing round it. The flame is positioned between electrodes that have a potential of several hundred volts across them. In the absence of any pollutants in the air being analysed, combustion of the hydrogen produces very few ions and there is almost no current between the electrodes. However, the presence of hydrocarbons in the analysed air produces ions in the flame and a net d.c. current between the electrodes. The detector does not identify the type of hydrocarbon present. However, this information can be obtained if the detector is used as part of a gas chromatography system (see later). A disadvantage of this type of detector is that it is quite expensive. However, it is very sensitive, and can detect pollutant concentrations of a few parts per million.

Gas chromatography

Gas chromatography is not in fact a single technique, but is rather a collective term used to describe a number of different though similar techniques that involve separating and then analysing a number of different pollutants in an air sample. A number of different gases can be measured, including carbon monoxide, hydrogen sulphide, nitrous oxide, and various hydrocarbons and chlorofluorocarbons (CFCs). The general principle of gas chromatography is to mix a sample of the air to be analysed with an inert carrier gas and to inject this into a chromatographic column, as shown in Figure 6.6. The column is filled with a nonvolatile substance called the stationary phase that selectively retards different components in the air sample. Both liquids and solids are used as the stationary phase in different gas-chromatography techniques. The different pollutants in the air sample emerge from the column at different times, according to the degree of retardation in the column. The inert gas serves to assist the flow of the pollutants through the column, but does not suffer any retardation itself. Pollutant-specific detectors at the output of the column measure the quantity of each pollutant in the sample. There is usually a chart recorder at the output of the instrument that produces a graphical output called a chromatogram (shown on the right-hand side of Figure 6.6). This has a series of spikes along the

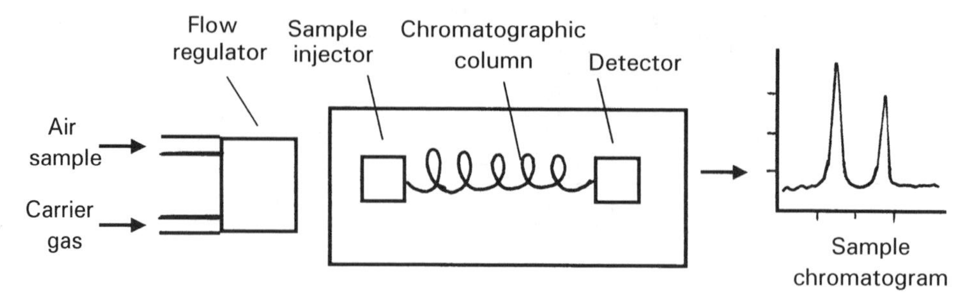

Figure 6.6 *Gas chromatography measurement.*

time axis. The identity of each gas detected is indicated by its position along this time axis (influenced by the degree of retardation for the gas in the column), and the height of each spike indicates the concentration of that particular constituent. Various forms of gas detector are used in gas chromatography, such as the flame ionisation detector and the thermal conductivity sensor.

Gas chromatography produces good results, but is somewhat expensive and requires skilled maintenance. One particular limitation is that it only detects and quantifies gases that are volatile. Volatility, and therefore performance, is improved by increasing the operating temperature, but care has to be taken to keep the temperature below that at which constituents in the air sample start to decompose. Usually, operational temperatures up to about 300 °C are satisfactory. Whatever operational temperature is chosen, it is important to maintain this at a constant value, because the temperature influences the shape of the chromatogram obtained, and hence the deductions made about the chemical composition of the air sample.

Mass spectrometer

The mass spectrometer is a sophisticated instrument that, in consequence, is very expensive. It also requires skilled maintenance. However, it is particularly suitable for on-line monitoring, because of its high speed of response. Also, it has the particular advantage of being able to detect many different pollutants simultaneously. In addition, it has a particularly wide measurement range, from pure gas (100%) down to a few parts per million.

The principle of mass spectrometry is to direct an electron beam at a sample of gas contained in a tube. The beam ionises the gas, producing positively charged molecules. Charged molecules of each different gas in the sample have a specific and characteristic mass–charge ratio. The charged molecules are then sorted according to their mass–charge ratio, either electrically or magnetically.

In electrical sorting, the charged molecules pass down a tube containing four electrodes arranged as opposite pairs. A voltage of specific magnitude and frequency, chosen according to the specific gas to be detected, is applied alternately to opposite pairs of electrodes. The field due to this voltage deflects most molecules, and only those with one specific mass–charge ratio pass down the tube to a detector at its end, where the magnitude of the charged particles detected corresponds with the concentration of one gas in the sample matching one specific molecule mass–charge ratio. All the different component gases in a sample can be quantified by repeating this process on new samples with different voltage magnitudes/frequencies corresponding with each gas to be detected.

In the alternative magnetic sorting, the stream of charged molecules passes down a tube and is deflected by magnetic plates placed either side of the tube on to a series of Faraday collectors that produce output currents. The angle of deflection of the molecules differs according to their mass–charge ratio. Thus, each Faraday collector only detects particles of one particular mass–charge ratio, and so the output current magnitudes from the collectors correspond with the concentration of each gas in the sample.

Ion mobility spectrometer

This is a relatively new type of instrument that has similar operating principles to the mass spectrometer. A sample of gas is introduced into a long tube. The sample is subjected to a radioactive beta source at the entrance to the tube, which converts some molecules to negative ions whose mass is characteristic of their parent gas. An electric potential is applied to the ions, and their resultant velocity down the tube varies according to their mass. Ions from different constituent molecules in the gas arrive at a collecting electrode at different times, according to their mass. The output from the electrode consists of a varying current whose time spectrum corresponds with the constituents in the gas sample.

These devices have a wide measurement range from 0.01 ppm up to 100 ppm, but are unfortunately very expensive. They can be used to measure a variety of ionisable organic and inorganic compounds. They are used particularly in a hand-held form for detecting toxic gases. Further information can be found in Reference 8.

Intelligent array sensor

This sensor is included within this chapter, because, although it is currently too expensive (£25 000 typical cost) for use in environmental monitoring applications, its price is likely to fall rapidly in the future. In operation, it emulates the human nose by identifying gases from their characteristic odours. In consequence, it is often known by the name of *electronic nose*.

Most techniques of gas analysis and measurement covered in this chapter can quantify odours consisting of single gases. However, very few techniques can deal effectively with the presence of multiple gases. The intelligent array sensor does have this ability, and it is therefore likely to become widely used in environmental monitoring applications once its cost falls to reasonable levels.

The intelligent array sensor is an electronic system that has an array of different input sensors. Each sensor in the array is sensitive only to one particular gas or to a small group of gases with similar chemical characteristics. The sensors used are commonly either organic conducting polymers[9] or metal oxide semiconductors. Both of these have a very fast response, with an output in the form of a resistance change according to the magnitude of each gas detected. If the pattern of resistance changes in the sensors is recorded when a range of known gases or gas mixtures of various concentrations are applied to the sensors, a pollutant/response data set can be built up. This can then be applied to a pattern-classifying neural network, as described in more detail in the papers referenced,[8,10–11] which can subsequently be used to identify and quantify unknown pollutant gases.

Despite its current high cost, the intelligent array sensor is currently used in some medical applications[12] as a diagnostic tool for identifying odours emanating from and characteristic of particular bacteriological infections. It is also used in various quality-control functions in the food industry, such as monitoring cheese ripening processes and controlling fermentation and microwave cooking procedures. This usage is providing useful operational experience, and environmental applications are likely to appear once the cost of these devices falls.

Mass-sensitive detector

A mass-sensitive detector consists of a coated piezoelectric crystal in which the oscillation frequency is changed by mass changes in the coating when it is exposed to a pollutant. Different organic and inorganic compounds can be detected according to the material used for the coating.

Piezoelectric badges

Piezoelectric badges have similar dimensions to a credit card, and monitor the exposure of workers over a given period of time (often one eight-hour shift) in workplaces where there is a possibility of toxic gases being generated. Badges are very sensitive and can measure toxic gases in concentrations of only a few parts per billion. Normally, each worker carries his/her own personal badge, usually worn near the face. However, if required, this technology can be used in a nonpersonal detector mounted at one or more fixed locations in a workplace. Each badge is sensitive to only one specific gas. Hence, if there is the possibility of exposure to several gases, the worker concerned must wear a separate badge for each gas.

The active part of a badge consists of a solid gel containing a catalyst and reagent that is mounted on a polypropylene film having piezoelectric properties. Exposure of the gel to gas causes a colour change in it, and the final intensity of colour indicates the amount of gas that the badge has been exposed to over the monitored period. This quantity measurement is made by putting the badge in a reader at the end of the period of exposure. The reader exposes the badge to light from a red, light-emitting diode, and the gel absorbs some of the light according to the intensity of its colour. The light absorption causes a heat rise in the gel, which in turn causes a piezoelectric voltage to develop across the polypropylene film as the heat causes it to expand. This voltage provides the output reading from the instrument. Commercial applications for measuring styrene and other gases are described in the referenced paper.[13]

Chemiluminescent analysers

The chemiluminescent analyser is a specialised device that only measures nitric oxide, but it has the advantage of having a wide measurement range for gas concentrations between a few parts per billion and several per cent. It can also measure nitrogen dioxide if this is first converted to nitric oxide by a reduction catalyst such as stainless steel or molybdenum.

The mode of operation can be explained as follows. A gas sample is mixed with ozone produced within the instrument, and the gas/ozone mixture is then reduced to an absolute pressure around 25 mbar. Infrared light is emitted during the ensuing chemical reaction, as ozone and nitric oxide combine to form nitrogen dioxide. Measurement of the magnitude of the light generated indicates the concentration of nitric oxide in the sample.

Paper tape colorimetric sensor

A paper tape impregnated with a reagent is driven slowly across the polluted air stream. The tape changes colour or darkens according to the concentration of gas present. By choosing a suitable reagent, this technique can be used to measure hydrogen sulphide and ammonia concentrations, but accuracy is poor.

References

1. Wark, K., Warner, C.F. and Davis, W.T., 1998, *Air Pollution: its Origin and Control* (Addison Wesley Longman, Reading, Massachusetts).
2. Osborn, P.D., 1989, *The Engineer's Clean Air Handbook* (Butterworth, Severoaks, Kent).
3. Griffin, R.D., 1994, *Principles of Air Quality Management* (Lewis Publishers (CRC), Boca Raton, Florida).
4. Colville, R.F., 1987, Sampling for process analysis, *Measurement and Control*, March 1987, 19–26.
5. Clarke, A.G., 1998, *Industrial Air Pollution Monitoring* (Chapman and Hall, London).
6. Gaisford, W.C. and Rawson, D.M., 1989, Biosensors for environmental monitors, *Measurement and Control*, **22**, 183–6.
7. Universal Sensors Ltd, 2000, 'Biosensors', *http://intel.ucc.ie/sensors/universal/index.html* (last accessed June 2003).
8. Persaud, K.C., Khaffat, S.M. and Pisanelli, A.M., 1996, Measurement of Sensory quality using electronic sensing systems, *Measurement and Control*, **29**, 17–20.
9. Persaud, K.C, Barlett, J. and Payne, P., 1992, Design Strategies for Gas and Odour sensing, in Dario, P., Sandini, G. and Aebisher, P. (eds), *Robots and Biological Systems*, Nato ASI Series (Berlin: Springer-Verlag), pp. 579–602.
10. Gardner, J.W., Hines, E.L. and Pang, C., 1996, Detection of vapours and odours from a multi-sensor array using pattern recognition, *Measurement and Control*, **29**, 172–8.
11. Harwood, D., 2001, Something in the air, *IEE Review*, **47**, 10–14.
12. Osmetech PLC, 2003, *www.osmetech.com* (last accessed June 2003).
13. Piezoptic Ltd, 2000, *http://www.itl.co.uk/www/piezoptic* (last accessed June 2003).

7

Quantification and Effects of Water Pollution

7.1 Sources and Forms of Water Pollution

Almost all water pollution is man-made, with the primary sources being industrial processes, farming and the processing of human waste in sewage works. A small amount of pollution also comes from natural sources, such as the decay of wild animals, but this is insignificant compared with man-made sources.

Pollution of water can take many forms, such as:

- temperature change;
- colour change;
- odour;
- excess acidity or alkalinity;
- metallic contamination;
- contamination by organic materials;
- other suspended solids in water.

7.1.1 Dispersion of pollutants

The effects of pollution in flowing water are mitigated as the distance increases from the point of discharge of the pollutant into the water. However, it is dangerous to assume any particular pattern or rate for the dispersion of pollutants. It might reasonably be assumed that a pollutant discharged into a river would be dispersed quite rapidly, and that pollutant concentration would therefore decrease to a low value within a short distance of the discharge point. Ideally, dispersion of pollutants in river water should occur in three ways: laterally, vertically and longitudinally.

ISO 14000 Environmental Management Standards: Engineering and Financial Aspects. Alan S. Morris.
 ISBN 0-470-85128-7

However, such perfect dispersion rarely occurs in practice, and relatively high concentrations of pollutant can persist in a river in particular areas.

In the case of industrial pollutants that originate in premises adjacent to domestic housing, the effluent from the houses has a beneficial diluting effect on the pollutants from the industrial process if a common sewer is used. However, because domestic sewage flow is much reduced between the hours of 6 p.m. and 7 a.m., industrial pollutants can cause a serious problem if discharge occurs during the night-time hours. To avoid this problem, there is often a requirement that industrial effluent is stored overnight and released at a uniform rate during the daytime.

7.2 Consequences of Water Pollution

The consequences of water pollution depend very much on the nature of the pollutant, its concentration and the rate of dilution (if discharged into flowing water). What is acceptable in terms of pollution levels also depends on the use to which the water affected is to be put. For toxic pollutants, an important criterion is whether the water is to be drunk by either humans or animals, in which case only very small concentrations of toxic pollutants are permissible. However, toxic pollutants also have to be strictly controlled to protect biological life in water. Apart from the obvious consequences of toxic pollutants, other pollutants also have various adverse effects, as discussed below.

7.2.1 Dissolved oxygen content

Oxygen does not actually dissolve in water in the sense that it reacts chemically with it. Rather, the term 'dissolved oxygen' refers to the physical distribution of oxygen molecules in water. Oxygen enters water both by photosynthesis and by absorption from the atmosphere. The content due to photosynthesis arises as a by-product of the action of aquatic plants and algae. Additional oxygen enters flowing water from the atmosphere, particularly where the water flow is disturbed by obstructions and waterfalls.

Fish depend for their survival on the supply of dissolved oxygen in water, which they take in through their gills. Hence, any decrease in the level of dissolved oxygen damages fish and even kills them if the oxygen in the water decreases below a critical level. Pollutants affect the concentration of dissolved oxygen in water in a number of ways, as discussed more specifically in the paragraphs below.

7.2.2 Water temperature

The discharge of effluents into rivers can cause a significant increase in the mean temperature of the flowing water. Temperature increases have several adverse effects. Firstly, temperature rises increase the oxygen requirement of fish, and large temperature increases prove fatal. For example, a temperature rise of 10 °C at least doubles the oxygen requirement. However, even very small increases in temperature inhibit the breeding activity in fish. Another problem is that the concentration of dissolved oxygen in water decreases as the temperature rises, because of increased oxygen loss from the water surface. Furthermore, the increased activity of aquatic plants and organisms at higher water temperature also increases the oxygen demand and causes a further reduction in the amount of dissolved oxygen in the water.

7.2.3 Water colour

Apart from its aesthetic appearance, which may be objectionable to humans, colour in water absorbs light and reduces the growth rate of weeds. Because weeds provide food, this can have serious adverse affects on various forms of aquatic life. Coloured water is also unsuitable for use in certain industrial processes.

7.2.4 Odours in water

Unless they are of great enough magnitude to be objectionable to humans, odours in water have little direct consequence either on humans/animals that drink it or aquatic life that lives in it. However, odours indicate the presence of pollutants whose existence may otherwise be undetected. In consequence, the detection of an odour usually leads to further analytical tests to identify the pollutant that is causing the odour.

7.2.5 pH value of water

Biological life in water is affected if the pH value falls outside the optimum range of 6.7 to 8.8, and ceases altogether if the pH value falls below 4.0 or exceeds 9.5. If pH values are outside the optimum range, then they can also cause a large increase in the toxicity of metallic contaminants in water, especially lead, copper, zinc, chromium and nickel.

7.2.6 Metallic contamination in water

Metallic contaminants are extracted and stored by marine animals, particularly shellfish. This can result in relatively high concentrations of arsenic, copper and other metals when marine animals are eaten by humans, with consequent dangers to human health.

7.2.7 Contamination of water by organic materials

The primary problem with organic pollutants is that they absorb oxygen. This affects fish, as noted earlier, and can kill them if oxygen depletion is severe enough. Biological action on organic solids can also give rise to the emission of hydrogen sulphide, which is highly toxic. In addition, organic pollutants can have a stimulating effect on the growth of micro-organisms in water. The micro-organisms involved include viruses, bacteria, fungi, algae and protozoa. These can affect human and animal health to a greater or lesser extent, and their concentration in water is a very important quality parameter that has to be monitored closely.

7.2.8 Other suspended solids in water

Apart from the problems noted earlier in respect of particular kinds of suspended solids, all types of suspended solid cause a number of additional problems. Some inorganic pollutants stimulate the growth of micro-organisms in the same way as organic compounds. Suspended solids also give water a cloudy appearance that is aesthetically unattractive to humans, and the reduced transmission of light through it has the same affect as colour in water (see Section 7.2.3). Even nontoxic solids can kill fish by choking their gills. Also, in high concentrations, suspended solids can smother aquatic plants, thus affecting fish indirectly by disturbing the food chain.

Particular solids such as nitrates, phosphates and potash have a stimulating effect on aquatic plants, and cause them to emit increased levels of oxygen into water during daylight. However, at night, the plants become absorbers of oxygen. This cyclical nature of the oxygen content of water can have a serious effect on fish activity, and causes particular complaints in fishing waters.

7.3 Water Sampling in Rivers

The normal expectation is that control procedures implemented by companies to prevent the discharge of damaging amounts of pollutants into watercourses work successfully to prevent the occurrence of significant water pollution problems. Thus, water sampling and testing to detect pollution is usually carried out by bodies like river authorities, etc., rather than by the companies that are liable to cause pollution. In spite of this, it is advantageous for industrial companies to have some knowledge of the sampling and testing procedures that are likely to be applied by river authorities and similar bodies, particularly because any deviation from good sampling and testing practice by such authorities is likely to provide a good defence against any legal proceedings that may be brought against alleged polluters. It may also be advantageous in a few cases for companies to carry out their own testing, particularly where penalties for pollution are severe, since self-testing can often detect the failure of control schemes and the onset of pollution problems before a serious situation develops. However, such decisions have to be taken with proper regard to the cost of sampling and testing, as discussed below.

Water sampling may take place either routinely at periodic intervals of time or, alternatively, at random times in response to actual or suspected pollution incidents. In either case, due regard must be made to the manner in which pollutants are dispersed. In some cases, pollutants are well dispersed, and a single sample will suffice. However, it has been mentioned earlier that the distribution of pollutants within water is often not uniform. In this case, to ensure that the pollutant content in the sample accurately reflects the mean pollutant concentration in the water sampled, the samples taken for analysis must be taken from many points, so that the total sample is as representative as possible of the water sampled. In the case of water flowing in rivers or other forms of channel, this means taking samples from several points across the width, at several depths and at several points in the direction of flow.

If sampling is to take place on a routine, repetitive basis, the sampling programme must be designed with regard to the purpose of sampling and the measurement parameters required. For example, the requirement may be to measure maximum or minimum concentrations of pollutants, or to detect changes or trends. The necessary care with which samples are collected from multiple points to ensure representativeness in the sample must also take the accuracy requirement into account, and this applies equally to single samples as well as periodic samples. This care is needed, since the sampling cost is dependent on the number of samples taken, and a sampling regime that produces greater accuracy than what is required incurs unnecessary cost. In many cases, a small increase in accuracy can incur quite a large increase in cost, principally because of the greater sophistication of the equipment needed to collect samples.

If legal action in response to a pollution incident may be a possibility, it is essential that anyone collecting samples that may possibly be used in evidence is properly qualified and follows correct sampling procedures. It is also necessary that vessels used for transporting samples prior to analysis are correctly designed and ensure that oxygen levels in the sample are not altered during transit.

One frequent problem affecting the collection of samples is that a significant amount of time often elapses between a problem such as dead fish being reported and the necessary expert sampler arriving at the scene to collect samples. This means that, if the discharge of pollutants is a one-off occurrence or only occurs intermittently, the pollutant will have travelled a long way downstream and will have been substantially diluted by the time that the sample is taken. In such cases, samples can rarely be used in evidence, but it is often possible to infer the probable cause and source of pollution and win the polluter's cooperation in preventing further occurrences of the problem.

7.4 Testing of River Water for Pollution

7.4.1 Introduction

Self-testing by companies is often able to detect the onset of pollution problems before a serious problem develops. Unfortunately, the cost of pollution-monitoring schemes is relatively high, and therefore many companies are only motivated to apply them if the penalties for pollution are severe. In consequence, as noted in the last section, water sampling and testing procedures are more commonly applied by river authorities than by the companies who cause pollution. However, even if companies are not involved in water testing themselves, it is useful for them to have some knowledge of the sampling and testing procedures applied by river authorities, because any deviation from good sampling and testing practice is likely to provide a good defence against legal proceedings brought in response to pollution of watercourses.

7.4.2 General principles

One comment that pervades all aspects of testing is that speed is essential. Testing must be carried out as soon as possible after the sample is collected, since many pollutants are unstable and change their composition after collection, by reactions that involve either oxidation or chemical reduction. Testing at the site of sample collection is therefore the ideal, although this is rarely possible, because of practical and economic reasons. If rapid testing and analysis is not possible, then the sample must be treated prior to transportation from the scene of collection in order to fix the concentrations of any unstable pollutants in the sample. If the fixing is to be achieved by appropriate chemical treatment, the nature of the pollutants must be known such that the only unknown parameters are the concentrations of the pollutants. However, cooling the sample will always reduce the rate at which pollutants change chemically, and this is recommended if more sophisticated treatment is not possible or desirable for economic reasons. Keeping light away from the sample is also desirable, since light can trigger or speed up some chemical reactions.

7.4.3 Measurement techniques

All pollutant measurement procedures are normally carried out by taking a sample of water (see Section 7.4.2) and carrying out analysis in a laboratory. A number of different measurement techniques exist that can be summarised under the following headings:

- general, nonspecific tests;
- gravimetric measurement of solid content;
- volumetric analysis;
- colorimetric analysis;
- pH measurement;
- dissolved oxygen measurement;
- temperature measurement;
- biosensors;
- odour analysis;
- microbiological parameter measurement.

General, nonspecific tests

One general, nonspecific test for poisonous pollutants in water is known as the *Daphnia test*. This involves adding the aquatic animal *Daphnia magna* to a sample of water and maintaining it at a temperature of 16–18 °C for 24 hours. *Daphnia* is extremely sensitive to even very low concentrations of poisons, and therefore the test result is negative if the *Daphnia* survives for 24 hours. If the test is positive, further more specific tests are applied to identify the nature of the poison present.

Another general procedure for detecting toxins uses a test instrument that contains nitrifying bacteria within a small column, with an ammonia-detecting electrode at the outlet of the column. The nitrifying bacteria normally converts waterborne ammonia into nitrates, but even small quantities of toxins greatly reduce the ammonia-to-nitrate conversion activity, which is detected at the electrode. The test indicates toxic contamination, but does not identify the pollutant.

A further general, nonspecific test that detects the total organic content in water involves measuring the amount of oxygen needed to oxidise all organic pollutants in a sample.

Gravimetric measurement of solid content

This technique can only be applied in a properly equipped laboratory, because of the nature of the equipment required. The exact procedure followed depends on what form of measurement is required.

Total solids are measured by evaporating the sample and weighing the solids remaining. The equipment required consists of a drying oven and a balance accurate to 0.1 mg. The weighed solid can be separated into volatile and nonvolatile components by heating the evaporated sample to 500 °C. At this temperature, only nonvolatile components are left.

Suspended solids are measured by filtering the sample under vacuum through preweighed glass-fibre paper with a pore size of 0.45 μm. The paper is then dried and weighed. As for total solids, the volatile solid content can be removed by heating the filter paper to 500 °C.

Sulphate content in a sample can be determined by adding barium chloride. This results in a barium sulphate precipitate that can be filtered off, dried and weighed.

Volumetric analysis

This technique has to be applied in some form of laboratory, but the requirements are such that a mobile laboratory with a limited set of facilities is suitable if on-site analysis is required. Volumetric analysis involves adding a reagent to a fixed volume of a sample until some specific end-point to the resulting chemical reaction is reached. The end-point is determined by an appropriate chemical indicator that is chosen according to the type of pollutant to be detected. Volumetric analysis can be used to detect a number of different pollutants. It can also be used to measure pH, as described in the later section on pH measurement techniques.

Colorimetric analysis

Colorimetric analysis has to be carried out in some form of laboratory. This is usually housed in a permanent building, although a mobile laboratory is acceptable in some cases, as noted below. The technique involves adding a reagent to a water sample. The reagent reacts with the pollutant being measured to form a completely soluble product. The product forms a solution of a specific colour, with depth of colour indicating the concentration of pollutant.

Depth of colour is measured in one of four ways, which all involve determining the optical density of the solution by measuring the transmission of light through it. Because the transmission of light would be inhibited by any suspended solids, it is essential that the sample is carefully prepared and does not contain any suspended solids.

The first way to measure the depth of colour is to use Nessler tubes, which contain a set of standard solutions of various known concentrations. These are viewed along with the test sample against a white base, with the best match indicating the pollutant concentration. The accuracy of this method is quite poor, as determining the best match in colour between the test and standard solutions is a matter of judgement. Also, the colour of the standard solutions fade over time and new solutions have to be made up periodically. Particularly for this latter reason, the use of Nessler tubes is usually restricted to a fixed laboratory.

The second way is to use a set of coloured glass filters that are placed in front of a tube of distilled water. These are compared with the colour of the test sample in a tube alongside. Looking for the best match is again a matter of human judgement, but at least the coloured discs are stable in colour, removing one source of inaccuracy. This technique can be applied in a mobile laboratory if required.

The third way is to use an instrument known as a colorimeter or absorptiometer. This has a low-voltage lamp that passes a beam of light through the sample. The output light is detected by a photoelectric cell, and the optical density of the solution is displayed on a meter. Measurement sensitivity can be improved by putting a suitable colour filter in the light path inside the instrument. Before use, the instrument must be calibrated to the zero optical density of the test sample by introducing the sample without any reagent added. A calibration curve is also required that gives the optical density for known concentrations of pollutant in the sample.

The fourth way is to use an alternative instrument called a spectrophotometer. This uses similar principles to a colorimeter, but there is a prism inside the instrument that produces monochromatic light at a specific wavelength, thereby increasing measurement accuracy and sensitivity. As for the colorimeter, a calibration curve and determination of the zero optical density point are required.

pH measurement

The traditional way of measuring acidity and alkalinity is to apply volumetric analysis using a *chemical indicator* that changes colour according to the pH value. Litmus paper can be used to give an approximate measurement, but better accuracy is obtained with phenolphthalein (pink above pH 8.2; colourless below pH 8.2) and methyl orange (purple below pH 4.5 and green above pH 4.5). The neutral points at pH 8.2 and pH 4.5 respectively do not coincide exactly with the real neutral point of pH7, but the amount of acidity or alkalinity between pH values of 4.5 and 8.2 is very small, and so there is little error in practice in using these two indicators.

The *glass electrode* (see Figure 7.1(a)) is now the most common pH sensor and consists of a glass probe with a thin, pH-sensitive glass membrane at its tip. The probe contains two electrodes, a measuring one and a reference one, separated by a solid glass partition. Despite the name of the device, neither of the electrodes are in fact glass. The reference electrode is a screened electrode (often silver chloride), immersed in a buffer solution (usually potassium chloride at pH 7) that provides a stable reference e.m.f. that is usually 0 V. The measuring electrode is surrounded by a neutral solution contained within its own partitioned part of the glass probe. The glass membrane permits the diffusion of ions according to the hydrogen ion con-

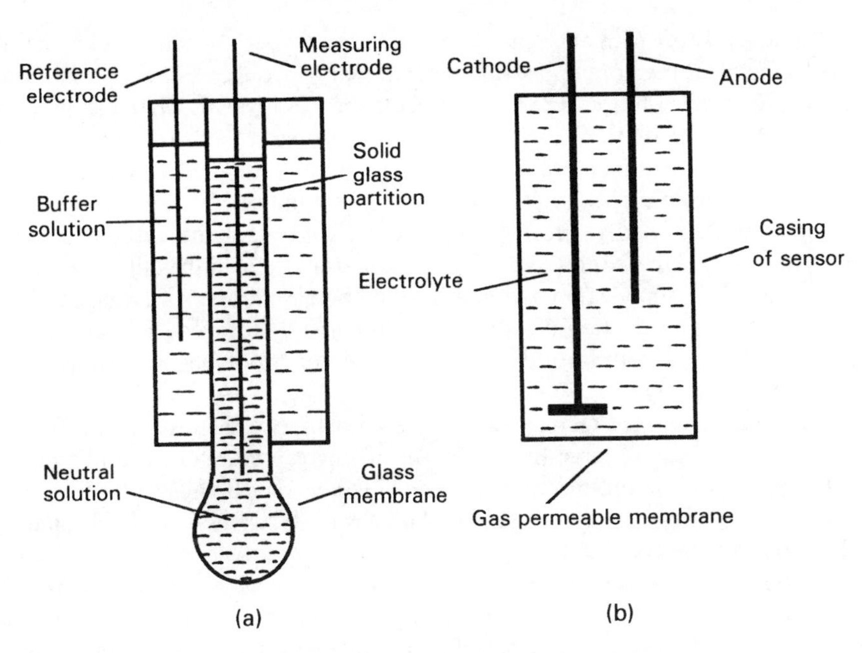

Figure 7.1 *Water quality measurement: (a) the glass electrode for pH measurement; (b) dissolved oxygen measurement.*

centration in the fluid outside the probe. The measuring electrode responds to the ion flow and generates an e.m.f. proportional to pH, that is amplified and fed to a display meter. The characteristics of the glass electrode are very dependent on ambient temperature, with both zero drift and sensitivity drift occurring. Thus, temperature compensation is essential. This is normally achieved through calibrating the system output before use, by immersing the probe in solutions at reference pH values. Sometimes, a temperature sensor is included within the electrode assembly, so that correct temperature compensation can be applied for all measurements.

Whilst being theoretically capable of measuring the full range of pH values between 0 and 14, the upper limit in practice is generally a pH value of about 12, because electrode contamination becomes a serious problem at very high alkaline concentrations, and also glass starts to dissolve at such high pH values. Glass also dissolves in acid solutions containing fluoride, and this represents a further limitation in use. If required, the latter problems can be overcome to some extent by using special types of glass.

Great care is necessary in the use of the glass electrode. Firstly, the measuring probe has a very high resistance (typically $10^9\,\Omega$), and a very low output, typically 60 mV if the pH of the measured sample differs by one from that of the buffer solution (e.g. sample at pH 8 and buffer solution at pH 7). Hence, the output signal from the probes must be electrically screened to prevent any stray pickup, and electrical insulation of the assembly must be very high. The assembly must also be very efficiently sealed to prevent the ingress of moisture.

A second problem with the glass electrode is the deterioration in accuracy that occurs as the glass membrane becomes coated with various substances it is exposed to in the measured solution. Cleaning at prescribed intervals is therefore necessary, and this must be carried out carefully, using the correct procedures, to avoid damaging the delicate glass membrane at the end of the probe. The best cleaning procedure varies according to the nature of the contamination. In some cases, careful brushing or wiping is adequate, whereas in other cases spraying with chemical solvents is necessary. Ultrasonic cleaning is often a useful technique, although it tends to be expensive. Steam cleaning should not be attempted, as this damages the pH-sensitive membrane. Mention must also be made about storage. The glass electrode must not be allowed to dry out during storage, as this would cause serious damage to the pH-sensitive layer.

Finally, caution must be taken about the response time of the instrument. The glass electrode has a relatively large time constant of one to two minutes, and so it must be left to settle for a long time before the reading is taken. If this causes serious difficulties, special forms of low-resistivity glass electrode are now available that have smaller time constants.

The *antimony electrode* is constructed similarly to the glass electrode, but uses antimony instead of glass for all parts, including the membrane. The normal measurement range is pH 1 to 11. It is particularly suitable for environments containing abrasive particles of fluorine-based fluids, which rapidly destroy glass electrodes. It is also more robust than the glass electrode, and can be cleaned simply by rubbing it with emery cloth. However, its time constant is very large, and its output response is grossly nonlinear. Because of this, its use is normally limited to environments where a glass electrode is unsuitable.

Finally, two further ways of measuring pH can be mentioned. The *fibre optic pH sensor*[1] measures the intensity of light reflected from the tip of a probe coated in a chemical indicator, whose colour changes with pH. The *fluorescence sensor*[1] uses the principle of fluorescence to measure pH by introducing fluoresceinamine into the measured fluid and measuring the pH-sensitive intensity of the fluorescence.

Dissolved oxygen measurement

All dissolved oxygen meters work on the basis of measuring the dissolved oxygen content in a sample electrochemically, in a cell as shown in Figure 7.1(b), containing an anode, cathode, electrolyte solution and a gas-permeable membrane that allows the diffusion of oxygen through it to the cathode. The current output from the device is proportional to the dissolved oxygen content in the sample. Alternative devices working on this principle differ mainly in the materials used for the electrodes and electrolyte.

The *Clark cell* (sometimes called a *polarographic cell*) uses a silver cathode, noble metal anode (gold platinum or palladium), with a potassium chloride or potassium bromide electrolyte. The cell is biased by a battery-supplied d.c. voltage of 800 mV applied between the anode and cathode. The diffusion of oxygen through the membrane to the cathode causes an imbalance of electrons in the electrolyte, and electrons are therefore emitted from the anode, producing an output current.

In use, the Clark cell suffers from a number of operational problems. Over time, silver chloride builds up on the anode, which therefore has to be cleaned periodically. The chemical reaction in the cell also causes a shift in the pH value of the electrolyte away from neutrality. Hence, the electrolyte has to be replaced frequently. The measurement time is also long, since a warm-up period of about ten minutes is necessary to allow potentials in the cell to reach steady-state.

The *galvanic probe* is an alternative to the Clark cell and uses a zinc or lead electrode for the anode and a gold or silver electrode for the cathode. Various electrolytes can be used, such as potassium chloride, potassium bromide, sodium chloride and sodium hydroxide. The galvanic probe has advantages over the Clark cell, in not requiring a bias voltage and not requiring a warm-up period before the output measurement can be read. It also requires much less maintenance, because the electrolyte does not deteriorate, and the deposit at the anode falls off instead of building up as a coat that has to be removed.

The suitability of the various dissolved oxygen probes available in different applications depends on which other gases permeate through the membrane from the water along with oxygen. Hydrogen sulphide attacks the silver cathode used in Clark cells, but does not attack the cathode (gold or silver) used in galvanic probes, because of the negative charge. Sulphur dioxide, because it converts to hydrogen sulphide at high pH values, causes similar problems in Clark cells. Ammonia, and also carbon dioxide at high concentrations, reacts with the anode in galvanic cells.

Water temperature measurement

The effect of a pollutant discharge on the water temperature in a flowing river can be calculated by measuring the temperature and volume flow rate of the pollutant before the point of discharge, and also the flow rate of the river. This enables a prediction to be made about the likely temperature rise in the river due to the pollutant discharge, and whether this is likely to cause a serious problem. These measurements and calculations can be carried out on a continuous basis if required.

Instrumentation systems to continuously measure the temperature of the water in rivers directly are rarely implemented. The more usual procedure is to measure the temperature manually as and when required. Such measurements are normally made with a standard liquid-in-glass thermometer. If the typical measurement uncertainty of around ±1% of these is inadequate, special high-grade instruments are available that offer measurement uncertainty as low as ±0.15%.

While not done commonly, it is possible to implement systems that measure the temperature of a river continuously. This requires a sensor that is anchored in the water, with signal cables carried on an umbilical cord. The sensor usually has to be protected by wire mesh to avoid damage by debris in the river. However, any mesh used must be frequently cleared of debris to avoid the temperature of the water trapped inside the mesh possibly becoming different to that of the rest of the water in the river. Suitable sensors for such a system include resistance thermometers, thermistors and silicon-diode based semiconductor devices.

Biosensors

Biosensors[2] are a relatively new development that can be used to measure both airborne and waterborne pollutants. Their principles have already been explained earlier in Chapter 6. Hence, little else needs to be said here, except to note their advantages compared with many alternative forms of sensing, particularly with respect to the measurement time. For example, cell-based biosensors can measure oxygen content in water within 15 minutes, whereas some conventional laboratory-based techniques can take five days.

Odour analysis

As noted earlier, odours emitted from water have no direct adverse consequences themselves, unless they are deemed objectionable to humans. However, they are a valuable indicator of the existence of pollutants that are causing the odour and may be causing damage to the ecosystem in the water.

Whether odours are objectionable is a very subjective question. The conventional approach to assessing whether odours are acceptable is to organise a human panel, who try to reach a consensus view about whether or not a particular odour is likely to be objectionable to the majority of people. However, this approach is very expensive, apart from only providing a subjective answer. Because of this, much effort is being expended in developing technological systems that can identify the nature of odours, quantify their magnitude and predict their acceptability or otherwise to humans.

In the discussion on gas measurement and analysis in the last chapter, many techniques were presented that can identify and quantify odours consisting of a single gas. Situations where odours emanate from a mixture of gases present greater difficulty, but the individual components can still be quantified by some of the techniques reviewed. However, translating such measurements into a prediction of human perception of the odour is a much harder task. Out of the techniques reviewed, the intelligent array sensor system (electronic nose) is the only one that offers a full solution to the problem of identifying and quantifying gases in an odour and also predicting the likely perception of the odour by humans. If used for this purpose, the neural network training set for the system would consist of the human reaction to various pollutant gases and gas mixtures at various concentrations.

Microbiological parameter measurement

Unacceptable micro-organism growth in water usually arises as a result of pollution by inorganic and organic substances that come from industrial sources. The size of most micro-organisms is such that even an optical microscope does not provide sufficient magnification to identify and quantify them, and therefore an electronic microscope normally has to be used. This involves the careful preparation of a slide from a water sample of known volume. However, as the main purpose of such micro-organism measurement is to assess the safety or otherwise of drinking water, such measurement is usually carried out by river authorities rather than by industrial companies. Therefore, no further discussion of the technique is appropriate here.

7.4.4 Instrument calibration

Finally, before leaving the subject of water sampling and testing techniques, mention must also be made of the importance of implementing proper calibration procedures. All measuring equipment, whether used for on-site or off-site analysis, needs periodic calibration to ensure that the measurements provided are accurate. Therefore, provision must be made in the design of analysis equipment for the introduction of calibration samples of liquid that contain known concentrations of the pollutant detected. More general guidance on the principles of calibration procedures can also be found in the earlier discussion in Chapter 4.

References

1. Colvile, R.F., 1985, On-line pH measurement and control, *Measurement and Control*, **18**, 396–9.
2. Gaisford, W.C. and Rawson, D.M., 1989, Biosensors for environmental monitors, *Measurement and Control*, **22**, 183–6.

8

Control of Air and Water Pollution

8.1 Air Pollution Control

A well-known adage is that prevention is better than cure. Thus, the starting point in controlling air pollution should always be to look at pollutant sources and try to minimise pollutant emission. Unfortunately, it is impossible in a book of this general nature to attempt to cover all methods of controlling emission for every type of pollutant from every type of pollution source around the world. Indeed, no single book could hope to achieve this, even if devoted specifically to the subject of pollution control. Hence, this chapter can only cover some of the most important principles in pollution control.

The starting point in implementing a pollution control system is to set targets for the extent to which pollutants are to be removed. Whilst it is desirable to minimise pollution to the lowest level that is technically feasible, due regard must be given to the economic cost of pollutant removal. If technical solutions to pollution problems were pursued regardless of cost, the financial well-being of a company would be threatened and some workers would have to be made redundant, even if the company avoided having to close down entirely. This need to balance environmental needs against economic reality is recognised by ISO 14001, and was discussed at length in Chapter 2. Thus, the targets set for pollutant emission to the atmosphere are usually less than what is technically achievable. Subsequently, when the EMS system is examined by auditors, the criteria for deciding whether the system satisfies ISO 14001 are to determine whether the pollution reduction targets set are reasonable, and whether any legislative pollution limits have been satisfied.

Once pollution reduction targets have been set, the EMS must prescribe a strategy for meeting the targets. A core part in this strategy is to ensure that efficient operation is maintained in all manufacturing and servicing activities that take place in a company. As pollution levels usually increase whenever there is a departure from normal operation, controls must be in place to prevent faults occurring and to detect

ISO 14000 Environmental Management Standards: Engineering and Financial Aspects. Alan S. Morris.
 ISBN 0-470-85128-7

any that do occur quickly. Risk analysis and reliability enhancement are very important in minimising the occurrence of faults, and these are discussed at greater length in Chapter 11. Statistical process control (SPC) is also a very useful tool in detecting the onset of faults, as discussed in Chapter 12. If SPC works properly, faults are detected as soon as they begin to occur, allowing prompt remedial action to be taken before any serious pollution has occurred as a result of the fault.

In the end, despite good process control and systems that can detect faults quickly so that emission of increased quantities of pollutants is prevented, the emission of significant levels of pollutant from many industrial processes is still unavoidable. In answer to this, various techniques are available that can remove both particulate and gaseous pollutants at the output of industrial processes.

8.1.1 Particulate pollutants

The diameter of particulate pollutants varies over a range between 0.01 μm and 100 μm. The large width of this range means that it is impossible to design a single piece of equipment that can efficiently remove particles across the whole range of diameters. Therefore, various types of equipment exist, with each one dealing with particles in a relatively small range of sizes.

Thus, before suitable equipment to remove particulate matter from flowing gases can be specified, the size range of the particles to be removed must be known. This requires expert knowledge of the process producing the pollutant particles, and a full discussion is therefore beyond the scope of this book. However, it is possible to give some general guidance. The ash particles that are typically produced by combustion processes have diameters in the range of 1 μm to 100 μm. Combustion also vaporises materials that subsequently condense into particles in the range of 0.01 μm to 1 μm. Metal-based dusts and fumes contain particles across the full range of sizes between 0.01 μm and 100 μm, whereas other processes produce particles in smaller ranges, e.g. cement dust (5–100 μm), fertilisers (10–100 μm), insecticides (1–10 μm) and paint pigments (0.1–10 μm). Equipment to remove particulate matter generally uses one of five principles: gravitational settling, centrifugal, electrostatic, filtration and washing. A brief discussion of these is given below, but readers requiring fuller details are referred elsewhere[1–3].

Gravitation settling

Gravitation settling is mainly suited to removing large particles in the size range above 10 μm. The procedure consists of directing the polluted gas through a duct and over hoppers that the particles fall into (see Figure 8.1(a)). For gravitational settling to work efficiently, the gas flow must have a low velocity and be nonturbulent.

Electrostatic precipitators

Electrostatic precipitators operate by passing gas between two electrodes and imparting an electrostatic charge to particulate matter in the gas, such that it migrates to one of the electrodes and is thus removed from the gas. Precipitators are made in

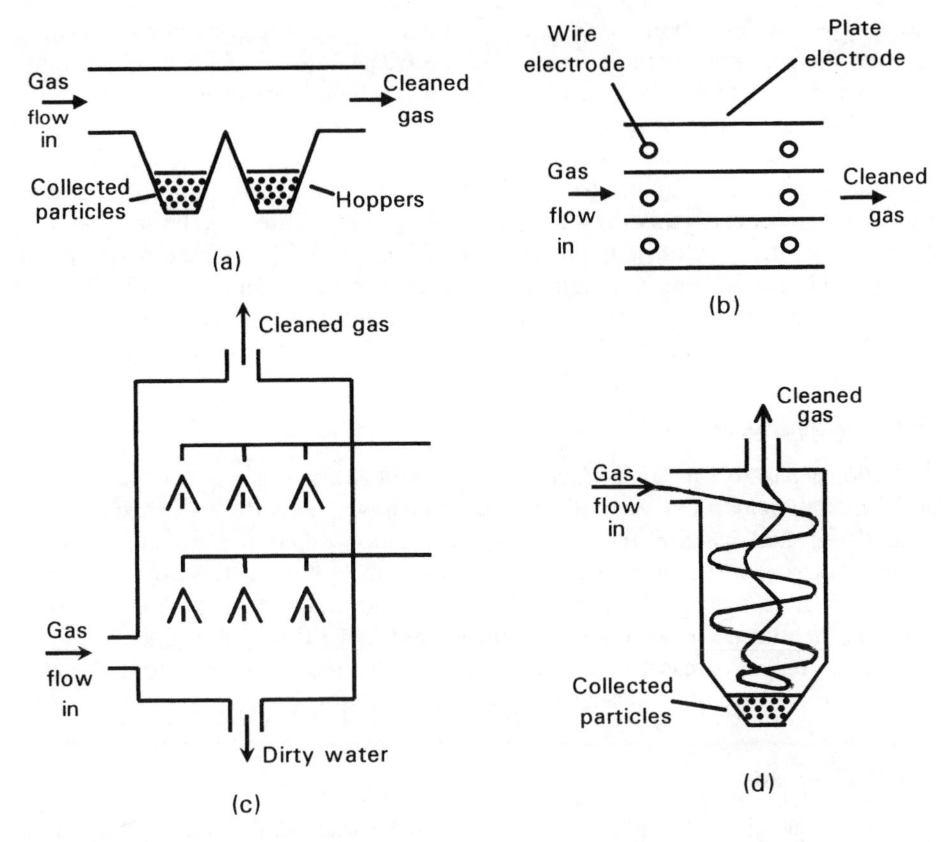

Figure 8.1 *Particulate removal systems: (a) gravitational settling; (b) centrifugal separator; (c) precipitation; (d) washing system.*

various forms, with a common arrangement being shown in Figure 8.1(b). Gas flows through tubes formed by a series of parallel metal plates that act as electrodes. Although Figure 8.1(b) only shows three tubes for convenience, a much greater number are actually used in many designs according to the flow rate and volume of gas to be handled. A series of wire electrodes are placed between the plates, and a high d.c. voltage (typically 50 kV) is applied between the wire electrodes and the parallel plates. This voltage creates a flow of electrons between the electrodes that causes the flowing gas molecules to form negative ions. These ions collide with particles in the gas and transfer their charge. The electrostatic charge on the particles then causes them to migrate to the plate electrodes, from where they are removed at regular intervals of time. Electrostatic precipitators are particularly suitable for removing particulate matter from high-volume gas flows. They can remove particles efficiently over a wide size range between 0.05 μm and 100 μm. They are also relatively cheap to operate and can withstand gas temperatures up to 650 °C and gas pressures up to 10 bar. The only limitation on the applicability of electrostatic precipitators is the nature of the particles in the gas. For efficient collection, these must have an electrical resistivity within a certain range. Low-resistivity particles, e.g. carbon-black dust, transfer their charge too easily to the plate electrode and fall off the plate back into the gas stream once their electrostatic charge is insufficient for them to adhere to the plate. High-resistivity particles cause problems for two reasons. Firstly, they insert a high-resistance layer between the electrodes, thereby inhibiting the electron flow and creation of ions. Secondly, back-ionisation can occur, in which air trapped in the layer of particles becomes ionised by the large voltage drop across the layer, producing positive ions that neutralise the negative ions. Unfortunately, many types of pollutant particles at the output of industrial processes have a high resistivity. However, in many cases, the resistivity can be lowered to a satisfactory value for precipitators by reducing the gas temperature and pressure and increasing its humidity.

Filtration separators

Filtration separators have a porous structure and consist of granular or fibrous material that traps particulate matter but allows gases to pass through. According to their design and material used in their construction, filters are available that can remove particles very efficiently across the whole range of diameters from 0.01 μm to 100 μm. Unfortunately, filters suffer from two main operational problems. Firstly, there is a significant loss of pressure in the gas as it flows through the filter. Secondly, this pressure loss increases over a relatively small period of use, requiring the filter to be either cleaned or replaced frequently.

Washing principles

Washing principles are applied in equipment that uses a liquid (normally water) to capture particles in a gas. Such equipment is often called a *wet separator* or a *scrubber* and comes in several alternative forms. In one simple form shown in Figure 8.1(c), the polluted gas flows upwards through a chamber in which a liquid is sprayed downwards through a multijet system. The liquid collects particles in the gas and then flows out at the bottom of the chamber. At some later stage in the process, the particles have to be removed from the dirty liquid if it is to be reused (or before it is discharged into a watercourse if the cleaning liquid used is water). Particles in the range between 0.1 μm and 20 μm can be removed by this method.

Centrifugal separators

Centrifugal separators direct gas tangentially into a cylindrical vessel (see Figure 8.1(d)). This causes the gas to flow in a spiral pattern around the inner surface of the cylinder. The centrifugal forces developed in the gas due to its circular motion deposit particles on the cylinder surface, from where they slide down into a hopper. When the gas hits the bottom of the cylinder, it is reflected upwards and flows within a narrow ascending spiral to an exhaust point at the top. Standard centrifugal separators only efficiently remove particles larger than 25 μm, but special designs can remove particles as small as 5 μm.

8.1.2 Gaseous pollutants

The six main techniques used to remove gaseous pollutants are solid adsorption, liquid solvent absorption, oxidation by combustion (both direct and catalytic), catalytic reduction, noncatalytic reduction and use of biofilters. In addition, some special techniques are available for removing chlorofluorocarbon (CFC) gases. Again, as for the case of equipment to remove particulate pollutants, a brief discussion of these various techniques is given below, but the reader requiring greater detail is referred elsewhere[1,3–4].

Solid adsorption

Adsorption involves passing polluted air over a solid material that collects pollutants on its surface. Collection may be either physical, using the intermolecular forces between the pollutant and the collecting surface, or it may be chemical, involving a chemical reaction between the pollutant and the collecting material. In the former case, the collecting surface can be periodically cleaned and reused. However, collection involving a chemical reaction usually requires that the adsorber be replaced periodically. Adsorption provides an economical way of removing pollutants that are only present in small concentrations. It is also very useful for removing pollutants that are not combustible (or where combustion is difficult).

Examples of adsorbent materials include charcoal (removes hydrocarbons), silica gel (removes water vapour) and zeolites. Zeolites are crystalline metal aluminosilicates. Their crystal structure is purpose-designed to remove particular pollutants, and they are often known as *molecular sieves*. Different crystal structures can remove various hydrocarbons, water vapour, ammonia, nitrogen oxides, sulphur dioxide, hydrogen sulphide and carbon dioxide.

Liquid-solvent absorption

Absorption involves bringing the polluted air into contact with a liquid that absorbs the pollutants. This process is often called *liquid scrubbing* and it is used to remove sulphur dioxide, hydrogen sulphide, nitrogen oxides and some hydrocarbons. Various designs exist for absorbing chambers. One common design has a vertical chamber, as shown in Figure 8.2, in which a clean liquid is pumped in at the top and flows down over an inert packing material inside the chamber. The polluted air flows upward from the bottom to the top of the chamber and, as it does so, the pollutant particles in it are absorbed into the liquid. The dirty liquid then flows out at the bottom of the chamber. The packing material has a purpose-designed shape that aims to maximise the surface area of liquid exposed to the polluted air. The liquid pumped through the chamber is chosen according to the pollutant to be removed.

Combustion

Combustion is a useful alternative to adsorption or absorption in many situations, and can be used to remove various combustible pollutant gases, including carbon

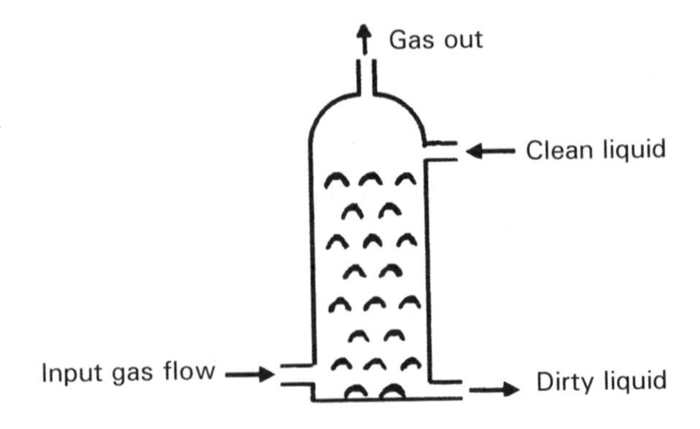

Figure 8.2 Liquid scrubbing system.

monoxide, hydrogen sulphide, cyanide gases and various types of hydrocarbon. It involves heating contaminated air in a combustion chamber for a given length of time. This causes oxidation of combustible pollutant gases, together with the emission of heat. The combustion unit is normally designed to create turbulence in the airflow, to ensure that the pollutants and air are thoroughly mixed, so that the oxidation time is minimised. If the combustion process is properly controlled, which principally means ensuring that the combustion temperature is high enough, the main products are water vapour and carbon dioxide. These are harmless to humans, although of course excessive emission of carbon dioxide is thought to be one of the contributory factors in global warming.

Unfortunately, depending on the nature and concentration of the pollutant to be removed, there is often insufficient oxygen in the air carrying the pollutant to achieve complete oxidisation and, in such cases, it is necessary to pump additional oxygen into the combustion chamber. It is also sometimes necessary to preheat the polluted air ahead of its entry into the combustion chamber, and/or inject additional fuel (such as natural gas) into the combustion chamber to raise the temperature and enhance the oxidation of pollutants.

Despite its attraction as an efficient means of removing many forms of gaseous pollutant, thermal oxidation of pollutants in air via combustion is expensive and consumes large amounts of energy, even if preheating of the air and additional oxygen or fuel are not needed. To minimise cost, heat exchangers are often incorporated into the combustion unit. These recover up to 75% of the heat generated during the oxidation process, but the addition of a heat exchanger has to be carefully considered, since they are expensive and can double the basic cost of the combustion unit. However, this cost is somewhat mitigated if there is a need to preheat the air before combustion, and the heat recovered from heat exchangers is used for this purpose.

Operational costs can also be reduced in many situations by applying catalytic oxidation, using catalysts like platinum, palladium and rhodium. A catalyst decreases the temperature at which oxidation occurs, and also reduces the length of time needed for combustion. Thus, energy costs are reduced, but of course the cost of the catalyst has to be added. As an example of the benefit of using a catalyst, benzene has to be heated to 800 °C for a time of 1 s to destroy 99% of it. However, with a catalyst, it only has to be heated to 230 °C for 0.24 s to destroy the same 99% proportion. Catalysts are therefore beneficial, but cannot be used in atmospheres that destroy catalysts, such as those containing halogen compounds. Compounds of phosphorus, silicon or sulphur and those containing heavy metals (zinc, lead, iron, etc.) also shorten the life of catalysts, and diminish the economic advantage of catalytic oxidisation.

Catalytic conversion

Catalytic conversion is primarily used just to remove nitrogen oxides. The procedure involves injecting ammonia into the output flue of a process. The gas mixture is then heated and passed over a catalyst, which causes a chemical reaction and leads to products of nitrogen and water vapour. The catalyst can consist of materials like platinum, palladium, vanadium oxide, titanium oxide and zeolites (crystalline metal aluminosilicates). The optimum temperature for the catalytic reaction varies according to the catalyst used, as given in Table 8.1.

Noncatalytic conversion

The main use of noncatalytic conversion is also to remove nitrogen oxides. It is an almost identical process to catalytic conversion, except that no catalyst is used. However, because no catalyst is used, the gas has to be heated to a higher temperature (typically 1000 °C). To avoid unnecessarily high energy costs, it is usually arranged for the cleaning plant to be located very close to the output of the chemical process that is producing the pollutant, since the temperature at this point is often already close to 1000 °C. Sometimes, noncatalytic conversion plant uses urea ($CO(NH_2)_2$) instead of ammonia as the reducing agent, although this produces an additional product of carbon dioxide as well as nitrogen and water.

Biofilters

Biofilters can only treat relatively low volumes of gas, but are increasingly popular for applications like the treatment of gases emitted from waste-disposal composting plants. Contaminated air is pumped to the base of the biofilter, which is typically 1 m deep and consists of a layer of soil, sand or compost supported on a bed of washed gravel, as shown in Figure 8.3. Odorous gases in the air are removed by a mixture of physical, chemical and biological processes. The biological processes are

Table 8.1 *Optimum temperatures for various catalysts*

Catalyst	Optimum temperature (°C)
Platinum Palladium	175–290
Vanadium oxide Titanium oxide	260–450
Metal aluminosilicate	450+

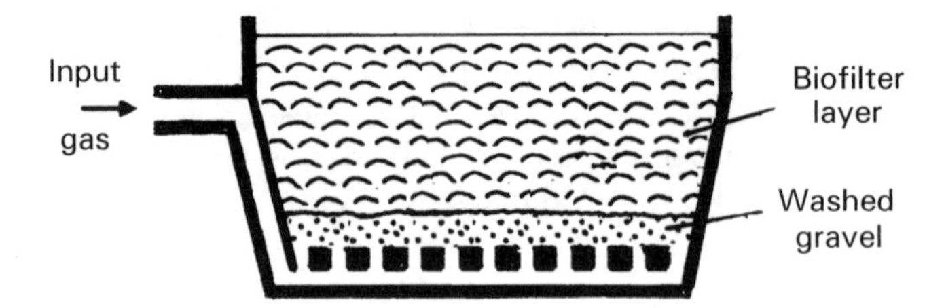

Figure 8.3 Structure of a biofilter.

performed by various bacterial and fungal micro-organisms that exist naturally in large quantities in the filter layer. The removal processes are inhibited if the filter dries out, so this is usually prevented by injecting moisture into the incoming air stream. The filter is self-sustaining, since nutrients within it allow generation of new micro-organisms as old ones are used up.

The bacteria and fungi in the filter convert odorous gases into carbon dioxide, nitrogen, sulphates and water, which can all be safely discharged into the atmosphere. This activity is enhanced by two physical processes. Adsorption causes the odorous gases to adhere to the surface of filter particles, and absorption causes the gases to be partly absorbed into the surface layer of filter particles. To optimise these physical processes, the filter must be sufficiently porous. In consequence, inert porous material is often mixed with the soil/sand/compost in the filter layer to increase its porosity. Further information on biofilters can be found in Reference 4.

Special techniques for chlorofluorocarbon (CFC) gases

Chlorofluorocarbon (CFC) gases deserve special mention. These are thought to be a primary cause of global warming, but they are difficult to remove by the techniques discussed earlier. The special techniques used instead are (1) passing the contaminated air through carbon filters (99% removal) or (2) treatment of the air in microwave plasma processing units (99.99% removal). Further details can be found in Reference 5.

8.2 Water Pollution Control

Water pollution can occur due to both unplanned and planned emission of pollutants. Unplanned emission covers all situations where pollution occurs as a result of accidental or unintended circumstances. On the other hand, planned emission involves the discharge of pollutants into a watercourse where the expected concentration in the water a short distance from the discharge point is such that harmful effects are not anticipated. Of course, this assumes that both the pollutant emission remains at or below some specified value and the volume of water flowing in the watercourse remains at or above some specified value.

Unplanned emission arises due to some fault condition or the breakdown of procedures that are intended to prevent pollutant emission. Most instances of pollution arise because of either the leakage of tanks holding liquids and powders or the spillage of liquids/powders during transfer from one container to another. In both these cases, pollutants enter the ground and, from there, eventually find their way into watercourses. To avoid leakage from tanks, regular inspection is required to detect corrosion and deterioration in seals, which are the main causes of leakage. Avoidance of spillage during liquid/powder transfer is largely achieved by establishing good working practices to ensure that due care is taken during transfer operations. However, well-designed containers that facilitate the transfer of liquids/powders from one to another also contribute to reducing spillage.

Somewhat less frequently, fault conditions in process plant lead to abnormally high pressures that cause seals to rupture and liquids to escape. Two steps can be taken to minimise this risk. Firstly, all seals should be regularly inspected, and replaced as necessary, as part of maintenance procedures. Secondly, process pressures should be monitored and an appropriate response made if pressure starts to increase abnormally. Statistical process control, as discussed in Chapter 12, is a very useful technique for obtaining the earliest possible warning of the start of an abnormal pressure increase.

In the case of planned discharge to a river, the usual method of control is to measure the quantity and concentration of pollutant before discharge to the river and, by concurrent measurement of the water flow rate in the river, to predict whether the concentration of pollutant in the river will exceed allowable levels. Appropriate action then has to be taken at an earlier point in the plant if there is a risk of exceeding allowable pollution levels in the river.

The correct procedure for sampling pollutants prior to discharge into a river is to take off a quantity of liquid in such a manner that the concentration of pollutant in the sample is representative of the pollutant concentration in the whole flow. As a general rule, the best place to take samples is at the output of a pump or in a vertical run of pipe. The flowing liquid is likely to be well mixed in either of these places. Horizontal pipe runs are generally unsuitable for sampling since fluids often flow in layers, and sampling just one layer will be unrepresentative, particularly because the density of the pollutant is likely to be different to that of the fluid carrying it, causing it to congregate at either the top or bottom of the pipe. Samples must be taken at periodic intervals and analysed. If this procedure is to give accurate measurements, it is very important that a proper procedure is in place to ensure efficient cleaning of the sampling apparatus between samples being taken.

The material used for the vessel that collects samples must also be chosen carefully, since some materials absorb certain types of pollutant. As a general rule, noble metals (platinum, etc.) and stainless steel exhibit low absorption for most materials, whereas mild steel, cupro-nickel alloys and most plastics absorb significant proportions of some pollutants.

References

1. Griffin, R.D., 1994, *Principles of Air Quality Management*, Lewis Publishers (CRC), Boca Raton, Florida.

2. Osborn, P.D., 1989, *The Engineer's Clean Air Handbook*, Butterworths, Sevenoaks, Kent, UK.
3. Wark, K., Warner, C.F. and Davis, W.T., 1998, *Air Pollution: its Origin and Control*, Addison-Wesley Longman, Reading, Massachusetts.
4. Williams, T.O. and Miller, F.C., Biofilters and facility operations, Part II, *Biocycle*, **33(11)**, 75–8.
5. Farmer, A., 1997, *Managing Environmental Pollution*, Routledge, London.
6. Mitsubishi, 2001, *www.mhi.co.jp/efin/a2000/annual8.htm* (last accessed June 2003).

9

Noise, Vibration and Shock Pollution

Noise, vibration and shock are forms of pollution that only affect the humans subjected to them. They do not have any effect on the physical environment itself, and are therefore not considered in many environmental management systems. Nevertheless, they can cause great annoyance to humans and, in some cases, particularly in respect of noise, they are subject to legislative limits. It is therefore appropriate to dedicate a chapter in this book to these subjects.

9.1 Noise

Sound energy is emitted from many sources in both industrial and nonindustrial environments and is generally known by the name of *noise* if it is undesirable or annoying in any respect. Even a low level of noise can cause great annoyance to the people subjected to it, and high levels of noise can actually cause hearing damage. Apart from annoyance and possible hearing loss, noise in the workplace causes communication difficulties, and also causes loss of output where the persons subjected to it are involved in tasks requiring high concentration. Extreme noise can even cause material failures through fatigue stresses set up by noise-induced vibration. The risk of this is particularly high with a jet engine or a rocket motor.

Various items of legislation exist to control the creation of noise. Court orders can be made against houses or factories in a neighbourhood that create noise exceeding a certain acceptable level. In extreme cases, where hearing damage may be possible, health and safety legislation comes into effect. Such legislation clearly requires the existence of accurate methods of quantifying sound levels.

Airborne sound can be defined as a periodic variation in air pressure about the atmospheric mean, and may be composed of either a simple or a complex waveform.

ISO 14000 Environmental Management Standards: Engineering and Financial Aspects. Alan S. Morris.
 ISBN 0-470-85128-7

Many primary sources of noise exist in industrial situations. Common primary sources are interacting parts such as gear teeth and bearings, motion of pistons between end-stops in hydraulic and pneumatic actuators, air and gas jets, combustion processes and vibrating components. However, noise from primary sources is frequently reflected from various boundaries in the environment. Each such boundary becomes a secondary noise source and often results in a standing wave, such that sound is nullified in some positions but reinforced in others.

Sound is usually measured in terms of the *sound pressure level*, SPL, which is defined as:

$$SPL = 20\log_{10}(P/0.0002) \text{ decibels (dB)}$$

where P is the r.m.s. sound pressure in μbar. The quietest sound that the average human ear can detect is a tone at a frequency of 1 kHz and SPL of 0 dB. At the upper end, an SPL of around 120 dB causes discomfort, and 144 dB causes physical pain. The human ear has the greatest sensitivity to sound at a frequency of 4 kHz. Hence, the SPL that a human will readily tolerate is a minimum for sound at 4 kHz, but increases substantially for lower and higher frequencies. Normally, human tolerance to noise is specified in terms of the *loudness level*, which is a subjective term that defines the magnitude of sound as judged by the listener.

Loudness level is expressed in units of *phons*. The relationship between phons and the SPL varies according to the sound frequency. At 1 kHz, the loudness level in phons is equal to the SPL in dB, so that a loudness level of 50 phons equals an SPL of 50 dB. At the frequency of 4 kHz, where human hearing is most sensitive, the same loudness level of 50 phons as perceived by the human ear is produced by a SPL of 42 Hz. At frequencies higher and lower than 4 kHz, the SPL required to produce a loudness level of 50 phons is greater than 42 dB. For example, at 50 Hz, a SPL of 72 dB is required to produce a loudness level of 50 phons.

9.1.1 Noise measurement

Sound is usually measured with a *sound meter*. This essentially processes the signal collected by a microphone, as shown in Figure 9.1. The microphone is a diaphragm-type pressure-measuring device that converts sound pressure into a displacement. The displacement is applied to a displacement transducer (normally capacitive, inductive or piezoelectric type) which produces a low-magnitude voltage output. This is amplified, filtered and finally gives an output voltage signal that can be recorded. The filtering process has a frequency response approximating that of the human ear, so that the sound meter 'hears' sounds in the same way as a human ear. In other words, the meter selectively attenuates frequencies according to the sensitivity of the human ear at each frequency, so that the sound-level measurement output accurately reflects the sound level heard by humans.

The sound meter is a useful tool for investigating complaints about noise pollution. In its normal mode of operation, it gives an output that is compensated for the nonlinear sensitivity of the human ear and therefore gives a fair measurement of the long-term degree of noise pollution existing. However, besides this facility to

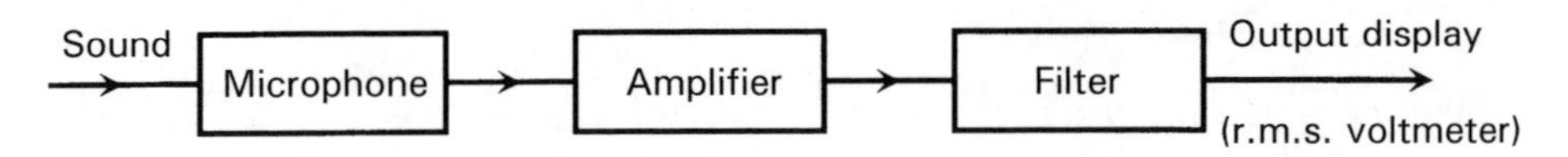

Figure 9.1 Elements of a sound meter.

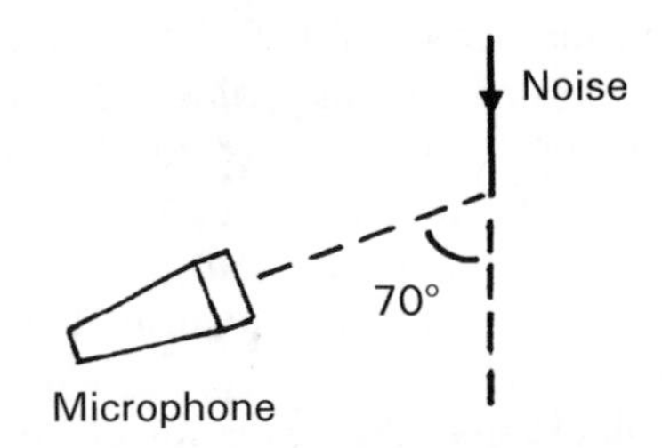

Figure 9.2 Recommended angle of incidence between microphone and noise source for a sound meter.

continuously record noise loudness levels, many sound meters also perform additional functions such as measuring the average loudness and peak loudness over a period of time. Average loudness is measured with the meter in integrating mode, with the summed loudness divided by the time that it is integrated over. The display of peak loudness, captured via a peak-hold facility in the meter, is important in the case of transient noises such as the blows of a forging hammer that are difficult to measure accurately from continuous recordings of noise levels, since such noise sources are characterised by very large magnitude noise that decays back to an ambient level over a very short space of time.

In practice, sound meters have to be used with care, particularly with regard to the relative angle between the microphone and the noise source. The microphone acts as an obstacle to the sound waves from the noise source, which increases the sound pressure on the microphone. Consequently, the output reading exaggerates the SPL and compensation has to be applied. The effect is quite significant for high sound frequencies, but becomes negligible at low frequencies. An angle of incidence of 70 ° between the microphone and noise source is often recommended, as shown in Figure 9.2, and some sound meters apply automatic compensation for the apparent pressure increase, assuming this 70 ° angle of incidence. However, yet further difficulties are met if significant reflected noise sources exist. In this case, the angle of 70 ° may have to be reduced substantially to avoid a direct path (0 °) between the secondary noise source and the microphone.

Calibration of the measurement system is also required. This involves observing the output reading of the sound meter when sounds with a known SPL are directed at it. Suitable standard sources of noise are available commercially, although care has to be taken in matching the acoustical properties of the calibration sound source with those of the microphone in the sound meter. Not every type of sound source is suitable for every microphone, and therefore appropriate advice should be sought from the sound meter manufacturer.

9.1.2 Noise analysis

Whilst the sound meter adequately demonstrates whether noise control legislation has been contravened in any given situation, it is of no use as a diagnostic tool to identify noise sources, since it does not indicate the primary frequency component in noise. This information is necessary for diagnosis, since different noise sources have different frequencies. In order to identify the noise source frequency, spectrum analysis has to be carried out on the signal captured by the microphone. Better accuracy in analysis is normally obtained if the microphone is pointed directly towards the suspected noise source (angle of incidence = 0°), rather than at the usual angle of 70°. The optimum angle of incidence is normally achieved by varying the microphone direction until a peak reading is obtained on the sound meter.

Spectrum analysis can be carried out by either analogue or digital means. Analogue analysis involves passing the noise measured by a microphone through a set of band-pass filters that each determines the amplitude of noise at the centre-frequency of the filter pass-band. A commonly used set of filters has centre frequencies at 31.25, 62.5, 125, 250, 500, 1000, 2000, 4000, 8000 and 16000 Hz. This set is known as a full-octave analyser and provides measurement of noise amplitude at 10 frequencies. If finer spectrum analysis is required, one-third octave or one-tenth octave filter-sets can be used that produce amplitude measurements at 30 and 100 different frequencies respectively.

Digital spectrum analysis involves applying the analogue voltage output from the microphone to an analogue-to-digital converter to change it to digital form. Then, a digital signal processing technique called the fast Fourier transform can be applied that produces a frequency/amplitude plot for the noise.

Determining the direction that noise is coming from can also be valuable in identifying the noise source. The human ear is very poor in this task, but an instrument called a sound intensity meter is able to do this quite well.

9.1.3 Noise reduction

The first step in noise reduction is to identify the source(s) generating the noise, for which spectrum analysis is generally helpful. Once a noise source has been identified, attempts can be made to either reduce its magnitude, or else shift the noise to a different frequency that the human ear is less sensitive to. Unfortunately, this is not always possible and, even where it is possible, the expert knowledge of acoustic consultants is generally needed to achieve success.

Once attempts to either reduce the amplitude of noise emitted from a source or else shift it to a less annoying frequency have been exhausted, the only remaining option to reduce annoyance is to try and absorb as much of the noise that is emitted as possible. Substantial noise reduction can be achieved by surrounding noise sources with a box-type structure that is constructed from sound-absorbing materials formed into wedges. This approach can be extended to secondary-source problems where noise reflections at boundaries in the workplace cause standing waves and amplification of the noise in particular locations. Such noise peaks can be reduced by covering the reflecting boundaries with sound-absorbing material. Alternatively, it might be possible in some cases to remove the boundaries that are causing the troublesome reflections.

9.2 Vibration

Various industrial operations are liable to generate vibrations, particularly the operation of machinery when there is some unbalance in moving parts. A vibration is a time-varying displacement at relatively high frequency that has a continuous form and a nature that is repetitive at least to some extent. In some forms of vibration, the waveform may be sinusoidal, although it can be quite complex in other types of vibration. Humans are very susceptible to vibration and can detect, via their fingertips, vibrational oscillations as small as 0.3 microns (0.3 μm) in amplitude at a frequency of 300 Hz. At frequencies above and below 300 Hz, sensitivity is less, but even relatively small oscillations can still cause annoyance, reduced working ability and possible health damage. In most companies, vibration only affects personnel inside the company and there are no external environmental effects. However, small companies on industrial estates may encounter problems where vibrations are transmitted from their operations to neighbouring premises, which then becomes an environmental problem.

9.2.1 Measurement of vibration

Vibration of a structure consists of linear harmonic motion in which the displacement x from the equilibrium position at any general point in time t is given by:

$$x = X_p \sin(\omega t) \tag{9.1}$$

where X_p is the peak displacement from the equilibrium position and ω is the angular frequency of the oscillations. Thus, vibration is characterised by two parameters, the peak displacement or intensity of vibration X_p and the frequency of vibration ω.

There are two common methods for measuring the frequency of vibration. The first method involves using a tool consisting of a hollow cantilever beam constructed in two parts, whereby one part can slide inside the other and thus alter the length of the beam. Graduations are provided along the inner tube, so that the effective length of the beam can be read off as it slides inside the outer tube. If the end of beam is placed against the vibrating structure and then its length is varied, resonance will be excited for a particular length, and the end of the beam will then oscillate with a significant amplitude. The length of the beam at this resonant point then indicates the oscillation frequency of the vibrating structure. The alternative technique is to use a stroboscope. The stroboscope is an instrument that produces light pulses at a variable frequency. If a stroboscope is directed at a vibrating structure and the frequency of light pulses is varied, there will be a particular frequency of pulses at which the structure apparently stops vibrating. At this point, the frequency of light pulses is equal to the frequency of the vibration.

The peak displacement or intensity of vibration can be measured in several ways. If equation (9.1) is differentiated with respect to time, an expression for the instantaneous velocity v of the vibrating structure at any general point in time t is obtained as:

$$v = \mathrm{d}x/\mathrm{d}t = -\omega X_{\mathrm{p}} \cos(\omega t) \tag{9.2}$$

Then, differentiating (9.2) with respect to time gives an expression for the acceleration A of the structure at any general point in time t as:

$$A = -\omega^2 X_{\mathrm{p}} \sin(\omega t) \tag{9.3}$$

Equations (9.1) to (9.3) show that the intensity of vibration X_{p} can be measured in terms of either displacement, velocity or acceleration. It is apparent from the equations that displacements are large at low frequencies, and that either displacement or velocity transducers are therefore theoretically best for measuring vibration at such frequencies. Accelerometers would only appear to be preferable at high frequencies. However, there are considerable practical difficulties in mounting and calibrating displacement and velocity transducers, and so they are rarely used in practice. Consequently, vibration is usually measured by accelerometers at all frequencies.

The form of accelerometer used for vibration measurement has to be chosen carefully. Equation (9.3) shows that the peak acceleration A_{p} occurs when $\sin(\omega t) = -1$ and can be written as:

$$A_{\mathrm{p}} = \omega^2 X_{\mathrm{p}}$$

This square-law relationship between the peak acceleration and the oscillation frequency means that the magnitude of the peak acceleration to be measured can be very high. In addition, both the frequency of oscillation and the magnitude of displacement from the equilibrium position in vibrations have a tendency to vary randomly, which further constrains the choice of accelerometer. Furthermore, consideration must be given to the fact that attaching an accelerometer to the vibrating structure will significantly affect the vibration characteristics if the structure has a low mass. The effect of such 'loading' of the measured system can be quantified by the following equation:

$$A_{\mathrm{l}} = A_{\mathrm{b}} \left(\frac{M_{\mathrm{b}}}{M_{\mathrm{b}} + M_{\mathrm{a}}} \right)$$

where A_{l} is the acceleration of the structure with the accelerometer attached, A_{b} is the acceleration of the structure without the accelerometer, M_{a} is the mass of the accelerometer and M_{b} is the mass of the structure. To minimise this system loading effect, the accelerometer used to measure vibration must have a low mass.

The traditional form of accelerometer, as shown in Figure 9.3, consists of a mass suspended by a spring and damper inside a casing that is rigidly fastened to the body

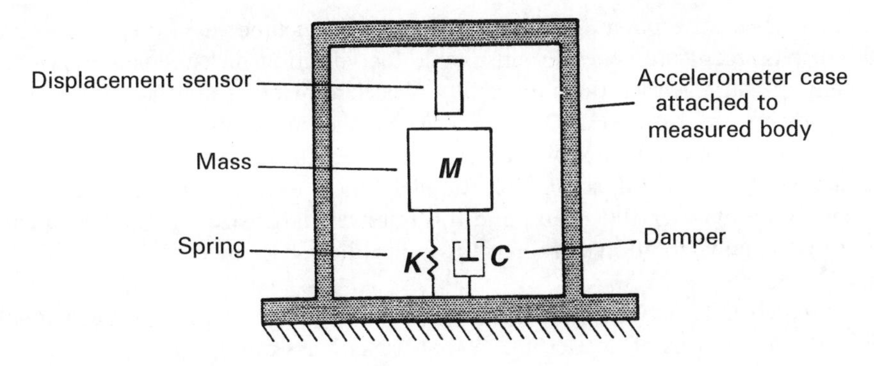

Figure 9.3 Schematic representation of traditional form of accelerometer.

undergoing acceleration. Any acceleration of the body causes a force, F_a, on the mass, M, given by:

$$F_a = M\ddot{x}$$

This force is opposed by the restraining effect, F_S, of a spring with spring constant K, and the net result is that the mass is displaced by a distance x from its starting position, such that:

$$F_S = Kx$$

In steady-state, when the mass inside is accelerating at the same rate as the case of the accelerometer, $F_a = F_S$ and so:

$$Kx = M\ddot{x} \quad \text{or} \quad \ddot{x} = (Kx)/M$$

This is the equation of motion of a second-order system, and, in the absence of damping, the output of the accelerometer would consist of nondecaying oscillations. A damper is therefore included within the instrument, which produces a damping force, F_d, proportional to the velocity of the mass M given by:

$$F_d = C\dot{x}$$

This modifies the equation of motion to:

$$Kx + C\dot{x} = M\ddot{x}$$

The acceleration value is obtained by measuring the displacement of the mass. The various types of accelerometer that are commercially available differ mainly in the technique used to measure the mass displacement, although some differences also occur in the type of spring element and form of damping used. Accelerometers that use either a resistive potentiometer or a variable-inductance device to measure the mass displacement are generally unsuitable for vibration measurement, since they can only measure accelerations up to 50 *g*. Accelerometers using the linear variable displacement transducer (LVDT) to measure the mass displacement are more suitable, since they can measure accelerations up to about 700 *g* with a typical inaccuracy of ±1% of full scale. Unfortunately, these have a typical physical size of 125 cm^3 and a mass of 100 grams, and this relatively large size and mass may cause problems in many vibration measurement applications, because of the system loading effect.

Fortunately, alternative forms of accelerometer also exist that have much smaller size and mass. Forms of accelerometer that use either strain gauge or piezoresistive sensors to measure the mass displacement can measure accelerations up to 200 g with an inaccuracy of ±1%. In these, the sensor serves as the spring element as well as measuring mass displacement, thus simplifying the instrument's construction and resulting in a much smaller size (typically 3 cm^3) and mass (typically 25 grams). Piezo-electric type accelerometers, where the mass displacement is measured with a piezo-electric crystal, can measure even larger accelerations up to 1000 *g*, with an inaccuracy of ±1%. Commonly used materials for the crystal, which also acts as the spring and damper within the instrument, are lead metaniobate, lead titanate, lead zirconate and barium titanate. Despite the piezoelectric accelerometer having a larger physical size and mass (typically 15 cm^3 and 50 grams) than strain gauge and piezoresistive types, it is the most common type of accelerometer used in vibration measurement, because of its capability of measuring the high-frequency oscillations that typically occur in vibrations.

As well as an accelerometer, a vibration measurement system requires other elements to translate the accelerometer output into a recorded signal, as shown in Figure 9.4. The three other necessary elements are a signal-conditioning element, a signal analyser and a signal recorder. The signal-conditioning element amplifies the relatively weak output signal from the accelerometer, and also transforms the high output impedance of the accelerometer to a lower impedance value. The signal analyser then converts the signal into the form required for output. The output parameter may be either displacement, velocity or acceleration, and may be expressed as either the peak-value, r.m.s.-value or average absolute value. The final element of the measurement system is the signal recorder.

All elements of the measurement system, and especially the signal recorder, must be chosen very carefully to avoid distortion of the vibration waveform. The bandwidth should be such that it is at least a factor of ten better than the bandwidth of the vibration frequency components at both ends. Thus its lowest frequency limit should be less than or equal to 0.1 times the fundamental frequency of vibration,

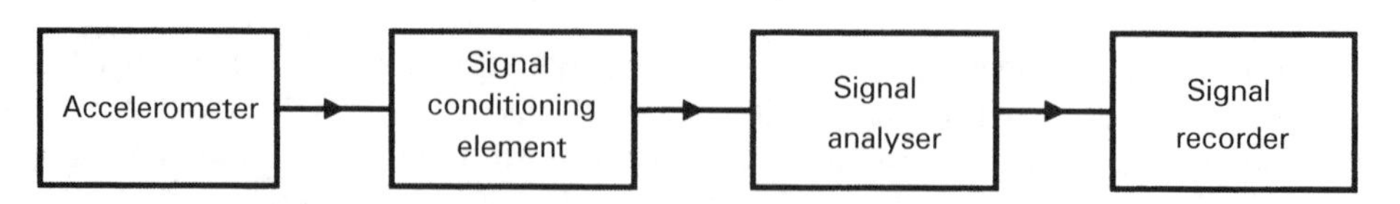

Figure 9.4 *Elements of vibration measurement system.*

and its upper frequency limit should be greater than or equal to ten times the highest significant vibration frequency component.

All elements in the measurement system require calibration, but the element that requires most frequent attention is the vibration transducer (normally a piezoelectric accelerometer). At the lowest level of calibration, the output of the vibration transducer is compared with that of a standard accelerometer. In this procedure, both transducers are mounted on a piece of equipment that is able to excite vibrations at a range of frequencies and amplitudes. The standard instrument is most commonly a piezoelectric accelerometer, as this has the best stability and frequency range. If the vibration transducer itself measures acceleration, as is commonly the case, the outputs of the two transducers can be compared directly. Otherwise, the outputs have to be converted to the same forms using equations (9.1–9.3) above. Calibration at yearly intervals is usually recommended.

Reference standard accelerometers are similarly used in comparison procedures at all intermediate levels of calibration. Only at the very highest levels of calibration are absolute methods used. The most common absolute methods of calibration involve interferometric techniques which give measurement uncertainty levels of around ±0.5%. Such facilities are only usually found in national standards laboratories.

9.2.2 Control of vibration

Ideally, the best way of dealing with vibration is to identify the source and then try to eliminate the cause of the vibration. This may be possible in cases where a vibration is due to an imbalance in a rotating or reciprocating component. This is a common solution to vibrations in motor vehicles caused by radial variations in the mass of the wheels and tyres, in which weights are added around the circumference of the wheel to correct the imbalance. However, this elimination-at-source approach is not possible in many situations, and therefore other ways of dealing with vibration have been developed, as discussed below.

Vibration isolation

The basic principle of vibration isolation is to mount the vibrating structure on a spring-damper system whose natural frequency is substantially lower than the frequency of vibration in the structure. This system, which is shown schematically in Figure 9.5, substantially reduces the transmission of vibrations into the floor below the structure. A spring-damper system with a natural frequency between 5 and 10 Hz is suitable for most types of vibrating structure. Mountings made from cork or rubber are frequently used, as such materials act as a combined spring and damper, and simplify the construction of the isolator.

The spring constant in the mounting has to be chosen according to the mass of the vibrating structure. The ideal stiffness is chosen according to the formula:

$$K = 0.04Mf^2 \text{ kg/cm} \tag{9.4}$$

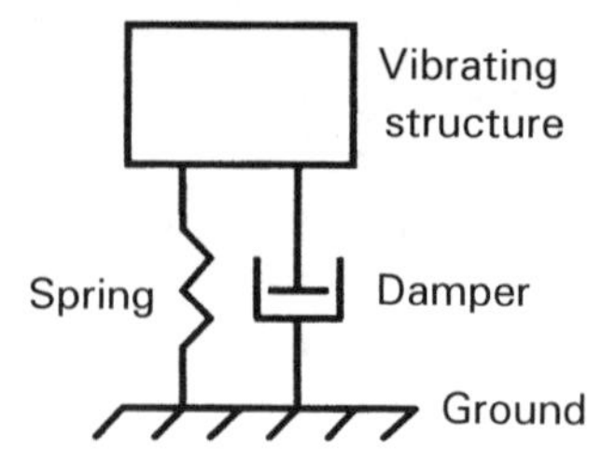

Figure 9.5 Schematic representation of vibration isolation system.

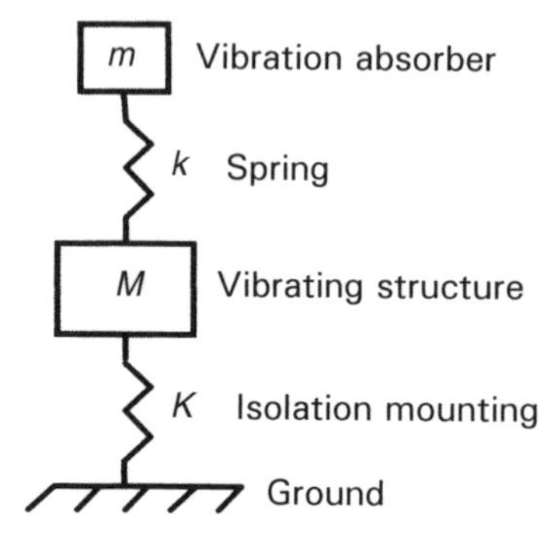

Figure 9.6 Schematic representation of dynamic vibration absorber system.

where M is the mass of the vibrating structure and f is the resonant frequency of the mounting. If the mounting has a resonant frequency of 10 Hz and the vibrating structure has a mass of 100 kg, the necessary spring constant in the mounting calculated according to (9.4) is 400 kg/cm. In practice, rather than balance the vibrating structure on a single mounting, it is more common to support it on four mountings positioned at each corner of the structure. In this case, each mounting only carries one-quarter of the weight of the structure and therefore, using the above values of M and f, a spring constant of 100 kg/cm would be required for each.

Dynamic vibration absorber system

Isolation is the simplest approach to vibration control, and should therefore be considered first. However, it is not suitable in every situation, and an alternative approach is to design a dynamic vibration absorber system. This involves a system of the form shown in Figure 9.6. In addition to a vibration-isolating mounting system with a spring constant K, the vibrating structure of mass M is connected to a vibration absorber system of mass m and spring constant k. The resonant frequency of the absorber system is tuned to the frequency of the vibrating structure. In theory, this absorbs the vibration of the structure and eliminates transmission of vibrations into the ground below. Unfortunately, this only works if the structure is vibrating at either a single frequency, or else within a narrow band of closely spaced frequencies.

Stiffening the vibrating structure

Vibrations in structures like beams and materials in sheet form can often be reduced by stiffening the structure in some way. This has the effect of shifting the vibration to a higher frequency that is not excited during the normal operation of the system that the structure is part of. However, shifting the vibration frequency of one component sometimes has the effect of exciting serious vibrations in other components in the system. If this happens, alternative solutions to the original vibration must be sought.

Vibration damping

One way of applying damping is to arrange for either the vibrating structure itself, or else a protrusion from it, to slide against a control surface with pressure applied across the interface, as shown in Figure 9.7. Friction at the interface between the vibrating structure and the control surface produces a damping force that inhibits vibration. This technique is known as *interface damping*.

Alternatively, damping can be achieved by adding *viscoelastic material* such as mastic to the vibrating structure. This is a common method of controlling vibration in metal sheets. Various viscoelastic materials are available that each give optimum damping over a particular frequency range. Thus, the material used has to be chosen carefully, according to the vibration frequency to be damped. The viscoelastic material can be applied in one of three ways, as shown in Figure 9.8. Firstly, the structure can be designed as a sandwich so that it is constructed as a structure of two thin sheets with a layer of viscoelastic material sandwiched in between. Secondly, the viscoelastic material can be sprayed on to one surface of the structure. Thirdly, the material can be sprayed on to both sides of the structure. Over certain frequency ranges, sandwich construction gives the best damping. However, over other frequency ranges, a single layer of material on one surface of the vibrating structure gives better damping. In either case, increasing the thickness of the viscoelestic layer tends to widen the frequency range over which good damping is achieved. With regard to the third option of having viscoelastic material on both sides of the vibrating structure, this is sometimes used where it is difficult to get sufficient material to adhere on one side of the structure. However, practical tests have indicated that a single coat on both sides is less effective in vibration suppression than spraying a double-thickness coat on just one side.

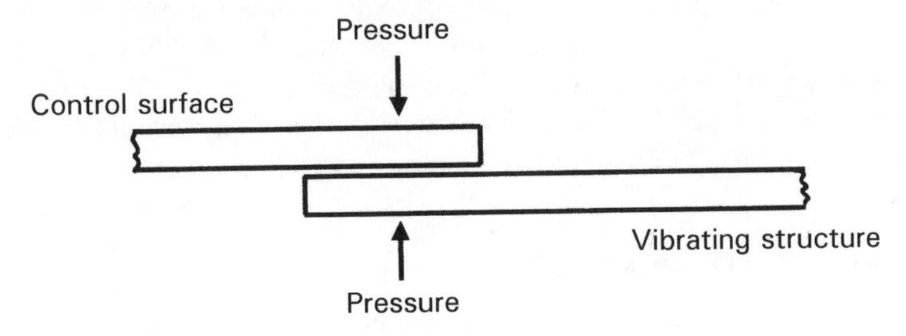

Figure 9.7 *Vibration suppression by interface damping.*

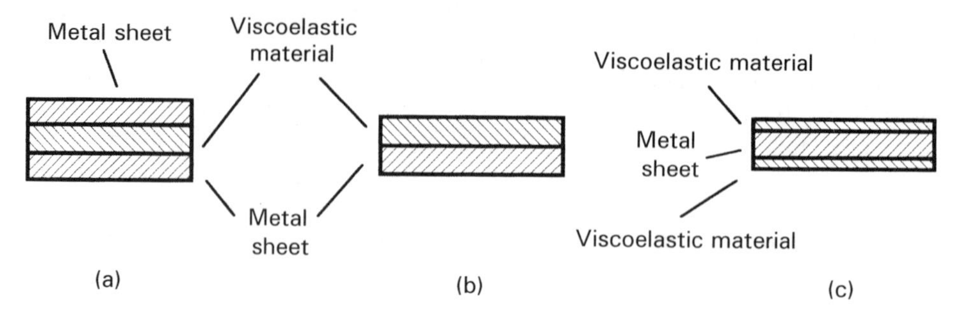

Figure 9.8 Damping with viscoelastic material: (a) sandwich construction; (b) viscoelastic material sprayed on one side; (c) viscoelastic material sprayed on both sides.

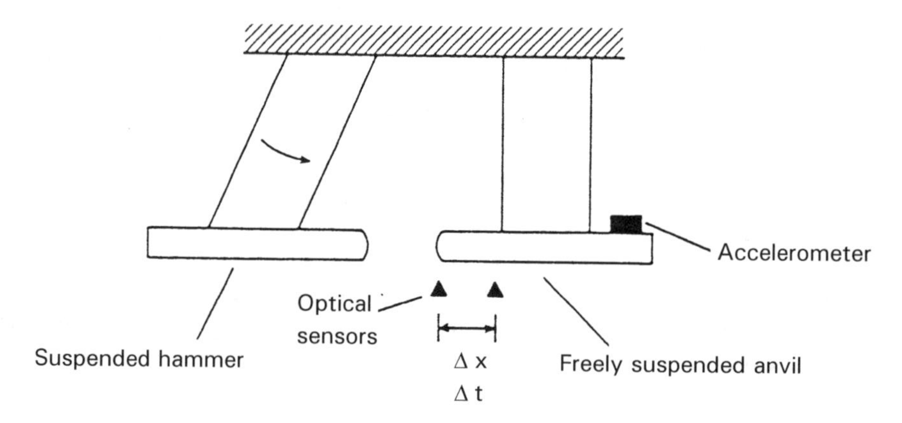

Figure 9.9 Shock machine. From Morris (1997) Measurement and Calibration Requirements, © John Wiley & Sons, Ltd. Reproduced with permission.

9.3 Shock

Shock describes the result of a type of motion where a moving body is suddenly brought to rest, usually because of an impact or collision. Shocks are very annoying in themselves. However, in addition, they also usually excite a vibration that decays in magnitude over time but causes annoyance until it has decayed. Shocks are very common in industrial situations and characteristically involve large-magnitude deceleration (e.g. 500 g) that lasts for a very short time (e.g. 5 ms). Thus, unlike vibration, which is usually continuous, shock is effectively a single event. Although in practice shocks can often be repetitive, the frequency of repetition is normally low enough to allow the effect of each shock to be treated as a single event.

9.3.1 Measurement of shock

An instrument having a very high-frequency response is required for shock measurement, and for this reason, the piezoelectric accelerometer is commonly used.

Again, other elements are required for analysing and recording the signal, as shown in Figure 9.4.

Shock calibration requires calibration of all the elements used in the measurement system. Calibration of the piezoaccelerometer is carried out using a shock machine of the form shown in Figure 9.9. The accelerometer is mounted, together with a standard accelerometer, on the anvil of the machine, and records of the shock signals from the two transducers are recorded on a transient recorder and compared. If optical sensors are used to measure the time δt taken for the anvil to move a fixed distance δx, an absolute measurement of acceleration can also be obtained according to the following equation:

$$A = \int_0^{\delta t} a(t)\mathrm{d}t = \delta V = \frac{\delta x}{\delta t}$$

where $a(t)$ is the instantaneous acceleration at time t, and δv is the change in velocity of the anvil as it moves through δx. Traceability to fundamental standards is then provided by the procedures described in the last section on vibration measurement.

9.3.2 Control of shock

The normal way of controlling shock is to design a spring-damper isolation system similar to that used to isolate vibrations. This isolates the shock from the ground. In this case, the resonant frequency of the isolator should be designed to be less than or equal to $1/10T$, where T is the time duration of the shock pulse.

10

Waste Management

As almost all environmental pollution originates from waste produced in manufacturing operations and servicing activities of companies, a rigorous waste management policy is a key component in a successful environmental management system. However, as well as the environmental benefits, good waste management can lead to considerable cost savings, and thus significantly increase the profits of a company. It has been estimated that the average cost to companies of the waste that they produce is 4.5% of their turnover. If just one-quarter of this waste is avoided, the saving is around 1% of turnover. In companies where the profit margin is only 5% of turnover, such a saving actually amounts to a 20% increase in profits.

First of all, a waste management policy should examine all sources of waste, and should set targets for waste reduction. Reducing the creation of waste in this way leads to cost savings in terms of the raw materials lost in waste, and in the time and energy expended in making products that are subsequently discarded because they are of substandard quality. The second stage of a waste management policy is then to look at ways of disposing of unavoidable waste in the most cost-effective way.

10.1 Waste Reduction

As noted above, even modest reductions in the quantity of waste produced can have significant cost benefits for a company. The rest of this section therefore discusses a number of important steps in developing an effective waste reduction strategy. In addition to this, further information is now available on a number of Internet sites such as Reference 1.

The starting point in a waste reduction strategy is to identify where waste is being produced. Walking round the company and making a flow chart of material and waste flows is a good starting point. Asking employees about where waste is

ISO 14000 Environmental Management Standards: Engineering and Financial Aspects. Alan S. Morris.
 ISBN 0-470-85128-7

occurring and how it might be reduced also yields useful information. Common sources of waste include offcut materials, use of disposable plastic or paper cups instead of reusable ceramic ones, fluid leaks, unnecessary use of paper, and loss of solvent materials through evaporation when the lids are left off their containers.

Having identified all waste sources, the next step is to analyse how and why the waste is produced. Appropriate action can then be taken to reduce the magnitude of waste generation by setting waste reduction targets that give priority to the worst problems. Of course, it will never be possible to eliminate all waste sources. For example, if circular pieces are being cut from square sheets of wood, some waste is inevitable. However, once waste sources have been identified, some brainstorming should be able to make a significant reduction in the amount of waste produced.

10.1.1 Mass balance

A more rigorous way of identifying sources of waste is a technique known as the *mass balance.* The mass balance concept is based on the physical principle that mass can neither be created nor destroyed. Thus, the amount of waste created in a process can be calculated by measuring the total weight of finished product at the output of the process and subtracting this from the total weight of input materials to the process. This difference is often known as the *mass loss* or *mass balance difference*. Apart from this use in identifying the waste arising in a single process, the mass balance principle is also used in several other ways. In one such use, it is applied to track the total flow of raw materials, products and waste across large regions in a country[2]. In another usage, it is applied in some European countries to control the disposal of hazardous waste within a legislative framework[3].

However, for the purposes of this book, the mass balance principle will be explained in terms of its use for identifying waste in a particular operation of a company. An example involving the manufacture of wooden coffee tables will be used to illustrate the principles involved. The product consists of four legs glued on to a top, as shown in Figure 10.1. After assembly, each table is finished by spraying a layer of varnish on to it. The wood for the tops is supplied in sheets that are 5 m long, 400 mm wide and 20 mm thick. The supplier has already planed the wood to a smooth finish on all surfaces. The legs are cut from 6 m lengths of 50 mm by 50 mm section wood. This is also supplied with all surfaces planed smooth. The varnish used is supplied in 10 litre drums, and the glue in 5-litre drums. The process was monitored over the production of 1000 tables, and the following masses were measured:

Raw materials used	Mass used (kg)
Wood for legs (286 lengths):	1561.6
Wood for tops (200 lengths):	2912.0
Varnish (10 drums):	132.5
Glue (five drums):	40.0
Total	4646.1

The Mass of the finished product (1000 tables) is 4441.6 kg, and hence, the mass loss (4646.1 – 4441.6): 204.5 kg.

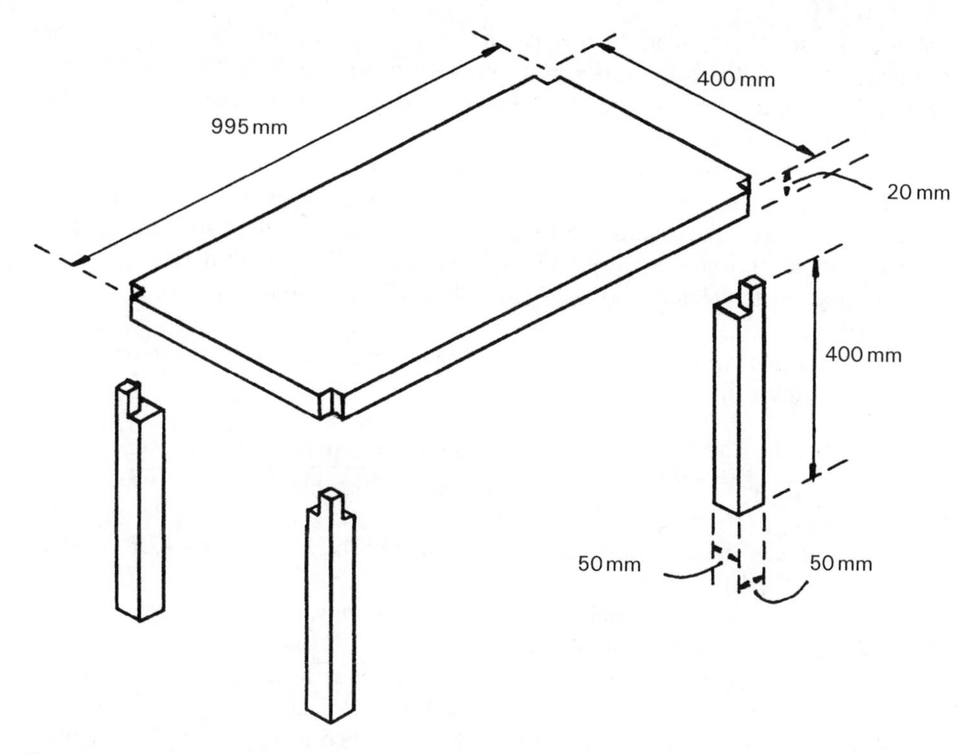

Figure 10.1 Application of mass balance to identify waste in manufacturing tables.

Once the mass loss has been calculated, some detective work is necessary to identify all the places where material is being lost as waste. For this particular example, it was possible to quantify almost 90% of the waste as follows:

Mass of offcuts remaining from 6 m lengths after cutting legs:	95.3 kg
Calculated mass of wood cut out from top and legs to make joints:	72.8 kg
Mass of 10 empty varnish drums (including any remaining varnish):	6.3 kg
Mass of four empty glue drums (including any remaining glue):	2.6 kg
Mass of part-finished drum of glue:	1.1 kg
Glue left on applicators:	0.4 kg
Estimate of varnish spilled, plus that cleaned out from spraying system:	1.3 kg
Total waste accounted for:	**179.8 kg**

The remaining waste (24.7 kg) can be accounted for by the material turned to sawdust during cutting operations.

Detailed analysis of waste in this way often suggests ways of reducing waste. In this particular example, the two largest sources of waste arise out of cutting the legs to length and cutting away material to make the joints. Without designing a completely different way of achieving rigid joints that avoids cutting away wood to form the joints, there is little that can be done to reduce the latter source of waste. However, wastage in cutting the legs can be achieved by revising the length of wood sections obtained from the supplier. In this particular case, negotiation with the supplier resulted in 5 m lengths being supplied instead of 6 m lengths. This reduced the offcut from each length from 361 mm to 167 mm, and reduced the mass wasted from 95.3 kg to 51.5 kg (allowing for buying 334 of the 5 m lengths instead of 286 of the 6 m lengths). Weighing the empty drums also revealed that some varnish and glue was being left on the sides. By introducing a simple instruction for workers to use a rubber scraper to remove and use the material (varnish or glue) remaining in the drums, most of this waste was eliminated. As well as the environmental benefit of reducing the waste created, the company achieved substantial cost savings through these changes.

The success in eliminating waste term is sometimes quantified by the *mass balance yield*. This is given by:

$$\text{mass balance yield} = \frac{\text{mass of input material}}{\text{mass of product}}$$

The mass balance yield is a very useful way of comparing the performance of plants that produce a similar product.

Unfortunately, the mass balance approach is not appropriate for every situation. The approach usually works well when all input and output materials involved are dry. However, the approach can get quite complicated if water is added or removed (by drying) from the process, or if chemical reactions are involved. If water is involved, all water added or removed must be quantified in some way. For chemical reactions, all products and gaseous emissions have to be quantified.

10.1.2 Waste reduction mechanisms

Unfortunately, as with almost all other aspects of environmental management systems, no general formula for waste reduction can be devised that will be generally applicable. Every situation is different, and so each case must be analysed separately, and a case-specific waste reduction strategy must be devised. To assist in generating such a strategy, some important waste reduction mechanisms that are applicable in many situations are set out below.

Product design

As well as reduction in waste generated during manufacturing operations and servicing activities, much can be done to assist the environment by good product design that minimises the quantity and toxicity of waste created by the product itself when it comes to the end of its useful life. As far as possible, products should be designed to achieve the following:

- Have a long life, so that product disposal and any waste generated happen as infrequently as possible.
- Use as much recyclable material in the product as possible.
- Avoid toxic materials in the product.
- Use the minimum amount of packaging.

Recycling materials

Recycling schemes make a significant contribution to waste reduction and also generate useful income into companies who operate them. Technology now exists that allows many products to be produced from recycled materials at an economic cost. For example, aluminium cans can be made from recycled aluminium. Newspapers, cardboard boxes and paper towels can be made from recycled paper. Also, glass bottles and other glass products can be made from recycled glass. The starting point for recycling schemes is the collection of recyclable materials. The separated material is then usually passed on to specialist companies who process the waste and then sell it as a raw material to companies who make new products with it.

As well as reducing the amount of material that is disposed of via landfill sites or by other means, recycling has other environmental benefits. Firstly, the extraction process for many raw materials (for example, bauxite to produce aluminium) creates a lot more pollution than processing recycled materials into a useable form. Secondly, manufacturing operations using recycled materials normally use less energy than when using new materials. Companies that use recycled materials also usually gain economic benefits, not only in reducing the cost of energy used but also in the cost of raw materials, since recycled materials are generally cheaper to buy than new materials.

10.2 Waste Disposal

As mentioned earlier, a proper waste strategy should firstly ensure that the amount of waste produced is minimised, and secondly ensure that any waste that is produced is recycled as far as possible. However, it is inevitable that some waste will still remain that has to be disposed of. Most countries have strict legislation about proper waste disposal and studying this should be the starting point in the development of a company's waste disposal strategy within an EMS. Legislation usually covers all aspects of waste disposal, including both nonhazardous waste and hazardous waste. Apart from the summary below, further information can be found in Reference 4.

10.2.1 Nonhazardous waste

Nonhazardous waste can be disposed of by either composting, landfill or combustion. Of these, composting is the most attractive from an environmental point of view, although potential problems of nuisance and pollutant emission need to be addressed carefully. The relative merits of each method of disposal are discussed further below.

Composting

Composting is a relatively cheap method of disposal for nonhazardous waste. It consists of the controlled biological decomposition of organic matter by microorganisms (bacteria and fungi) into a stable solid material. The necessary microorganisms are normally present naturally in the materials to be composted, and special additives are usually unnecessary. Various organic materials are suitable for composting, including food waste, paper waste, sewage biosolids, and plant and tree waste. The product of composting can be used for various purposes, such as improving the nutrient content and structure of soil for farmers and gardeners, enhancing the capability of soil to inhibit plant diseases, reducing the toxicity of pollutant-contaminated land, and stabilising soil to reduce erosion. For optimum efficiency, the composting process needs to be a fairly large-scale operation. Hence, it is more often carried out by municipal authorities rather than by private companies. However, it is sometimes economically viable for large companies to set up their own composting facility if they generate significant quantities of waste material that is suitable for composting.

One of the main costs in composting is sorting the incoming material to remove noncompostable material such as metal, glass and plastic. However, some of this cost can be recovered if the material extracted is sent to recycling units. Sorting is normally done by passing the material along a conveyor belt and using human labour to pick off unwanted items. This is costly for municipal authorities, but less of a problem for private companies since they have better control over the content of the incoming material if it is generated within the company, meaning that much less unsuitable material has to be removed.

The main parameters that have to be controlled in composting are: temperature, oxygen level and moisture level. Proper control of these parameters is necessary to accelerate the decomposition process, but also to minimise nuisance and pollution due to the process. If the pile of composting material is large enough to ensure that most of the material is away from the surface of the pile and therefore well insulated, the decomposition process itself usually generates sufficient heat to create an adequate temperature in the composting material. Therefore, heating the composting material by artificial means is not normally required. Nevertheless, some heating or cooling is sometimes required to maintain the optimum temperature for decomposition, which is 45–59 °C. At this temperature, most weed seeds in the material are killed, and the micro-organisms are at optimum efficiency. At lower temperatures, the micro-organisms become less active, whereas at higher temperatures, some very beneficial micro-organisms are killed.

Both aerobic and anaerobic micro-organisms exist. Aerobic ones require oxygen, whereas anaerobic ones do not. Aerobic micro-organisms are preferable because they decompose organic material 10 to 20 times faster than anaerobic ones. Also, anaerobic decomposition produces more nuisance, because greater quantities of hydrogen sulphide, amines and methane are produced. Thus, to ensure that aerobic decomposition predominates, the oxygen level must be maintained. In many cases, periodic turning of the compost pile injects sufficient new oxygen to replenish that which has been used up. This also has an additional benefit of turning the material on the outside of the pile into the inside, thus ensuring that it is subjected to the temperature rise necessary to decompose the material in it. However, it is sometimes necessary to inject additional oxygen by blowing air into the pile, particularly if the moisture content in the pile is high.

One of the biggest problems with composting is the emission of odorous gases, which can cause much public annoyance. To minimise complaints, composting plants should be sited as far as possible from significant centres of human population, and with due regard to the prevailing wind direction. If odours are only carried to centres of human population by particular wind directions, then the composting material can be covered to allow emissions to build up inside until the wind direction changes and the odorous gases can be released. Where odour emission is particularly serious, composting can be performed in fully enclosed vessels, where conditions for decomposition are kept optimum and gases emitted are applied to a biofilter or other form of treatment plant before release to the atmosphere. Unfortunately, this solution significantly increases the cost of the composting operation. A cheaper solution is simply to have a cover over the composting material from which the emitted gas is extracted and treated.

Much can also be done to limit odours by careful control of the composting process. Odours are increased when the oxygen level reduces and anaerobic decomposition predominates, thus increasing emission of hydrogen sulphide and amines. Thus, extra oxygen must be pumped into the pile if necessary, to prevent this happening. Moisture control is also necessary, since moisture increases the oxygen requirement, and a cover is therefore usually provided over the pile to prevent rain increasing the moisture content. The carbon–nitrogen ratio in the decomposing material is also important, with a C:N ratio between 25:1 and 35:1 being optimum. A ratio greater than 35:1 slows down decomposition, whereas less than 25:1 causes increased emission of ammonia. The optimum ratio is normally achieved by mixing different kinds of input material in a suitable proportion. Typical C:N ratios for different types of waste are given in Table 10.1.

Besides odour emission, the other potentially serious problem with composting is seepage of organic material, PCBs (from treated wood), pesticide residues and nitrates out of the pile and into the ground below, from where these pollutants can get into watercourses. Much of this potential pollution can be avoided by carefully screening the material that is put into the compost pile, and nitrate production can be minimised by carefully controlling the C:N ratio of the composting material. Controlling the moisture content of the pile also limits the amount of pollutants that seep downwards out of it. Of course, pollution can be prevented entirely by using a sealed vessel for the composting process, but at increased cost as already noted. Seepage of pollutants into watercourses can also be prevented by putting the composting material on a concrete platform with gutters around the edges to collect liquid that seeps out of the pile. This liquid can be collected and sent to a water-treatment plant.

Table 10.1 *C:N ratio of different waste materials*

Material	C:N ratio
Wood	700:1
Paper	170:1
Dead (dried) leaves and weeds	90:1
Fresh leaves and weeds	35:1
Grass clippings	15:1
Food scraps	15:1

Finally, one further possible hazard with composting is that decomposing organic material can ignite spontaneously when its temperature exceeds 93 °C, if the moisture content is between 25% and 45%. In practice, the temperature can only attain 93 °C if the pile exceeds 4 m in height. Thus, ignition can be avoided fairly easily by keeping the pile less than 3 m in height. Further good practice is to monitor the temperature and turn the pile if its temperature reaches 60 °C.

Combustion

Combustion is a controlled burning process that reduces the volume of waste by up to 90%. Provided that the temperature achieved is sufficiently high, combustion breaks down harmful chemical compounds into less harmful ones and also kills bacteria. When combustion is complete, the residual ash is harmless (although this must be checked) and can be safely disposed of by landfill or composting. A particular advantage of combustion from an environmental point of view is that energy can be recovered, which also has an economic benefit. This recovered energy is often used in heating schemes.

Both private companies and municipal authorities operate combustion units. Their main operational limitation is that toxic gases are produced that must be carefully controlled. Common methods of control are to apply the gases to either a liquid scrubber or a filter (see Chapter 8).

Landfill

Landfill sites are fairly widely available and readily accept waste at a modest charge, provided that the necessary documentation is completed that declares that none of the waste is hazardous. Unfortunately, disposal in landfill sites is open to abuse by people disposing of hazardous waste that is declared as nonhazardous. This clearly damages the environment, since waste inevitably seeps from landfill sites into watercourses. Hence, there is a trend towards legislation that discourages landfill disposal by imposing disposal taxes that make landfill disposal more expensive and encourages more recycling of materials and alternative disposal mechanisms.

Most of the problems with landfill can be avoided if they are sited sensibly, designed correctly and operated properly. Siting should be away from geographical faults and floodplains, and also as far away from known watercourses as possible. The bottom and sides of the area to be filled should be lined with a geomembrane or plastic sheet that is reinforced with 60 cm of clay. In spite of this, the area around it should be tested periodically by drilling boreholes to confirm that waste has not escaped.

Odour and rodent infestation are two potential problems with landfill sites. However, good operating practice, where the waste is frequently compacted and covered with a few centimetres of soil, minimises these problems. Methane (a 'greenhouse' gas) is also generated in landfill sites, but schemes now exist where the methane is collected and recycled as an energy source. Such recycling of methane also has economic benefits.

10.2.2 Hazardous waste

Hazardous waste is material that is corrosive, poisonous, radioactive, explosive, ignitable or harmful in any other way. It may exist in either solid, liquid, gas or slurry form. To avoid any ambiguity about whether a particular kind of waste is hazardous or not, for example if it is only mildly corrosive, legislation usually lists the specific types of waste covered. Such a list is normally quite long, and it is common to find legislation listing over 500 specific substances that are controlled.

Although it might be thought that only large companies would generate significant quantities of hazardous waste, in fact quite small businesses often generate sufficient hazardous waste to come within legislative control. For example, photograph-processing centres, decorators (especially in regard to paint and paint-thinners), dry cleaners (for clothes) and car repair companies all generate hazardous waste. Although some legislation about safe hazardous waste disposal also extends to householders, this is very difficult to enforce. However, companies operating an EMS should recognise this potential source of pollution by their customers in the general public, and respond to it by providing free disposal facilities for waste arising out of their products. This leads to schemes like disposal of old car batteries and used motor oil. Companies can also minimise pollution by their customers by reducing the amount of hazardous materials contained in the products that they sell to the public. For example, water-based products are much less harmful to the environment than solvent-based ones.

Legislation requires that all hazardous waste be treated to reduce its toxicity, after which it can be disposed of by composting, combustion or landfill. The disposal of untreated hazardous waste in landfill sites is normally prohibited, although there is often a dispensation to allow acceptance of small amounts of genuinely domestic hazardous waste. Treatment of hazardous waste can take place either on-site or off-site, although off-site treatment requires that a secure transportation system be in place to avoid pollutant escape. The specific treatment applied depends entirely on the type of waste involved. It is impossible to give specific guidance in a book of this nature, and this is an area where advice must be sought from appropriate experts.

Legislation governing the treatment of hazardous waste normally excludes techniques that merely dilute it, since this does not chemically change the toxic elements. Likewise, simply storing the waste indefinitely in sealed containers as an alternative to treatment is not allowed. However, temporary storage is allowed as long as the storage is secure and the only purpose of storage is to allow waste to build up until the quantity is sufficient for treatment to be more economic. However, the frequency with which the waste is emptied from the storage tank and treated must be documented and agreed.

References

1. Envirowise International, 2003, *Practical Environmental Advice for Businesses* (*www.envirowise.gov.uk*) (last accessed June 2003).

2. Forum for the future, 2002, *UK Mass Balance Project* (*www.massbalance.org*) (last accessed June 2003).
3. Katholieke Universiteit Leuven, 2001, *Use of the Mass Balance* (*www.chem.kuleuven.ac.belsafety*) (last accessed June 2003).
4. Environmental health and safety online, 2002, *Hazardous Waste Disposal*, 2002 (*www.ehso.com*) (last accessed June 2003).

11

System Reliability and Risk Assessment for Environmental Protection

In general terms, reliability and risk assessment describe procedures that identify abnormal incidents and accidents that might arise in some activity and quantify the likelihood of their occurrence. The discussion below summarises the main principles involved, but more comprehensive treatments on reliability and risk assessment can be found if required in the further reading suggested[1–5,7]. If several texts are read, it will become apparent that there are no universally agreed definitions of what the terms 'reliability' and 'risk assessment' mean in more specific terms, and there is much inconsistency in the way that these two terms are used. In some circles, the word 'hazard' is used to describe any abnormal incident or accident, and the word 'risk' is used to quantify the probability of a particular hazard occurring. In other circles, the word 'risk' is merely used to describe the hazard itself. In yet another usage in the design and analysis of safety systems, the word 'risk' combines two components, the probability of a hazardous incident occurring, and also the severity of the consequences of the incident.

The word 'reliability' also suffers some variability in the way that it is defined, although there is rather less variation than in definitions of risk. The usual definition of reliability is that it is the probability that a system will operate without faults for a given period of time, where the word 'fault' is used to denote any hazard as defined above. However, variation in the definition occurs according to whether or not the severity of the fault is taken into account. Also, reliability is sometimes defined in quasi-absolute rather than probabilistic terms, such as in terms of the mean-time-between-failures.

For the purposes of ISO 14001, it seems sensible to give the possibility of environmental pollution the same importance as the possibility of human injury when

ISO 14000 Environmental Management Standards: Engineering and Financial Aspects. Alan S. Morris.
 ISBN 0-470-85128-7

abnormal incidents occur. Thus, the definitions given below of the various risk-related terms that have been adopted in this book for environmental protection purposes are similar to those used for safety systems.

11.1 Definitions

Hazard: This is any possible abnormal incident or accident in some manufacturing operation or activity of a company that might cause environmental pollution.

Risk: This is a measure of the probability of occurrence and the severity of adverse effects arising from a particular hazard.

Risk assessment: This is a measure of the probability of each hazard occurring.

Risk analysis: This describes a combination of three related processes, assessing what hazards might arise, calculating the probability of each hazard occurring, and evaluating the consequences.

Risk management: This involves considering each identified hazard in turn, considering what options are available to reduce the likelihood of the hazard occurring and to respond if it does occur, and evaluating each possible action in terms of its costs and benefits.

11.2 Identifying Hazards

The origin of hazards can be divided into four categories, hardware faults, software faults, management failures and human mistakes. As almost all activities and manufacturing operations of companies involve hardware, software, and a managed human labour force, there is a potential for all four types of hazard to occur in most situations. The starting point for hazard identification is to draw a flow chart of all activities and then categorise each one according to the type of hazard that might arise. Past experience and historical information are valuable aids in this, and therefore the personnel carrying out this task need to have good understanding and knowledge about the processes for which the hazard analysis is being carried out.

11.2.1 Hardware failures

Hardware failures affecting an activity can be divided into three groups: critical failures, major failures and minor failures. Critical failures are those that cause total stoppage of an activity, and possibly the emission of a significant amount of pollution. Major failures are those where the activity is seriously downgraded but can continue to some extent, although often with an increased level of pollution. Minor failures are those where an activity is affected, but not to a great extent, and any increase in pollution is only slight. Deciding whether a hardware failure is critical, major or minor is part of risk analysis.

Identifying all the hardware faults that might arise is a difficult task and requires the expert knowledge of engineers who are intimately acquainted with the plant in question. However, even such expert knowledge cannot predict everything that might happen, although it should be able to predict most likely causes of failure. Some common types of hardware fault are listed below:

- variation in fuel supply;
- variation in cooling system;
- failure of power supply;
- mechanical failure of components;
- failure of seals (often due to excess pressures);
- complete or partial failure of pollutant removal systems;
- atmospheric effects (e.g. temperature, rain);
- failure of important measuring instruments.

The incidence of many of the types of fault in the list above can be reduced by good engineering practice, in avoiding use of components over a certain age, protecting components from hostile environments, and not subjecting them to large temperature and pressure changes. However, as with so many aspects of environmental management, the process of fault prediction is case-specific in many respects, and requires sound knowledge of the plant in question and good engineering judgement.

11.2.2 Software faults

As the number of computerised elements in manufacturing systems and other activities of companies have grown, the incidence of failures due to a software fault has grown in parallel. Every piece of software used must be regarded as a potential hazard. However, unlike hardware, software is unchanging. Therefore, if software has been used once in any situation without problems, it will always work in that situation, provided that everything remains exactly the same. Of course, a computer output can suddenly change in circumstances where there has not apparently been any change, but that could only be due to a hardware fault in the computer.

Software faults can arise when some untested combination of data is applied to a software program. Such incidents are notoriously difficult to predict, since it is impossible to test software rigorously and prove that no errors exist in it. However, the application of good software engineering principles when designing software can greatly minimise the risk of it containing errors. Software quality and software fault prediction are described in detail in[6].

During the hazard assessment exercise, all situations where there is some change should be regarded as a possible source of software failure. These include situations where a new part of software is used or a different range of data values are to be input into software. Otherwise, if there is no change, there will not be any software failures in software that has previously operated satisfactorily.

11.2.3 Management failures

Many hazards arise from organisational failures within a company that are due to poor management practices. More effective management can therefore greatly reduce these types of hazards. Some common examples of situations arising from poor management that lead to hazards are:

- breakdown in communications between staff;
- lack of incentives for staff to find or report problems;
- conflicts between management and staff;
- staff covering up mistakes and defects;
- slowness in staff correcting defects;
- non-compliance of staff with recommended maintenance procedures;
- inadequate data collection due to unenforced inspection policy.

11.2.4 Human mistakes

Human mistakes are probably the most difficult of all types of hazard to predict, because of their high degree of randomness. Even the most highly trained and conscientious human worker is liable to make mistakes from time to time, but it is impossible to predict when this might happen. In spite of this, much can be done to reduce the frequency and effect of human mistakes by taking the following measures:

- Ensuring that workers have appropriate training for their job, with refresher courses as necessary.
- Providing adequate rest breaks to ensure that workers do not suffer from fatigue.
- Avoiding excessive hours of work.
- Providing comfortable working conditions, particularly in terms of temperature and humidity level.
- Providing adequate lighting.
- Strictly enforcing a no-alcohol policy both before and during a shift of work.

11.3 Risk Assessment

Although risk management requires an objective assessment of the likelihood of occurrence of each hazard, this requirement is very difficult to satisfy, since there is always some degree of subjectivity in risk assessment in practice. Objectivity in risk assessment can only be achieved on the basis of historical information and a presumption that a past frequency of occurrence will continue in the future. However, as company activities and operations continue to evolve, some change is inevitable, and the presumption that what has happened before will happen again no longer holds. Thus, historical information can form the starting point in objective risk assessment, but some subjectivity is introduced when deciding how changes in the processes analysed will change the hazard risks.

11.3.1 Fault tree analysis

Fault tree analysis is a common approach to risk assessment. The starting point in this is to identify all the possible initiating events. An initiating event is any hazard that might arise and possibly cause pollution. An event tree is then drawn for each initiating event, showing all possible consequences of that event, as shown in Figure 11.1. The probability of each consequence in the event tree is then assessed. Reliability analysis (see Section 11.6) is often a useful aid in this.

The number of possible sequences of consequences following an initiating event can potentially be quite large. As shown in Figure 11.1, there are 2^N possible

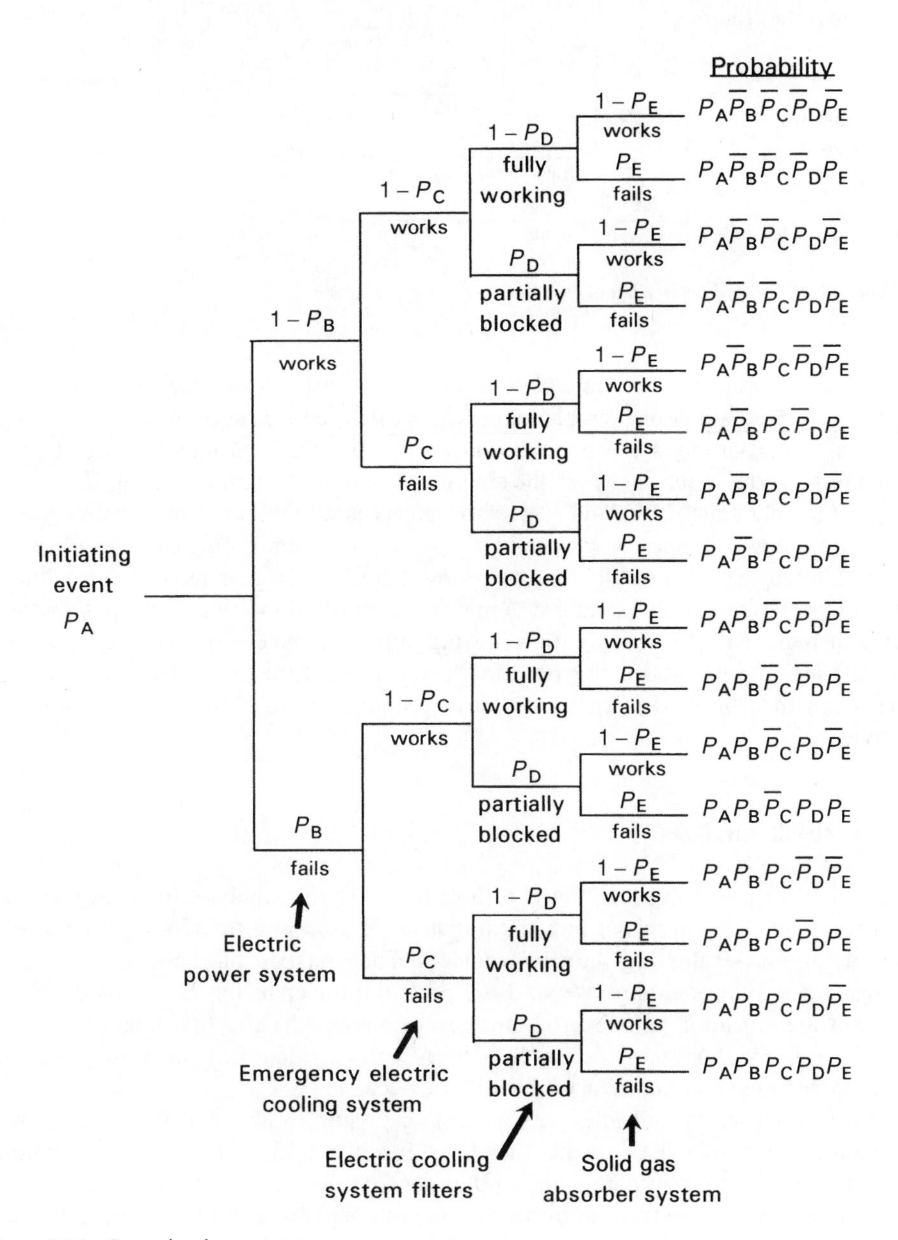

Figure 11.1 Example of an event tree.

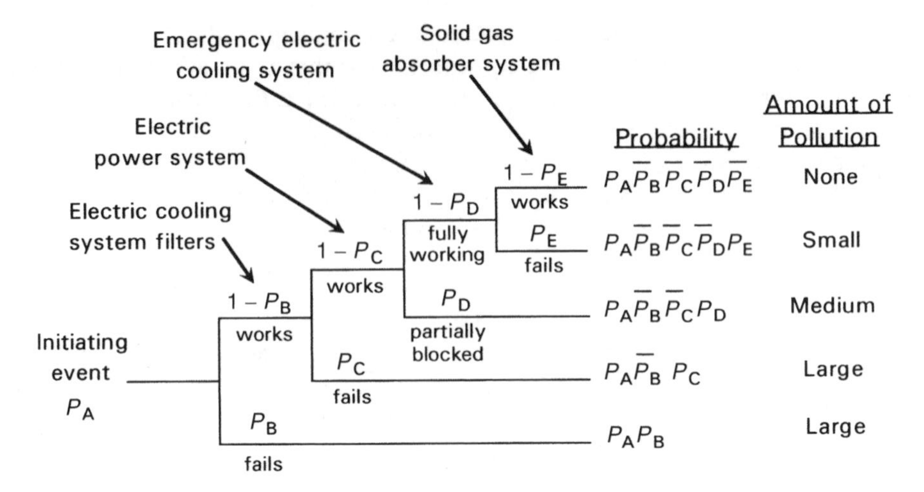

Figure 11.2 Example of a 'pruned' event tree.

sequences if there are N possible consequences arising from the initiating event. However, the number of possible sequences is often reduced substantially when engineering considerations are applied, allowing substantial 'pruning' of the tree. In Figure 11.1, the sequence where the emergency cooling system works but the electric power supply fails, is invalid, since the emergency cooling system cannot work without a power supply. Also, a solid gas absorber system is only capable of removing pollutant gases in small concentrations. Thus, any large emission of gas due to failure of the emergency cooling system will cause pollution, because the gas absorber cannot remove it. Thus, any path following failure of the emergency cooling system always leads to pollution. The resulting 'pruned' event tree is shown in Figure 11.2. This can then be used in risk analysis to evaluate the resulting pollution for each sequence.

11.4 Risk Analysis

As stated in the definitions given in Section 11.1, risk analysis involves the three processes of assessing what hazards might arise, assessing the risk of each hazard occurring and evaluating the consequences. The first two of these processes have already been considered above, and the potential for error has been noted. Firstly, hazard identification will never identify every possible hazard that might arise. Secondly, even for those hazards that have been correctly identified, some subjectivity is inevitable in calculating their probability of occurrence.

With this background, it is perhaps not surprising to find that the consequence-evaluation part of risk assessment is also subjective in many respects. For example, if a hazard causes a discharge of a pollutant into air, the consequence depends on factors like wind speeds, wind directions and atmospheric conditions: some assumptions can be made about these variables, but evaluation of the environmental consequences is necessarily subjective. Likewise, quantifying the effects of pollutant discharge into water requires assumptions to be made about the water flow rate and the pollutant dispersion characteristics. In the case of pollution by noise or vibration, assumptions have to be made about the highly subjective human perception of these quantities. Thus, psychological as well as mathematical considerations come into evaluating the consequence of hazards.

In spite of the subjective aspect of risk evaluation, some degree of quantification can often be achieved by scoring techniques in which the consequences arising from each hazard are given a numerical value. This is well established in safety system analysis in the form of SIL (safety integrity level) values.

Although objective measures can be made about the probability of pollution arising from various hazards and the sequences of consequences that can follow, the amount of subjectivity involved in such calculations means that accuracy is often quite low. As a result, pollution probabilities are often just expressed in subjective terms like 'small', 'medium' and 'high'. The right-hand side of Figure 11.2 shows such subjective evaluations of pollution probability for each of the event sequences shown. The sequence where everything works leads to no pollution risk. However, the path where the emergency cooling system works, but the gas absorber fails, will lead to a small amount of pollution, since the emergency cooling system is unable to provide quite enough cooling to prevent the creation of a small amount of pollutant gas, and this will not be removed if the gas absorber fails. The path where the filters are partially blocked leads to a medium amount of pollution since the solid absorber can only remove gas in small concentrations. Finally, if everything fails, the amount of pollution will be large. Such subjective evaluations of pollutant probability can be expressed in the form of a histogram, as shown in Figure 11.3.

If the environmental consequences of a fault are minor, a decision may be taken to continue with the activity affected whilst the fault is being rectified. In that case, some estimate of the time that will be taken to correct the fault is needed. One approach to this called the *three-point estimate* involves making three predictions

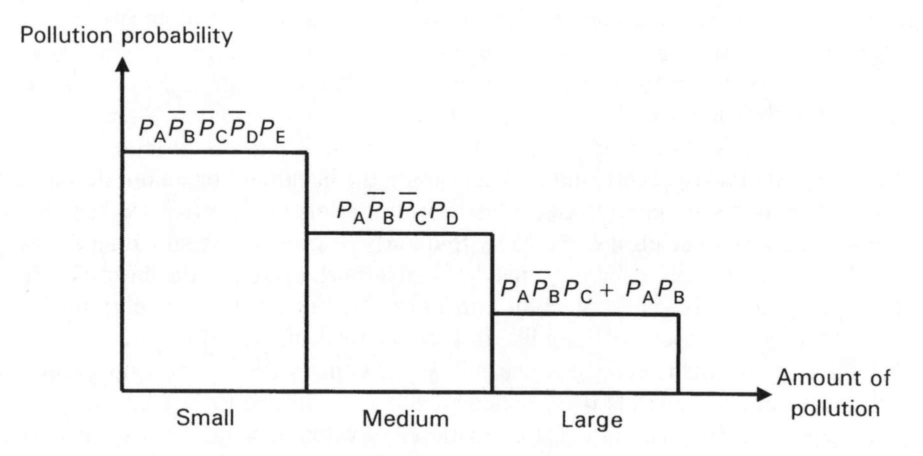

Figure 11.3 *Histogram of pollutant probabilities.*

about the time needed; an optimistic one, a 'most likely' one and a pessimistic one. The estimated time is then calculated according to the following formula:

$$\text{estimated time} = \frac{\text{optimistic time} + (4 \times \text{most likely time}) + \text{pessimistic time}}{6}$$

As an example, suppose that an optimistic time is three hours, the most likely time is six hours and a pessimistic time is 24 hours. Then the best estimate of the time needed to correct the fault is $[3 + (4 \times 6) + 24]/6 = 8.5$ hours. The above formula for calculating the time needed to correct the fault is based on $\pm 3\sigma$ limits within a beta distribution and means that, for the values given, 99.7% of the faults will be corrected within 8.5 hours (NB σ represents the standard deviation of the fault correction times).

11.5 Risk Management

For each hazard identified, the various ways of reducing the likelihood of the hazard occurring, or responding if it does occur, must be considered. Each possible action must then be evaluated in terms of its costs and benefits, with the aim of reducing environmental damage at an acceptable cost.

The likelihood of hazards occurring can generally be reduced by increasing the reliability of manufacturing plant and other systems. Inevitably, there is a cost in doing this. One common technique, as discussed in Section 11.6.7, is to identify the least reliable components in a system and duplicate them, so that one is kept on standby to replace the component in normal use if it fails. This can become quite expensive if every component with unsatisfactory reliability is duplicated, particularly because automatic switching systems are often needed to effect automatic transfer from the normal system component to the standby one. However, if the consequences of each hazard occurring are considered as well, it becomes possible to adopt a strategy of duplicating only those unreliable components whose failure leads to the greatest consequences. This makes the strategy much more cost effective.

11.6 Reliability Analysis

Reliability analysis is a useful aid in determining the likelihood of a fault developing in a system or system component. This is an important tool in risk assessment, as discussed earlier, to determine the likely frequency of environmental damage arising from faults. In addition, reliability analysis is an important part of the life cycle analysis of products whose environmental impact can be lessened by reliability improvements that increase their working life and reduce their impact while in use.

Reliability is formally defined as the ability of a component to perform its required function within the bounds of specified working conditions for a stated period of time. Factors such as manufacturing tolerances, quality variations in raw materials used, and differing operating conditions, all conspire to make the faultless operating life of a component impossible to predict. Such factors are subject to random variation and chance, and therefore reliability cannot be defined in absolute terms. The nearest one that can get to an absolute quantification of reliability are quasi-absolute terms like the mean-time-between-failures, which expresses the average time for which a component works without failure. Otherwise, reliability has to be expressed as a statistical parameter that defines the probability that no faults will develop in a component over a specified interval of time.

One immediate difficulty that arises when attempts are made to quantify the reliability of a component is defining what should be counted as a failure. Failures can generally be divided into three categories:

- *Critical failures* – where failure causes total loss of function in the component.
- *Major failures* – where failure causes a major loss of function in a component but it can still continue to be used to some extent.
- *Minor failures* – where failure leaves the component still able to perform its primary purpose, but with the loss of some convenience function.

In a product such as a transistor radio, failure of the speaker would be regarded as a critical failure, because that would render the radio totally useless. Failure of a noise suppression circuit within it would probably be counted as a major failure, as it would make the radio difficult to listen to, although some programmes might be heard if the user listened carefully. A minor failure might be failure of the tone adjustment control, which would impair the quality of sound reproduction by a small degree but programmes could still be heard reasonably well.

It is probably becoming apparent that classification of failures into these three categories is not necessarily straightforward. Many types of failure lie on the borderline between categories, and which side of the border they are placed is a matter of personal judgement. Nevertheless, deciding what should be counted as a serious failure is a prerequisite in reliability analysis. Once this has been decided, reliability can be quantified either in quasi-absolute or probabilistic terms.

11.6.1 Reliability quantification in quasi-absolute terms

Whilst reliability is essentially probabilistic in nature, it can be quantified in quasi-absolute terms by the mean-time-between-failures and the mean-time-to-failure parameters. It must be emphasised that these two quantities are only average values calculated over a number of identical examples of a component. The actual values for any particular sample of a component may vary substantially from these mean values.

The *mean time between failures* (MTBF) is a parameter that expresses the mean number of failures which occur in a component over a given period of time. For example, suppose that the history of a robot manipulator is logged over a one-year (365-day) period and the time intervals in days between faults occurring that require repair are as follows:

11 23 27 16 19 32 6 24 13 21 26 15 14 33 29 12 17 22

The mean interval is 20 days, which is therefore the mean-time-between-failures. A simpler way to calculate the MTBF is to merely record the total number of failures that occur over a given time interval. For N failures over a time interval t, the failure rate θ is given by: $\theta = N/t$. The MTBF is the reciprocal of θ, i.e.: MTBF = $1/\theta$.

The *mean time to failure* (MTTF) is a parameter that is associated with components which are discarded at the first failure, because it is either impossible or uneconomic to repair them. It expresses the average time before failure occurs, calculated over a number of identical components. Consumer goods such as cheap transistor radios are examples of products that cannot be repaired economically and are therefore discarded at the first major or critical failure. For example, suppose that a batch of 20 radios is put through an accelerated-wear test (in which a long period of typical use is simulated over a much shorter period), and the simulated life in years before serious failure for each radio is as follows:

7 9 13 6 10 11 8 9 14 8 8 12 9 15 11 9 10 12 8 11

The mean of these 20 numbers is 10. Therefore the simulated mean-time-to-failure is 10 years.

For many components, MTBF and MTTF figures are both relevant. The nature of a component often means that minor repairable faults will occur at various points in time during its use, but then, when it reaches a certain age, a catastrophic failure will occur. Over its working lifetime, the frequency of faults is quantified by the MTBF parameter, and the estimated working life before irredeemable failure occurs is quantified by the MTTF parameter.

A further reliability-associated term of importance is the *mean-time-to-repair* (MTTR). This expresses the average time needed for repair of a component, calculated over a number of typical faults that are likely to occur in it. Returning to the example of the robot manipulator, suppose that the time taken in hours to repair a particular fault on 18 occurrences was:

4 1 3 2 1 9 2 1 7 2 3 4 1 3 2 4 4 1

The mean of these values is three: therefore the mean-time-to-repair was three hours.

The relative importance of the mean-time-to-repair parameter varies according to the component to which it is applied. When applied to a critical element in a production process, the mean-time-to-repair is of equal importance to the mean-time-between-failures. What really matters in this case is the proportion of the total available production time that is lost whilst the critical element is inoperative. Alternatively, if production continues whilst the fault is being rectified, it is the time when environmental damage may be occurring. Clearly, repair time is of equal importance to the frequency of fault occurrence. Often, an element whose MTBF is low but whose MTTR is also low will be preferable to an alternative element where the MTBF is a little higher but the MTTR is a lot higher.

However, this argument about the relative importance of the MTBF and MTTR parameters assumes that elements have to be repaired immediately on failure. Where critical elements in production processes do not have a great cost, it is normal practice to keep spare elements on standby to replace elements on the production process as they fail. In this case, the MTTR is of very little importance. The amount of lost production or environmental damage that occurs depends on how long it takes to remove and replace a failed element. In this mode of working, the MTBF parameter assumes very much greater importance, because the total production time lost in a given interval of time is clearly proportional to the number of faults occurring over the interval. The MTTR would only become important if it became very large and interrupted the supply of standby elements.

The MTBF and MTTR parameters are often expressed in terms of a combined quantity known as the *availability* figure. This measures the proportion of the total time that a component is working, i.e. the proportion of the total time that it is in an unfailed state. The availability of a component is defined as the following ratio:

$$\text{AVAILABILITY} = \frac{\text{uptime}}{\text{total time}} = \frac{\text{MTBF}}{\text{MTBF}+\text{MTTR}}$$

Alternatively, the availability can be expressed as: $\text{AVAILABILITY} = \frac{1}{1+(\theta \times \text{MTTR})}$, where θ is the failure rate.

To minimise both production costs and also environmental damage, the aim must always be to maximise the MTBF figures and minimise the MTTR figure, thus maximising the availability. As far as the MTBF (and MTTF) figures are concerned, good design and high quality control standards during manufacture are the appropriate means of maximising these figures. Good design procedures, which mean that faults are easy to repair, are also an important factor in reducing the MTTR figure. However, many factors affecting the MTTR are outside the manufacturer's control to a large extent, because they are strongly influenced by customer practices. If a customer chooses to do their own repairs, the time taken to effect each repair is governed by the skill of the personnel they employ to do the work and the stocks of spare parts that they keep. If, on the other hand, repair work is entrusted to the manufacturer, then they can do much to reduce the MTTR by ensuring that their maintenance staff are well trained and motivated and respond quickly to breakdown calls. The manufacturer must also maintain an adequate stock of spare parts and have a means of ensuring that parts are delivered speedily as soon as they are requested. The existence of an efficient parts-delivery service is an important contributor to reducing MTTR figures, even where the customer does their own maintenance.

It is very important to stress that the MTBF, MTTF and MTTR figures discussed above are average values. The greater the number of products that the average is calculated over, the greater will be the accuracy of the figure derived. However, any particular sample of a product may have MTBF, etc. figures that differ significantly from the mean value.

11.6.2 Failure patterns

The pattern of failure in a product may increase, stay the same or decrease over its life. Many types of product exhibit all three patterns of increasing, constant and decreasing failure rate over some period of their working lives, and have a failure-rate/age curve of the form shown in Figure 11.4(a). This is frequently referred to as the *bathtub curve*.

Manufacturing defects in many products, such as substandard components, poor sealing against contamination, faulty assembly or bad connections, show up very early in their life. Thereafter, the rate of fault-incidence remains at a low and approximately constant level for a substantial period of time. Following this period, material fatigue and other ageing processes are common reasons for the failure rate to start to increase.

In the case of products that have a failure pattern of the form shown in Figure 11.4(a), of which electronic components are typical, it is normal practice for manufacturers to 'burn in' components until all of the components that are likely to fail in this early period of their life have failed. This should mean that, by the time that a product is delivered to the customer, it will have reached the stage where the rate of fault-incidence is constant. If this screening process is controlled correctly, then customers will never experience the part of the failure pattern where the rate of fault-incidence starts at a high level and decreases with time. The failure pattern of the product after delivery, as perceived by the customer, would therefore be as shown in Figure 11.4(b). Normal practice would be to replace the product either when its

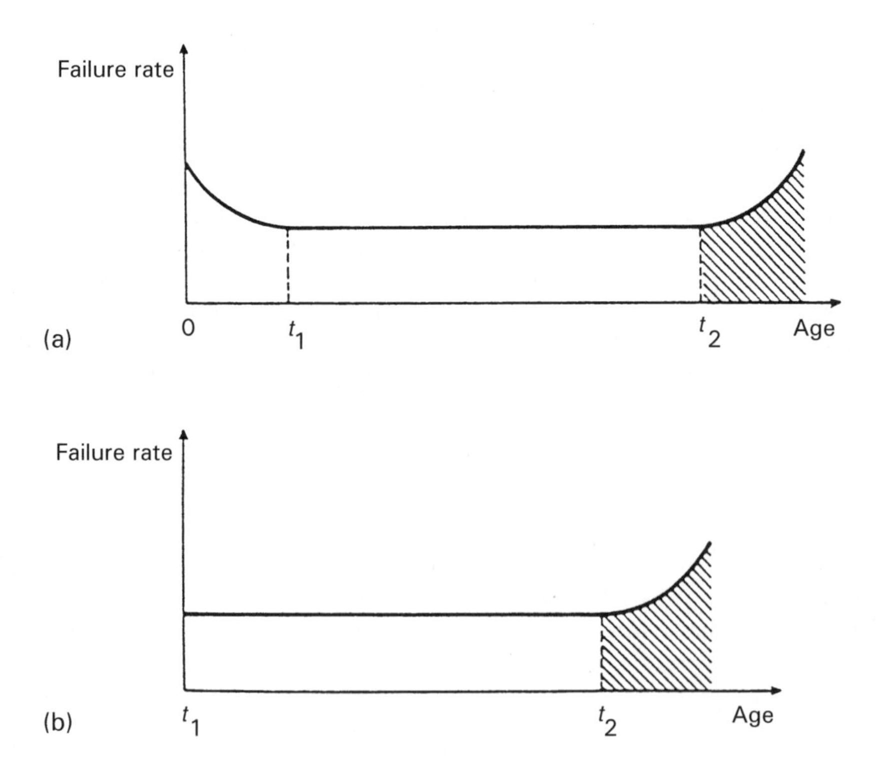

Figure 11.4 *'Bathtub' curves: (a) typical failure-rate/age characteristic of product immediately after manufacture; (b) target failure-rate/age characteristic when product is first used. From Morris (1991)* Measurement and Calibration Requirements, *© John Wiley & Sons, Ltd. Reproduced with permission.*

reliability reaches the right-hand shaded region in Figure 11.4 or shortly afterwards. This ensures that products are replaced before the rate of fault-incidence reaches a high level.

Complex systems that contain many different components typically exhibit a constant pattern of failure over the whole lifetime of the system. The various components each have their own failure pattern where the failure rate is increasing or decreasing with time. The greater the number of such components within a system, the greater is the tendency for the failure patterns in the individual components to cancel out and the rate of fault-incidence to assume a constant value.

11.6.3 Reliability quantification in probabilistic terms

The expression of reliability in quasi-absolute terms as the mean-time-between-failures has already been explained. If the average number of failures in a given time for a certain component is θ, the MTBF can be expressed as: MTBF = $1/\theta$.

In probabilistic terms, the *reliability* R_x of a component X is defined as the probability that the component will not fail within a certain period of time. The unreliability U_x is a corresponding term which expresses the probability that the component will fail within a certain time interval. The U_x and R_x are related by the expression:

$$U_x = 1 - R_x \tag{11.1}$$

For a given time interval t, U_x is related to θ (and hence to the MTBF) by the expression:

$$U_x = 1 - e^{-\theta t} \tag{11.2}$$

(It should be noted that this expression is only valid if the failure pattern is in the centre region of the bathtub curve, i.e. the failure rate is a constant. Also, the component must be in a working state at time $t = 0$, i.e. it must be delivered without faults.)

Examination of equation (11.2) shows that the unreliability is zero at time $t = 0$. Also, as t tends to ∞, the unreliability tends to a value of 1. This agrees with intuitive expectations that the value of unreliability should lie between values of 0 and 1. Another point of interest in equation (11.2), is to consider the unreliability when t = MTBF, i.e. $t = 1/\theta$. Then: $U_x = 1 - e^{-1} = 0.63$, i.e. the probability of a component failing after it has been operating for a length of time equal to the MTBF is 63%. Further analysis of equation (11.2) shows that, for $\theta t \leq 0.1$, $U_x \approx \theta t$. This is a useful formula for calculating (approximately) the reliability of a critical component which is only used for a time that is a small proportion of its MTBF.

11.6.4 Laws of reliability in complex systems

If a piece of equipment contains 100 integrated circuits, of which each has a mean failure rate of one failure per 10^6 hours, will the equipment operate for 1000 hours without failure? This is clearly an important question, but it is equally clear that it is impossible to give an answer in absolute terms. All that can be done is to express the likelihood or probability that the equipment will work for 1000 hours without failure. The reliability of a system containing multiple components that are either in series or in parallel can be expressed by applying the basic principles of probability[6].

Reliability of components in series

In many systems containing multiple components, the whole system fails if any one component within it fails. A good example of this is a system of fairy lights connected in series on a Christmas tree. One way in which the reliability of such a system of series components can be quantified is in terms of the probability that none of the components will fail within a given interval of time. Applying the joint probability rule[6], the reliability R_s of a system of n series components can be expressed as the product of the separate reliabilities of the individual components:

$$R_s = R_1 R_2 R_3 \cdots R_n \qquad (11.3)$$

In the case of n identical system components, equation (11.3) simplifies to:

$$R_s = (R_x)^n \qquad (11.4)$$

where R_x is the reliability of each component.

In the case of the integrated circuits mentioned at the start of this section, the probability of any individual chip failing within 1000 hours of operation is 0.1%. The reliability of each component is therefore (from equation (11.1)): 1.0 – 0.001 = 0.999.

As all of the transistors in the system are nominally identical, equation (11.4) is applicable, and the system reliability can be expressed as: $R_s = (0.999)^{100} = 0.905$.

Thus, there is a 90.5% probability that the system will operate for 1000 hours without failure.

Reliability of components in parallel

In many systems containing components connected in parallel, total system failure only occurs when all elements fail. Street lights are an example of such a system, where all lamps are connected in parallel on to the mains electrical supply. The street is only totally dark if all the lamps fail. For such systems, the system reliability, R_s, is given by:

$$R_s = 1 - U_s \qquad (11.5)$$

where U_s is the unreliability of the system. U_s is calculated in a similar manner to equations (11.3) and (11.4):

$$U_s = U_1 U_2 U_3 \cdots U_n \qquad (11.6)$$

or for identical system components:

$$U_S = (U_x)^n \qquad (11.7)$$

From equation (11.7):

$$R_s = (1 - U_s) = 1 - (U_x)^n = 1 - (1 - R_x)^n \tag{11.8}$$

As an example of the use of these equations, consider a lighting system containing four lamps connected in parallel. What is the probability of total system failure within the first 1000 hours of operation, given that the reliability of a single lamp over 1000 hours is 90%? All components in the system are identical, and therefore the system reliability can be calculated from equations (11.5) and (11.7):

$$U_x = 1 - R_x = 0.1. \quad \text{Hence,} \quad U_s = (0.1)^4 = 0.0001$$

Thus, the probability of total system failure (no light at all) over 1000 hours of operation is 0.01% or, expressed otherwise, the system reliability is 99.99%.

Binomial law of reliability

This law is applicable to systems of parallel components where some component failure can be tolerated, but system failure occurs before all individual components within it have failed. If R and U are the reliability and unreliability of n identical components connected in parallel in a system, then, by the binomial law of reliability:

$$(R + U)^n = 1 \tag{11.9}$$

Expanding equation (11.9):

$$R^n + nR^{n-1}U + \frac{n(n-1)}{2!}R^{n-2}U^2 + \cdots\cdots + \frac{n(n-1)(n-2)\cdots(n-r)}{r!}R^{n-r}U^r + U^n = 1$$

If the system does not fail unless all n components fail, then:

$$R_s = 1 - U^n$$

If the system fails when r components fail, then:

$$R_s = R^n + nR^{n-1}U + \cdots\cdots + \frac{n(n-1)(n-2)\cdots(n-r-1)}{(r-1)!}U^{n-1} \tag{11.10}$$

and since $R_s = 1 - U_s$,

$$U_s = \frac{n(n-1)(n-2)\cdots(n-r)}{r!}R^{n-r}U^r + \cdots\cdots + U^n$$

11.6.5 Availability of complex systems

The behaviour of complex systems containing many components is frequently described in terms of their availability rather than their reliability. This is because the availability figure is more directly relevant to production cost calculations and estimations of the likely environmental impacts. The equations to calculate the availability of systems containing series or parallel components have an identical form to those for calculating the corresponding reliabilities. Thus, for two series components:

$$A_s = A_1 \times A_2$$

and, for two parallel components:

$$A_s = A_1 + A_2 - (A_1 \times A_2)$$

where A_s is the net system reliability and A_1 and A_2 are the reliabilities of the separate system components.

11.6.6 Reliability calculations for sample systems

Example 11.1

The environment in a computer room is controlled by three identical air-conditioning units connected in parallel, as shown in Figure 11.5. The computer can continue to function if only one of the three units is working. If the reliability of each air-conditioning unit is 90% over 10 000 hours of operation, what is the overall reliability of the system?

For each unit, $R = 0.9$ and $U = 0.1$. Applying equation (11.10), with $n = 3$ and $r = 1$:

$$R_s = (0.9)^3 + [3 \times (0.9)^2 \times 0.1] + [3 \times 0.9 \times (0.1)^2] = 0.999$$

Thus, the overall system reliability is 99.9%, i.e. the probability that all three air-conditioning units will break down and leave the computer unable to run is only 0.01%.

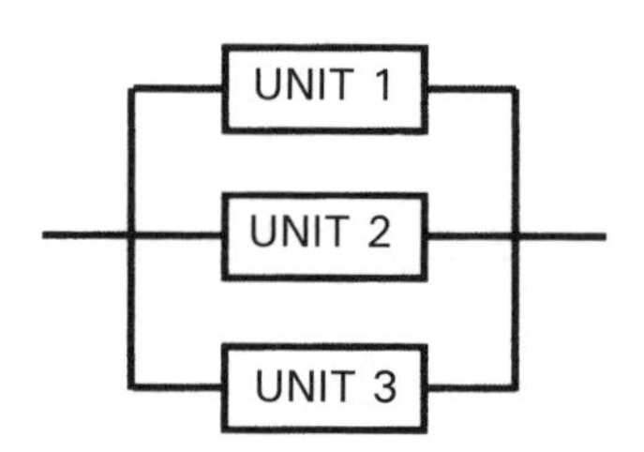

Figure 11.5 *Identical air-conditioning units in parallel.*

Example 11.2

In a machine shop, finished goods can be produced by either of two parallel routes A and B as shown in Figure 11.6. If the reliabilities of the machines are R_1, R_2, R_3,

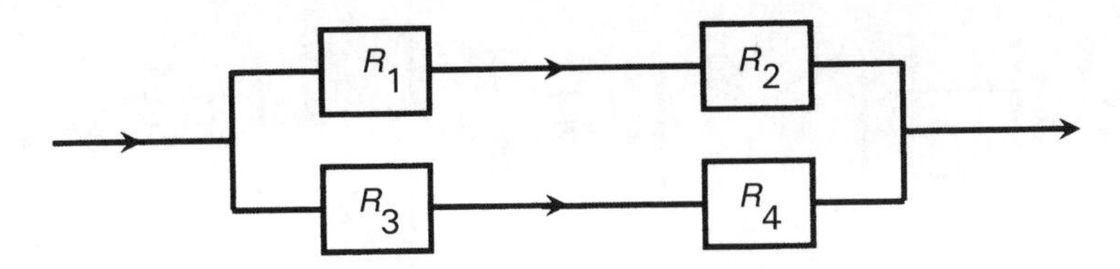

Figure 11.6 Production routes in parallel.

R_4, what is the total system reliability? (The system is only regarded as failed if neither of the two routes A and B is operational.)

From equation (11.3): $R_A = R_1 \times R_2$ and $R_B = R_3 \times R_4$

From equation (11.1): $U_A = 1 - R_A = 1 - R_1R_2$ and $U_B = 1 - R_B = 1 - R_3R_4$

From equation (11.5): $R_s = 1 - U_S$

From equation (11.6): $U_S = U_A \times U_B = (1 - R_1R_2)(1 - R_3R_4)$

Hence, $R_s = 1 - (1 - R_1R_2)(1 - R_3R_4)$

Example 11.3

Figure 11.7 shows a block diagram of some units in a production system that are connected in series. What is the overall system reliability if the reliabilities of the units for 1000 hours of operation are as follows?

$$R_1 = R_3 = R_5 = 0.99; \quad R_2 = R_4 = 0.90$$

Applying equation (11.3): $R_s = (0.99)^3 \times (0.90)^2 = 0.786$. Thus, the overall system reliability is 78.6%.

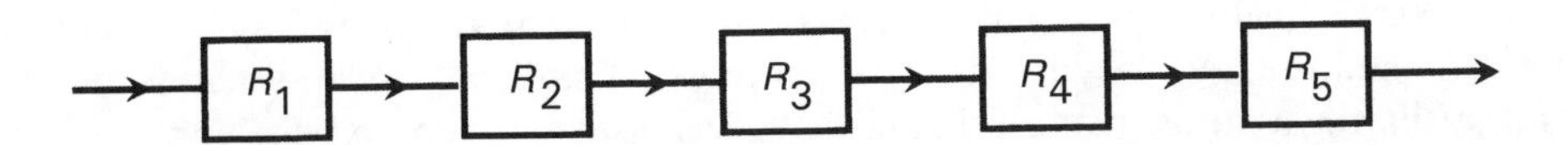

Figure 11.7 Production routes in series.

Example 11.4

Figure 11.8(a) shows a block diagram of the same production system referred to in example 11.3, but where the least reliable units in the system are duplicated. A switching system automatically transfers production to the standby unit when one of the duplicated units fails. If the values of R for each unit are the same as in example 3, what is the new system reliability?

The system is equivalent to the system shown in Figure 11.8(b), where R_A and R_B represent the net reliabilities of the pairs of identical production units.

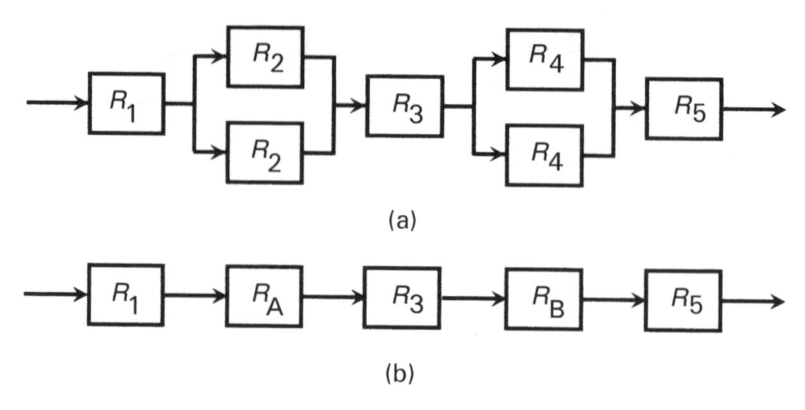

Figure 11.8 Improving reliability: (a) normal units in series with duplicated units; (b) alternative diagram for calculating the net efficiency of system (a) above.

From equation (11.8): $R_A = R_B = 1 - (1 - 0.90)^2 = 1 - 0.01 = 0.99$

Thus, $R_s = R_1 R_A R_3 R_B R_5 = (0.99)^5 = 0.951$. Therefore, the overall system reliability is now 95.1%.

11.6.7 Achieving high reliability

As a significant amount of environmental pollution arises as a result of faults in manufacturing systems, the reliability of manufacturing plant, and also of any special plant that is put in to reduce pollution, is of paramount importance. Several factors contribute towards achieving high reliability. Firstly, at the design stage of a system, all components included must be carefully chosen. They must be compatible with each other and suitable for use in the expected operating conditions. Secondly, good quality-control is necessary during the manufacture of products or construction of systems, to ensure that latent faults are not built in that will subsequently cause reliability problems. The more tightly that production parameters can be constrained to their target values, the less weaknesses will be inherent in the resultant product or system, and the less likely it will be to fail in use. Thirdly, if a new system is identified as having a failure pattern that initially decreases with time, a burn-in procedure should be followed, so that its failure pattern has reached steady-state before it is delivered to the customer.

Thus, sound analysis at the design stage, careful quality control during manufacture, and burn-in periods after manufacture where appropriate, will all ensure that the reliability of any particular system is the best that can practically be attained. However, in their subsequent use, many systems form part of a larger assembly, and the requirement then is to maximise the reliability of the whole assembly at minimum cost.

Several general rules about enhancing the reliability of an assembly containing several components become apparent if the examples in the last section (11.6.6) are studied. Firstly, it follows from example 11.3 (and also from equation (11.3)) that the greater the number of series components in a system, the less the overall system reliability tends to be. Therefore, attempts must be made at the design stage of such a system to include as few separate series components as possible. Having designed a system with as few separate series components as possible, example 11.4 shows that, by identifying the critical components in the system that have the worst reliability and duplicating them with identical components connected in parallel, the system reliability can be improved substantially. If this practice is followed, provision must be provided for replacing failed components by standby units. The most efficient way of doing this is via an automatic-acting switching system, but manual methods of replacement can also often work reasonably well. Similar principles of duplication apply in such areas as designing lighting systems. In a particular room, calculations may show that a single 80-watt fluorescent tube mounted centrally will provide adequate lighting. However, if reliability considerations are brought in, the designer will specify two 40-watt fluorescent tubes rather than a single 80-watt, one because this still provides some lighting even if one tube fails. The gain in system reliability is very much greater than the cost of achieving it because the energy consumption will be the same in either case and the only increased cost is the difference in cost between two 40-watt tubes and an 80-watt one. The general principle ensuing from this is that a system containing two or more identical components in parallel is usually more reliable than one that contains only a single component.

The principle of increasing reliability by placing components in parallel is often extended by deliberately putting more components in a system than it needs to function at 100% efficiency. This practice is known by the term *redundancy*. It is commonly applied in electrical circuits, where bad connections are a frequent cause of malfunctions. Where connections are made by plugs and sockets, it is often arranged that the same connection is made by two separate pairs of plugs and sockets. The second pair is redundant, i.e. it is not normally needed and the system can function at 100% efficiency without it, but it becomes useful if the first pair fails. A common example of this is the multiple earth connections to be found in a car, where, in good conditions, only one connection is theoretically necessary.

To summarise then, reliability considerations should be dominant at the design stage of a product or system. The design should be such that the product or system is as easy to create as possible. This will ensure that the incidence of faults during subsequent use of the product or system is minimised. High standards of quality control are of course a necessary condition for this as well. Finally, the general rule applies that, in a complex system, reliability is maximised when the number of series components is minimised, and when the number of parallel components that each perform the same function is as large as possible.

References and Further Reading

1. Dhillon, B.S. and Singh, C., 1996, *Engineering Reliability: New Techniques and Applications* (John Wiley & Sons, Chichester, UK).
2. Evans, J.R. and Olson, D.L., 1998, *Introduction to Simulation and Risk Analysis* (Prentice Hall, New Jersey).
3. Gray, S., 1995, *Practical Risk Assessment for Project Management* (John Wiley & Sons, Chichester, UK).

4. Haimes, Y.Y., 1995, *Risk Modelling, Assessment and Management* (John Wiley & Sons, Chichester, UK).
5. Kumamoto, H. and Henley, E.J., 1996, *Probabilistic Risk Assessment and Management for Engineers and Scientists* (IEEE Press, New York).
6. Morris, A.S., 1997, *Measurement and Calibration Requirements for Quality Assurance to ISO 9000* (John Wiley & Sons, Chichester, UK).
7. Vose, D., 2000, Quantitative Risk Analysis (John Wiley & Sons, Chichester, UK).

12

Statistical Process Control

When manufacturing plant is operating normally, the control procedures applied as part of the environmental management system usually ensure that any pollutants emitted are within defined limits and do not cause environmental problems. However, if a fault condition arises in the manufacturing plant, pollutant emission can rise significantly and lead to environmental damage. Most fault conditions are associated with changes in various operating parameters of the manufacturing process, and it would therefore seem to be a simple matter to just monitor these parameters with calibrated measuring instruments on a continuous basis (whilst correcting for any environmental errors, etc.), looking for any changes from the usual parameter values. Unfortunately, this is not as simple as it sounds, because many process parameters change value randomly within certain limits because of various random effects. Thus, in practice, almost every measurement will be away from the target value, and some mechanism is therefore needed to determine whether or not the deviations are within the range that can reasonably be expected.

Statistical process control provides this mechanism. It determines whether the variation in measurements of a process parameter is within the range to be expected according to the known random deviation characteristics of the parameter, or whether the deviation is outside the expected range and indicative of some problem in the manufacturing process. When applied correctly, statistical process control is an extremely valuable technique in minimising pollution arising from malfunctions in manufacturing systems. It is able to determine when things are starting to go wrong, before the stage is reached when the problem has grown to a sufficient magnitude where significantly increased pollution is emitted into the environment. This early-warning of the development of production problems also usually leads to significant savings in manufacturing costs, since early identification of process faults minimises

ISO 14000 Environmental Management Standards: Engineering and Financial Aspects. Alan S. Morris.
 ISBN 0-470-85128-7

the amount of defective product that is produced and subsequently discarded as being substandard.

In a book of this nature, it is inevitable that only an overview of the main principles of statistical process control can be covered. If required, more detailed information can be obtained from specialist texts on the subject, such as[1].

12.1 Conditions for Application of Statistical Process Control

Before statistical process control can be applied to a process variable, three conditions must be satisfied. Firstly, any systematic errors in the measurements must have been identified and either eliminated or compensated for. Secondly, the deviations of the measurements from the mean value must be truly random and Gaussian. Thirdly, the process must be under *statistical control*, which means that the mean and standard deviation of the measurements remain constant. Statistical process control can only be applied if, and only if, all of these three conditions are satisfied.

The condition regarding *systematic errors* is fairly easy to comply with. The various sources of systematic error were discussed at length in Chapter 4, and appropriate mechanisms were presented there for either eliminating them or compensating for them.

The condition about *Gaussian distribution* is rather difficult to test in a mathematically rigorous way. However, it is normally sufficient to plot some measurements on a histogram and confirm that they are symmetrically distributed about the mean value in a similar manner to that shown in Figure 4.1.

The final condition specified was about the process being under *statistical control.* This is verified by taking several sets of successive samples of a process parameter measurement and calculating the mean and standard deviation of each set. These will not be identically equal going from one set of measurements to the next, because each set only consists of a finite number of measurements out of an infinite data set (see the discussion on 'standard error of the mean' in Section 4.1.2 for further explanation), but differences from one data set to the next should only be small, and they should not show any trend, such as steadily increasing or decreasing values. The existence of statistical control is verified mathematically if the means of successive sets of measurements do not go outside the boundaries equal to three times the standard error of the mean. This is best explained by considering an example.

In a particular chemical production process involving an exothermic reaction, water-cooling is used to control the temperature, such that the correct products are obtained from the reaction. Suppose that several successive sets of twenty temperature measurements are taken and the mean and standard deviation (σ) are calculated for each, giving the following values:

Set 1: mean = 120.0; $\sigma = 2.05$
Set 2: mean = 119.5; $\sigma = 2.13$
Set 3: mean = 121.3; $\sigma = 2.02$
Set 4: mean = 120.4; $\sigma = 1.98$
Set 5: mean = 119.8; $\sigma = 2.08$

The first point to note is that the values of mean and standard deviation are not showing any increasing or decreasing trends. The standard error of the mean should now be calculated (See section 4.1.2):

$$\alpha = \sigma/\sqrt{n} = 2.13/\sqrt{20} = 0.48$$

(where n is the number of measurements in each data set)

$$\text{Hence, } 3\alpha = 1.44$$

The mean of the five successive data-set means can be calculated as:

$$\text{mean} = (120.0 + 119.5 + 121.3 + 120.4 + 119.8)/5 = 120.2$$

The maximum deviation of any of the five data-set means from this overall mean is (121.3 – 120.2) = 1.1. This is less than the calculated value for 3α of 1.44, showing that the process is under statistical control.

12.2 Principles of Statistical Process Control

Before proceeding further, it is useful to review some of the material on random errors presented in Section 4.1. Figure 4.2 shows that, for measurements subject to random variation within a Gaussian distribution, both very large and very small values extending towards plus and minus infinity are possible. However, the nature of the distribution is such that very large and very small values only occur rarely, and most measurements will fall reasonably close to the mean value. Statistical process control relies on this fact that most values will be close to the mean, and it is able to set a criterion for determining whether any particular measurement is inside or outside what could reasonably be expected according to random variation.

It was established in Section 4.1.2 that 99.7% of measurements lie within boundaries of $\pm 3\sigma$ either side of the mean of a set of measurements that are subject to random errors and part of a Gaussian distribution. This is normally used as the criterion for determining whether a particular measurement is inside the boundary of values that can reasonably be expected according to the known random variation characteristics. If any measurement falls outside the boundaries of $\pm 3\sigma$, there is only a 0.3% chance that such a large deviation is due just to random effects, and this is used to indicate that there is some abnormal condition in the production process which should be investigated and corrected.

Measurements that go outside the boundaries of $\pm 3\sigma$ usually indicate that some problem has already developed. However, in many production situations, it is desirable to have some prior warning that things are starting to go wrong before this stage is reached. Such prior warning is usually provided by establishing an additional check for values that go outside boundaries of $\pm 2\sigma$. For a set of measurements subject to random errors, 95.4% will be inside these boundaries and thus only 4.6% will be outside (as shown in Section 4.1.2). This is not a high enough probability to say that there is definitely a problem developing, but it can be used to initiate checks, especially if more than one measurement falls outside the $\pm 2\sigma$ boundaries within a short space of time.

Statistical process control of a production parameter is normally carried out in practice by plotting successive measurements on special charts. These are often known as *Shewhart charts*, after the American, Shewhart, who invented them in the 1920s. Three common types of chart are the *XBAR chart*, the *CUSUM chart* and the *RANGE chart*.

12.3 XBAR Chart (or MEAN Chart)

The *XBAR chart* is the one that is probably used most extensively. It is also sometimes referred to as a *MEAN chart*. The purpose of an XBAR chart is to show up measurements that differ from the mean value of the measurements by an abnormal amount in relation to what could reasonably be expected due to random errors. The mean of the measurements is commonly represented as $\overline{X}$ or XBAR, explaining why the chart is given the name '*XBAR chart*'.

In its most common form of implementation, an XBAR chart has the mean parameter value ($\overline{X}$) and the $\pm 2\sigma$ and $\pm 3\sigma$ boundaries drawn on it*. The $\pm 2\sigma$ boundaries are normally labelled as the upper and lower warning limits (UWL and LWL) and the $\pm 3\sigma$ ones as the upper and lower action limits (UAL and LAL). Referring back to the set of mass measurements given in Table 4.1, suppose that these are measurements in grams of the mass of breadcakes produced in a bakery. The mean of the measurements is 81.18 g, and the standard deviation is 0.30. The $\pm 2\sigma$ and $\pm 3\sigma$ boundaries are therefore ± 0.60 and ± 0.90 respectively and the warning and action limits can be calculated as follows:

$$\text{UWL} = 81.18 + 0.60 = 81.78; \quad \text{LWL} = 81.18 - 0.60 = 80.58$$
$$\text{UAL} = 81.18 + 0.90 = 82.08; \quad \text{LAL} = 81.18 - 0.90 = 80.28$$

An XBAR chart can be drawn for these mass measurements, as shown in Figure 12.1.

The basic principle of using an XBAR chart is to plot a succession of measurements of the process variable being monitored on the chart. If the process is properly under control, then all measurements plotted will remain within the boundaries marked by the LWL and UWL lines. This much is common to all statistical process control schemes. However, practice varies with regard to the response generated if measurements go outside the warning or action limits. The basic rules say that a measurement going outside the $\pm 2\sigma$ boundaries, as shown in Figure 12.2(a), should be taken as a warning that a fault may be developing in the production process, initiating checks that the sensors are working properly and carrying out other

* The formal standards on statistical process control described in BS 7785[2] and ISO 8258[3] specify slightly different boundaries of $\pm 1.96\sigma$ and $\pm 3.09\sigma$, which encompass 95.0% and 99.8% respectively of measurements that are varying only due to random effects.

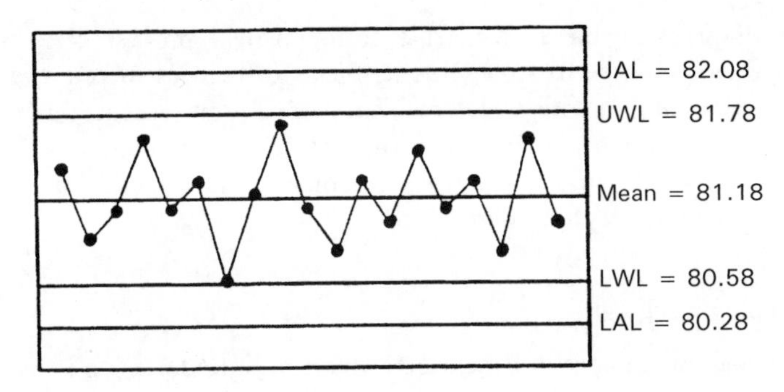

Figure 12.1 *XBAR chart for the 19 mass measurements given in Table 4.1. From Morris (1997)* Measurement and Calibration Requirements, *© John Wiley & Sons, Ltd. Reproduced with permission.*

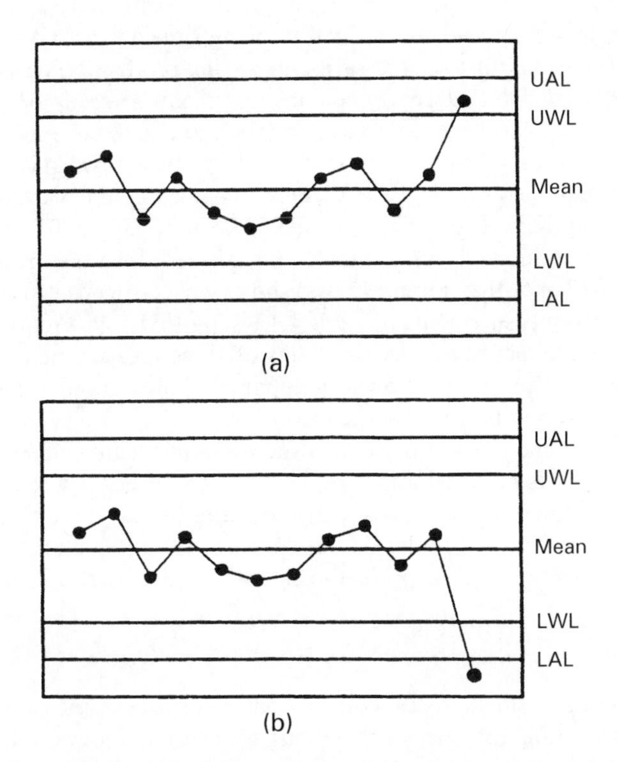

Figure 12.2 *Use of an XBAR chart to indicate process problems: (a) measurements going outside ±2σ limits; (b) measurements going outside ±3σ limits. From Morris (1997)* Measurement and Calibration Requirements, *© John Wiley & Sons, Ltd. Reproduced with permission.*

observations according to the nature of the process. Measurements going outside ±3σ limits, as shown in Figure 12.2(b), indicate that a definite fault has occurred, triggering remedial action which may include shutting the plant down if the fault cannot be readily identified and corrected.

However, in practice, these basic rules of statistical process control are often modified in various ways. To some extent, these differences in practice arise from differences in the sampling rate, which may vary from a few seconds for fast processes to as slow as one sample per day for some petrochemical processes. The sections below attempt to cover some of the ways in which the basic rules for responding to out-of-limit measurements are altered.

12.3.1 Sample averaging

If a single measurement goes outside the $\pm 2\sigma$ or $\pm 3\sigma$ boundaries on an XBAR chart, there is a finite probability (of 4.6% and 0.3% respectively) that the cause is a random deviation of the measurement and the instrument producing it rather than a fault in the monitored process. A measurement that exceeds $\pm 2\sigma$ or $\pm 3\sigma$ limits due to unusually large random effects is known as a *false alarm*. False alarms can be a serious problem, since they may lead to production being shut down to detect a fault that actually does not exist, thus unnecessarily increasing production costs.

The incidence of false alarms when measurements exceed these defined $\pm 2\sigma$ and $\pm 3\sigma$ boundaries can be greatly reduced if several successive measurements are averaged and the mean plotted on the control chart. If just two successive measurements are averaged, the probability of them both exceeding the warning limit of $\pm 2\sigma$ due to only random effects can be expressed as $0.046 \times 0.046 = 0.0021$ or 0.21%. If the average of three successive measurements is plotted, then the probability falls to $0.046^3 = 0.000097$ or 0.01%. Figure 12.3(a) shows what happens if the means of successive sets of two measurements in Table 4.1 are plotted, and Figure 12.3(b) shows a similar plot of the means of successive sets of three measurements. If figure 12.3 is compared with Figure 12.1, then it is apparent that averaging has the effect of moving the plotted points closer to the mean line.

Such plotting of the average of several successive measurements is clearly useful when a fast process is involved and samples are taken every few seconds. However, in the case of a slow process that is sampled infrequently, the technique is not of much use, because, if a real fault has developed in the production process, serious damage would result before the problem is identified and corrective action is taken.

12.3.2 Trends

The basic rules of statistical process control treat excursions of measurements outside $\pm 2\sigma$ limits as a warning that generates investigative action. However, if the process is changing rapidly, it may go on to exceed the unacceptable $\pm 3\sigma$ limits before there has been sufficient time to respond properly to such a warning and serious environmental pollution might result. This difficulty may be avoided if an additional rule of looking for trends is introduced into the control algorithm. A trend is a sequence of values that are either increasing or decreasing in value, due to events like a tool wearing. Figure 12.4 shows examples of such increasing and decreasing trends.

What has not been established so far is how many points there should be in a trend before action is taken. It is very important to get this right, otherwise false alarms may be generated. Starting from any measurement, P, the probability of the next

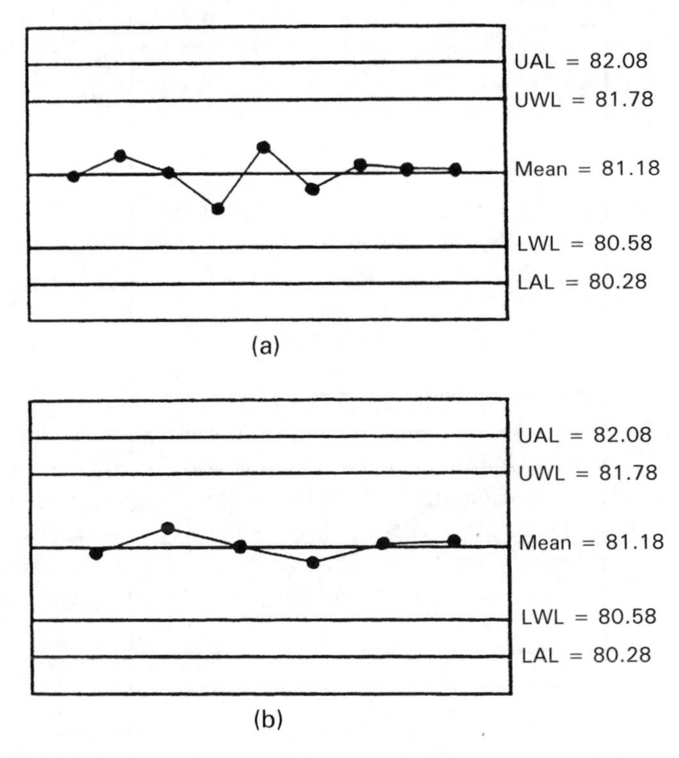

Figure 12.3 *Plotting the mean of successive measurements on an XBAR chart: (a) means of two measurements; (b) means of three measurements. From Morris (1997)* Measurement and Calibration Requirements, *© John Wiley & Sons, Ltd. Reproduced with permission.*

value being greater than P is represented by the parameter A in Figure 12.5 and the probability of it being smaller is represented by the parameter B. In general, A is not equal to B, and neither probability is easy to quantify. This makes it difficult to calculate the probability A^n or B^n that a succession of n values will increase or decrease consistently due only to random effects. However, because the sum $A + B$ is the total distance between the $+2\sigma$ and -2σ limits, i.e. equal to 0.954, there is a particular point where A and B are equal with a value of 0.477. This condition is approximately satisfied when the starting measurement P is close to the mean value. Therefore, starting from any particular point on the chart that is close to the mean value, there is approximately a 0.477 probability that the next measurement will be larger. The chance that the third measurement will be larger again is 0.477×0.477, i.e. 0.2275. The probability that this rising trend will continue just because of random variations to the fifth measurement is 0.052 or 5.2%. The probability of this trend of five measurements occurring because of random variations is similar to the chance (4.6%) of one measurement exceeding the $\pm 2\sigma$ boundaries. Such a trend containing five measurements is therefore taken as a warning of possible process problems. Continuing further, a trend continuing to the ninth measurement only has a 0.27% chance of occurring naturally with random variations. This is similar to the chance (0.3%)

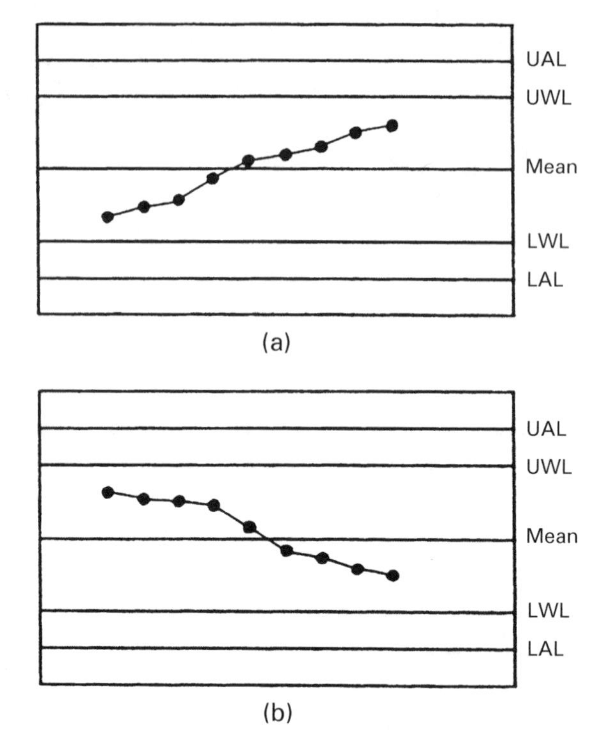

Figure 12.4 *Trends on XBAR chart: (a) increasing trend; (b) decreasing trend. From Morris (1997)* Measurement and Calibration Requirements, *© John Wiley & Sons, Ltd. Reproduced with permission.*

of a measurement exceeding the $\pm 3\sigma$ boundaries, and such a trend of nine measurements is therefore taken to indicate a problem with the process that must be investigated and corrected.

12.3.3 Runs

Like trends, runs can be very useful in predicting the onset of process problems before measurements start to approach the $\pm 2\sigma$ or $\pm 3\sigma$ boundaries. A run is a sequence of measurements that are consistently above or below the mean value, as shown in Figure 12.6. The probability of any particular measurement lying between the mean and the $+2\sigma$ boundary is 0.477 or 47.7% (because 95.4% of the measurements lie between $+2\sigma$ and -2σ). The probability of five successive values lying on the same side of the mean due to random effects is $0.477^4 = 0.052$ or 5.2%. This is almost the same as the 4.6% chance of a measurement lying outside the $\pm 2\sigma$ limits, and it is taken as a warning of possible process problems. The probability of nine successive measurements lying on the same side of the mean is only $0.477^8 = 0.0027$ or 0.27%. This is similar to the 0.3% probability of a single measurement lying outside the $\pm 3\sigma$ boundaries, and is therefore taken to indicate a problem with the process that must be investigated and corrected.

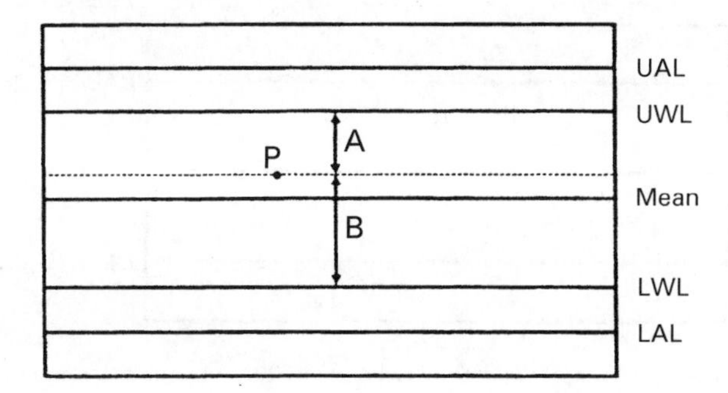

Figure 12.5 *Probability of next measurement being greater or smaller. From Morris (1997)* Measurement and Calibration Requirements, *© John Wiley & Sons, Ltd. Reproduced with permission.*

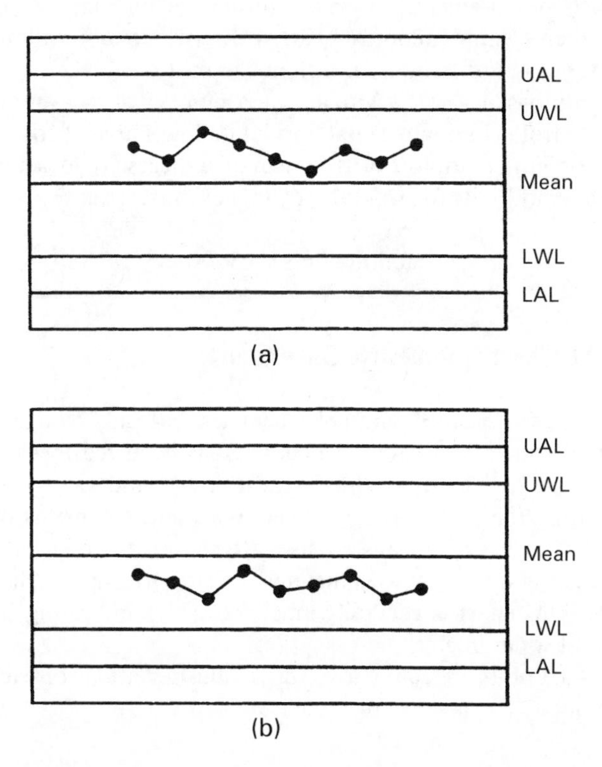

Figure 12.6 *Runs on XBAR chart: (a) measurements increasing; (b) measurements decreasing. From Morris (1997)* Measurement and Calibration Requirements, *© John Wiley & Sons, Ltd. Reproduced with permission.*

12.3.4 Simplified XBAR chart

If the statistical process control algorithm bases the decision about control action on two or more successive measurements, then it becomes unnecessary to have both warning and action limits. For example, if the decision is based on plots of the mean

Figure 12.7 *Simplified XBAR control chart. From Morris (1997)* Measurement and Calibration Requirements, *© John Wiley & Sons, Ltd. Reproduced with permission.*

of two successive measurements, or on a run/trend of nine increasing or decreasing measurements, then a simpler control chart with just $\pm2\sigma$ action boundaries and no warning boundaries is sufficient, as shown in Figure 12.7. If the mean of more than two measurements, or runs/trends with a greater number of measurements than nine, is used as the control criterion, then action limits lower than $\pm2\sigma$ can be used. The appropriate action limits for plots of the mean of n successive measurements that are equivalent to the $\pm3\sigma$ limits for single measurements are given by:

$$\text{action limits} = \pm3\sigma/\sqrt{n}$$

12.4 CUSUM Chart (cumulative Sum Chart)

If the mean value of a process parameter changes, but only by a small amount, an XBAR chart may fail to show that the change has occurred, because all measurements may continue to fall within the $\pm2\sigma$ warning boundaries for some time after the change. Plotting the mean of several successive measurements may be of some help in this situation, as such means are likely to show a run of mean values that are consistently either above or below the mean value drawn on the XBAR chart. However, a CUSUM chart is generally much better for indicating small changes in the mean value of a parameter.

A CUSUM chart plots the cumulative sum of the deviations of each measurement from the mean value. For a series of n measurements, x_i, $i = 1, \ldots n$, the cumulative sum is given by:

$$\text{CUSUM} = \sum_{i=1}^{n} d_i$$

Referring back to the set of 19 successive mass measurements given in Table 4.1, the CUSUM can be formed as shown in Table 12.1. Whilst the process is under statistical control, the CUSUM should remain close to zero and only make small excursions either side of zero. Whilst only random variations are taking place, negative deviations should occur in similar numbers to positive ones, and so the sum should be approximately zero. However, a potential problem can arise because of error in the mean value used to calculate the CUSUM. Referring back to the 19 mass measurements in Table 4.1, the mean is 81.178 947 if it is calculated to eight significant digits. This was approximated to a value of 81.18 for calculating the CUSUM. This means that there is an error of 0.001 053 (81.18 – 81.178 947) at every step in the CUSUM calculation. Thus, after calculating the CUSUM for two measurements, the error is 0.002 106, after three measurements it is 0.003 159, and after 19 measurements it is approximately 0.02. Clearly, the errors are cumulative and result in a value of the CUSUM that is significantly biased after a number of measurements. This necessitates periodically reinitialising the CUSUM to zero.

To demonstrate the value of CUSUM calculation, suppose that some change occurs in the production process after these 19 measurements have been taken and plotted, such that the mean increases from 81.18 to 81.28, and that a further ten measurements are then taken after the change. Table 12.2 shows these further ten measurements and the CUSUM calculation. If the XBAR chart in Figure 12.1 is continued by plotting these next 10 measurements, as shown in Figure 12.8, there is little indication that there has been a change in the process. However, if the full set of mass measurements is plotted on a CUSUM plot, as shown in Figure 12.9, the cumulative sum is seen to move randomly either side of zero for the first 19 measurements and then to move steadily upwards over the next ten measurements. This upward behaviour of the CUSUM plot is indicative of a change having taken place in the process.

Table 12.1 *CUSUM calculation for mass measurements given in Table 4.1*

Measurement	d_i	CUSUM	Measurement	d_i	CUSUM
81.4	0.22	0.22	80.8	−0.38	−0.18
80.9	−0.28	−0.06	81.3	0.12	−0.06
81.1	−0.08	−0.14	81.0	−0.18	−0.24
81.6	0.42	0.28	81.5	0.32	0.08
81.1	−0.08	0.2	81.1	−0.08	0.00
81.3	0.12	0.32	81.3	0.12	0.12
80.6	−0.58	−0.26	80.8	−0.38	−0.26
81.2	0.02	−0.24	81.6	0.42	0.16
81.7	0.52	0.28	81.0	−0.18	−0.02
81.1	−0.08	0.2			

Table 12.2 *Data values and CUSUM calculation for ten mass measurements following process change starting from CUSUM value after the 19 measurements in Table 4.1 as calculated in Table 12.1*

Measurement	d_i	CUSUM	Measurement	d_i	CUSUM
81.4	0.22	0.20	81.3	0.12	0.80
80.9	−0.28	−0.08	81.6	0.42	1.22
81.5	0.32	0.24	80.9	−0.28	0.94
81.7	0.52	0.76	81.7	0.52	1.46
81.1	−0.08	0.68	80.7	−0.48	0.98

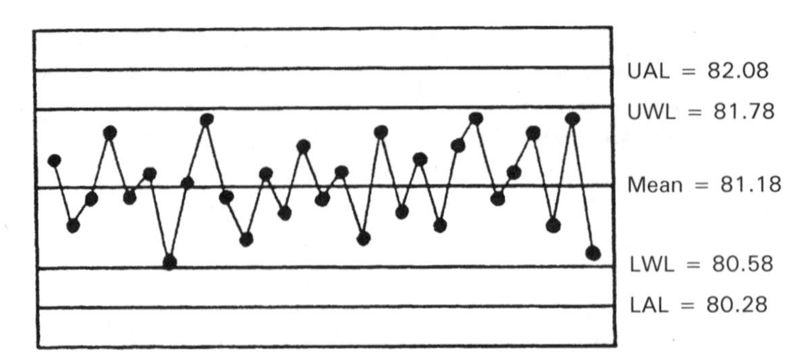

Figure 12.8 XBAR chart for 29 mass measurements. From Morris (1997) Measurement and Calibration Requirements, © John Wiley & Sons, Ltd. Reproduced with permission.

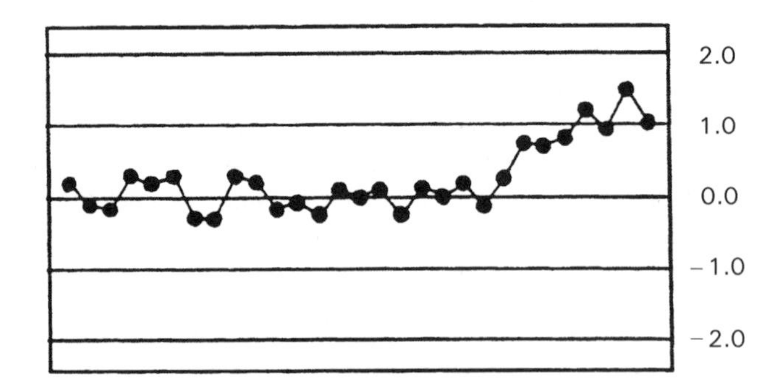

Figure 12.9 CUSUM chart for 29 mass measurements. From Morris (1997) Measurement and Calibration Requirements, © John Wiley & Sons, Ltd. Reproduced with permission.

The CUSUM chart is therefore much better than the XBAR chart for detecting small process changes. However, if a large change suddenly occurs that takes a measurement outside the $\pm 2\sigma$ or $\pm 3\sigma$ boundaries, this is indicated more quickly by an XBAR chart. This can be demonstrated if we return to the original set of 19 measurements in Table 4.1 and consider what happens if these are followed by a measurement of 81.8, which is just outside the $+2\sigma$ boundary of 81.78. The XBAR chart for the 19 + 1 measurements shown in Figure 12.10(a) immediately shows this excursion outside the upper warning limit. However, the corresponding CUSUM plot in Figure 12.10(b) fails to show this clearly. Thus, in practice, there is a strong case for plotting both XBAR and CUSUM charts in parallel.

12.5 RANGE Chart (R Chart)

The *RANGE chart* is often called an 'R' chart for simplicity. The chart plots the range of a set of several successive measurements of a parameter. The range of the measurements is simply the difference between the largest value and the smallest value

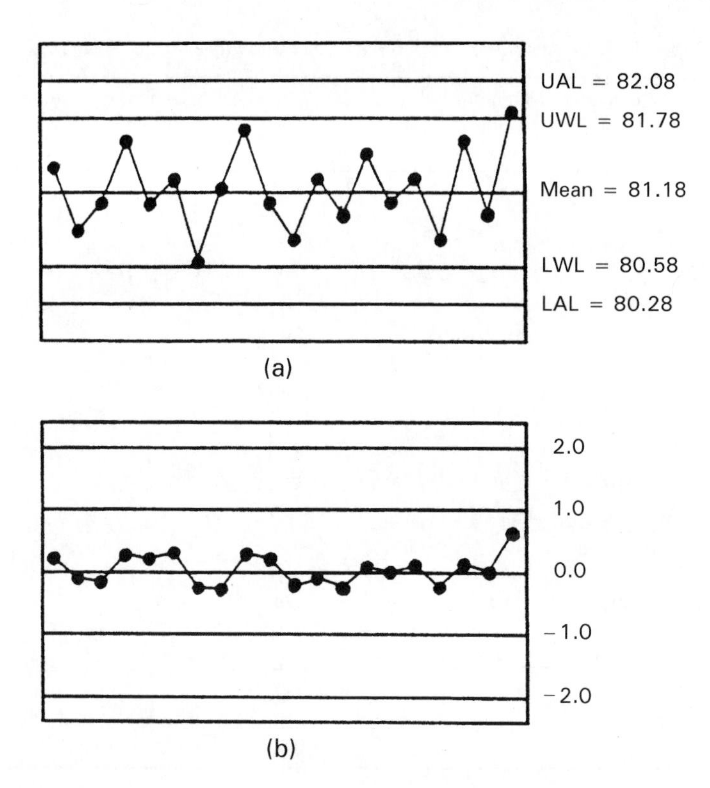

Figure 12.10 *Measurements exceeding +2σ limit: (a) plotted on XBAR chart; (b) plotted on CUSUM chart. From Morris (1997)* Measurement and Calibration Requirements, *© John Wiley & Sons, Ltd. Reproduced with permission.*

in the set. Such samples of several successive measurements are taken at regular intervals of time, and the range of each is plotted on the chart. The sample size for measurements to be plotted on an R chart can be as small as two, but it is more usual for the sample size to be at least four.

Table 12.3 shows some measurements of the thickness of paper produced in a paper mill. Four measurements are taken at a time, and this procedure is repeated every hour. The range of each set of four measurements is calculated in Table 12.3 and Figure 12.11 shows the corresponding R chart plotted.

The main use of an R chart is in identifying long term changes in a manufacturing process that cause the spread of measurements to change. The usual observation is an increase in spread, but a decrease can also sometimes occur, usually due to some malfunction in a measuring instrument. Whilst the process continues to operate correctly, the points plotted on the chart will remain within a constant band of values. However, any long-term trend for the points plotted on the chart to increase in magnitude is indicative of some change having taken place in the process that should be investigated. Likewise, any significant decrease in spread usually indicates that a measuring instrument is not working properly.

Table 12.3 *Paper thickness measurement and range calculation for successive samples*

Measurement (μm)	Maximum value x_1	Minimum value x_2 x_2	Range $x_1 - x_2$	Measurement	Maximum value x_1	Minimum value x_2 x_2	Range $x_1 - x_2$
97				96			
106				102			
104	106	97	9	104	104	96	8
101				101			
95				93			
99				97			
107	107	95	12	103	103	93	10
98				99			
102				106			
96				100			
101	102	96	6	102	106	94	12
99				94			
94				97			
105				101			
102	105	94	11	98	102	97	5
95				102			
100				106			
103				99			
97	103	97	6	96	106	96	10
101				102			

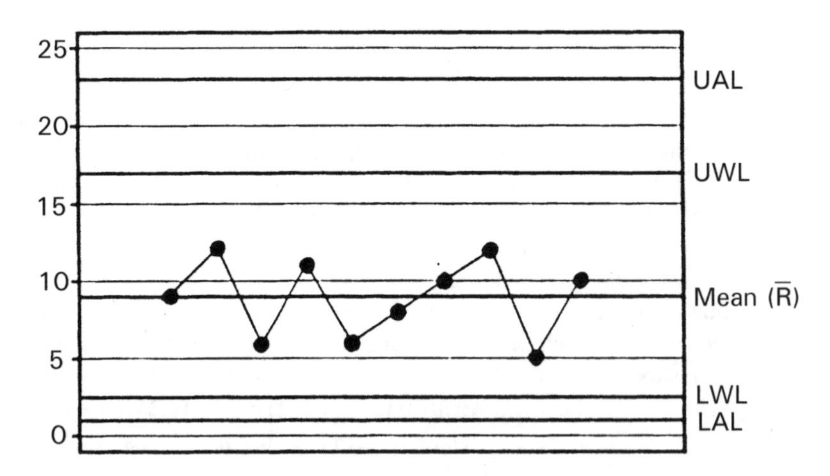

Figure 12.11 *Range (R) chart for paper thickness measurements. From Morris (1997)* Measurement and Calibration Requirements, *© John Wiley & Sons, Ltd. Reproduced with permission.*

12.5.1 Control limits for an R chart

Determining the control limits for an R chart is nothing like as straightforward as it is for XBAR and CUSUM charts, because the distribution of the ranges of samples is positively skewed, as shown in Figure 12.12. This means that the control limits are asymmetrical about the mean range.

The warning and action control limits are normally based on 97.5% and 99.9% probability limits (i.e. such that there is only a 2.5% and 0.1% probability respectively that any sample range outside these limits is due only to random effects). The actual values of the limits are calculated as:

$$\text{Limit} = K\overline{R} \tag{12.1}$$

where $\overline{R}$ is the mean range value and K is a constant. Values for K depend on the size of each sample and are derived from published tables, such as those in[1]. For a sample size (n) of four, the values of K are:

$$K_{\text{UAL}} = 2.57; \quad K_{\text{LAL}} = 0.10; \quad K_{\text{UWL}} = 1.93; \quad K_{\text{LWL}} = 0.29$$

The mean range value is calculated from:

$$\overline{R} = \left[\sum (R_1 + R_2 + \cdots + R_s)\right]/s$$

where R_1, R_2, etc. are the means of each successive sample, and s is the total number of sample range values used to calculate the mean sample range. Thus, from Table 12.3:

$$R = (9+12+6+11+6+8+10+12+5+10)/10 = 8.9$$

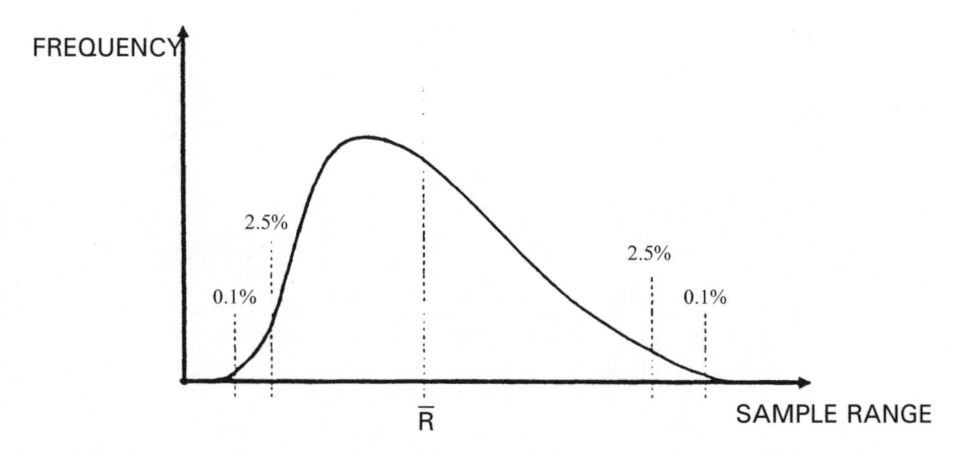

Figure 12.12 *Positively skewed limits on R chart. From Morris (1997)* Measurement and Calibration Requirements, *© John Wiley & Sons, Ltd. Reproduced with permission.*

Hence, using equation (12.1) above, the control limit values are:

$$\text{UAL} = 22.9; \quad \text{LAL} = 0.9; \quad \text{UWL} = 17.2; \quad \text{LWL} = 2.6$$

12.6 Summary of Control Charts

To operate statistical process control to the best advantage, it is usually necessary to plot all three types of control chart simultaneously, in order to gain the best advance warning of all the manufacturing process problems that may arise. When sudden changes occur, such as something breaking, the XBAR chart indicates this most rapidly. However, the other two types of chart are much better at showing up small changes in a process over a long period of time. Such small changes can either take the form of changes in the mean value of one or more process parameters, or alternatively take the form of a change in the spread of the measurements about the mean value. In some cases, both the mean value and spread can change together. For detecting changes in the mean, the CUSUM chart is best, whereas the RANGE chart is best for detecting changes in the spread of parameter values.

References

1. Oakland, J.S., 1996, *Statistical Process Control* (3rd Edn), (Butterworth-Heinemann, Oxford, UK).
2. BS 7785, 1994, *Shewhart Control Charts*, British Standards Institution, London.
3. ISO 8258, 1991, *Shewhart Control Charts* (International Standards Organisation, Geneva).

13

Monitoring Process Parameter Values to Minimise Pollution Risk

Environmental pollution often increases when process parameters vary from their usual value because of abnormal operating conditions. Such variations can be detected by statistical process control, as explained in the last chapter, allowing remedial action to be taken promptly before serious environmental pollution has occurred. However, a prerequisite for this is accurate measurement and recording of parameter values. ISO 14001 requires that such records are available to prove that the EMS system is working satisfactorily and to demonstrate that the company concerned is fulfilling its stated objectives in protecting the environment. Obtaining such records also has significant economic benefits, since parameter variations away from their target values invariably have an adverse effect on operating costs. The detection of variation in parameter values allows prompt remedial action to be taken.

The purpose of this chapter is therefore to review the main ways of measuring the common process parameters of temperature, pressure, fluid flow rate and liquid level. Variation of any of these can often have a significant environmental impact. As process parameter measurement systems for environmental protection purposes need to operate on a semi-continuous basis and automatically provide records of measurements, some kind of unmanned, computerised, data logging and analysis system is needed. Therefore, only measuring instruments that are suitable for input into such a system will be covered in this chapter, and indicators, gauges and handheld devices that require humans to observe and record measurements will be omitted. As well as this summary of measuring devices, appropriate ways of calibrating the various instruments involved will also be covered, so that the levels of measurement accuracy specified in the ISO 14001 EMS manual are guaranteed.

ISO 14000 Environmental Management Standards: Engineering and Financial Aspects. Alan S. Morris.
 ISBN 0-470-85128-7

13.1 Temperature Measurement

Temperature measurement related to environmental protection is needed for several purposes. Firstly, variations in the temperature at which chemical reactions take place can have significant environmental consequences, since temperature changes can affect the rate of chemical reaction and even the products of the reaction. Secondly, variations in river water temperature because of discharges into it can affect the health of fish, encourage algae growth and deplete the oxygen level in the water. Thirdly, temperature rises can lead to spontaneous combustion and the release of combustion products into the atmosphere, with consequent pollution. These are just a few examples of the situations where temperature measurement is necessary as part of an EMS.

The most common types of measuring device used in automatic temperature monitoring systems are thermocouples, resistance-change devices, radiation thermometers and semiconductor devices. As well as standard nonintelligent versions, these are also available in intelligent forms as discussed in Section 13.1.5. Other temperature measuring instruments like the liquid-in-glass thermometer, the bimetallic thermometer and the pressure thermometer (also known as the capillary thermometer and gas-filled thermometer) are not covered in this chapter, since these are merely temperature indicators. Very expensive devices like the quartz thermometer, the acoustic thermometer (ultrasonic thermometer) and fibre-optic temperature sensors are also excluded, since these are only used in a few special applications.

13.1.1 Thermocouples

Thermocouples are a very important class of device, as they provide the most commonly used method of measuring temperatures in industry. They consist of two metal wires that are connected at one end by either welding, soldering or, in some cases, just by twisting the wire ends together. The thermocouple is shown schematically in Figure 13.1(a), and in equivalent circuit form in Figure 13.1(b). Thermocouples depend for their operation on the thermoelectric effect. The principle of the thermoelectric effect is that an e.m.f. (voltage) is generated at the junction between different metals that is a function of the temperature of the junction. The general form of this relationship is:

$$e = a_1 T + a_2 T^2 + a_3 T^3 + \cdots + a_n T^n$$

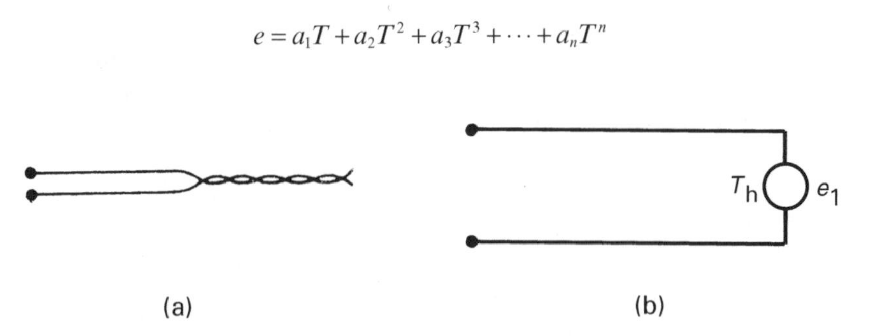

Figure 13.1 Thermocouple: (a) structure; (b) equivalent circuit.

This relationship is clearly nonlinear, and therefore is inconvenient for measurement applications. Fortunately, for certain pairs of materials, the terms involving squared and higher powers of $T (a_2T^2, a_3T^3$, etc.) are approximately zero and the e.m.f.–temperature relationship is approximately linear, according to:

$$e \approx a_1 T$$

Thermocouples are manufactured from various combinations of the base metals copper and iron, the base-metal alloys of alumel (Ni/Mn/Al/Si), chromel (Ni/Cr), constantan (Cu/Ni), nicrosil (Ni/Cr/Si) and nisil (Ni/Si/Mn), the noble metals platinum and tungsten, and the noble-metal alloys of platinum/rhodium and tungsten/rhenium. The particular combinations used as thermocouples are chosen, such that the e.m.f.–temperature relationship is approximately linear over a reasonable temperature range, as shown in Figure 13.2.

The five standard base-metal thermocouples are chromel–constantan (type E), iron–constantan (type J), chromel–alumel (type K), nicrosil–nisil (type N) and copper–constantan (type T). These are all relatively cheap to manufacture, but they

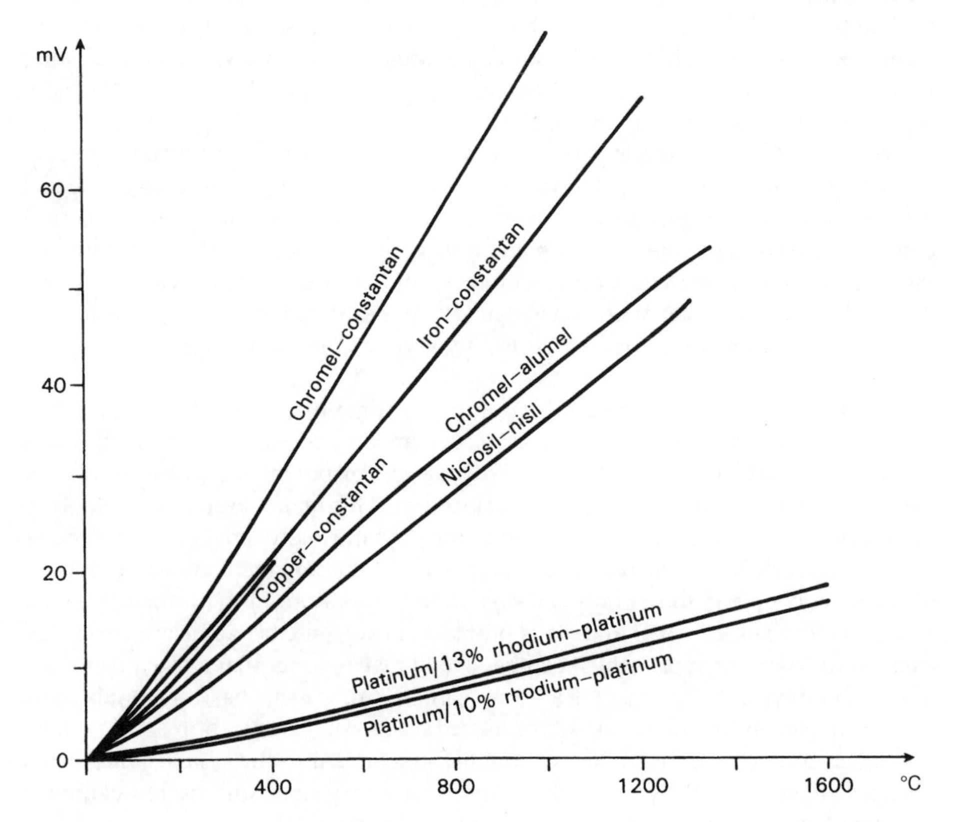

Figure 13.2 *E.m.f./temperature relationship for various common types of thermocouple. From Morris (1997)* Measurement and Calibration Requirements, *© John Wiley & Sons, Ltd. Reproduced with permission.*

become inaccurate with age and have a short life. Chromel–constantan devices give the highest measurement sensitivity of 80 μV/°C, with an inaccuracy of ±0.5% and a useful measuring range of –200 °C up to 900 °C. Unfortunately, whilst they can operate satisfactorily in oxidising environments, their performance and life are seriously affected by reducing atmospheres. Iron–constantan thermocouples have a sensitivity of 60 μV/°C, and are the preferred type for general-purpose measurements in the temperature range of –150 °C to +1000 °C, where the typical measurement inaccuracy is ±0.75%. Their performance is little affected by either oxidising or reducing atmospheres. Copper–constantan devices have a similar measurement sensitivity of 60 μV/°C and find their main application in measuring subzero temperatures down to –200 °C, with an inaccuracy of ±0.75%. They can also be used in both oxidising and reducing atmospheres to measure temperatures up to 350 °C. Chromel–alumel thermocouples have a measurement sensitivity of only 45 μV/°C, although their characteristic is particularly linear over the temperature range between 700 °C and 1200 °C, and this is therefore their main application. Like chromel–constantan devices, they are suitable for oxidising atmospheres, but not for reducing ones. The measurement inaccuracy for them is ±0.75%. Nicrosil–nisil thermocouples (type N) have similar characteristics and operating environment limitations to type K devices, with equally good linearity over the high-temperature measurement range, measurement inaccuracy of ±0.75% and measurement sensitivity of 40 μV/°C, but their long-term stability and life are at least three times better. A detailed comparison between type K and type N can be found in Reference 1.

Noble-metal thermocouples are much more expensive than base-metal types, but enjoy high stability and long life in conditions of high temperature and oxidising environments. They are chemically inert, except in reducing atmospheres. Thermocouples made from platinum and a platinum/rhodium alloy have a low inaccuracy of ±0.2% and can measure temperatures up to 1500 °C, but their measurement sensitivity is only 10 μV/°C. Alternative devices made from tungsten and a tungsten/rhenium alloy have a better sensitivity of 20 μV/°C, and can measure temperatures up to 2300 °C.

In order to make a thermocouple conform to some precisely defined e.m.f.–temperature characteristic so that accurate measurements are achieved, the metals used in thermocouples have to be refined to a high degree of pureness, and any alloys used have to be manufactured to an exact specification. This makes even base-metal thermocouples expensive and, in consequence, they are normally only a few centimetres long. It is clearly impractical to have the rest of the measurement system at one end of a thermocouple if this is only a few centimetres away from the hot environment being measured at the other end. This problem is overcome by connecting extension leads to the thermocouple, which can be several metres long. If the thermocouple is a base-metal type, the extension leads are made from the same basic materials as the thermocouple, but the materials are produced to a lower specification and are therefore much cheaper. This introduces two more junctions into the system, where small voltages are generated. However, these are usually very small and do not cause any significant measurement error.

In the case of a noble-metal thermocouple, the metals used are so expensive that it would be too costly to manufacture extension leads from the same materials, even if they were produced to a lower specification. In this case, compensating leads are used instead. These are extension leads made from base metals and alloys that have similar thermoelectric characteristics to the thermocouple materials, so that the measurement error introduced at the junction is minimised. An example of this is the use of nickel/copper–copper extension leads connected to a platinum/rhodium–platinum thermocouple.

The final equivalent circuit of a thermocouple connected to extension/compensating leads is shown in Figure 13.3. In addition to the voltage e_1 generated at the closed (hot) junction, and the voltages e_2 and e_3 generated at the junctions between the thermocouple wires and the extension leads, there are two further voltages e_4 and e_5 generated at the junction (known as the reference junction) between the open ends of the extension leads and the connection wires to the rest of the measurement system. The *law of intermediate metals* can be used to show that $e_4 + e_5$ is equivalent to the voltage e_1' that would be generated if the two thermocouple wires were connected directly together at the reference junction temperature T_r. If it is assumed that $e_2 \approx e_3 \approx 0$, this means that the final output voltage E measured at the open ends of the extension leads is the sum of e_1 and e_1'. To simplify the use of thermocouples, standard *thermocouple tables* are published that translate between the measured output voltage E and the hot junction temperature T_h. Table 13.1 shows a sample table for an iron–constantan thermocouple. Tables normally automatically compensate for the e.m.f e_1' at the reference junction according to the e.m.f generated at this junction when it is at 0 °C. Thus, such tables are only valid when the reference junction is at 0 °C, and this condition is met in practice by immersing the junction in an ice bath where the ice is just melting.

Unfortunately, it is not easy to maintain the reference junction temperature at 0 °C in many applications. In this case, the reference junction is placed in an environment maintained at some nonzero temperature by an electrical heating element. Correction then has to be made to the temperature indicated by the thermocouple tables, to allow for the nonzero reference junction temperature. The necessary correction can be calculated by applying the *law of intermediate temperatures*, which states that:

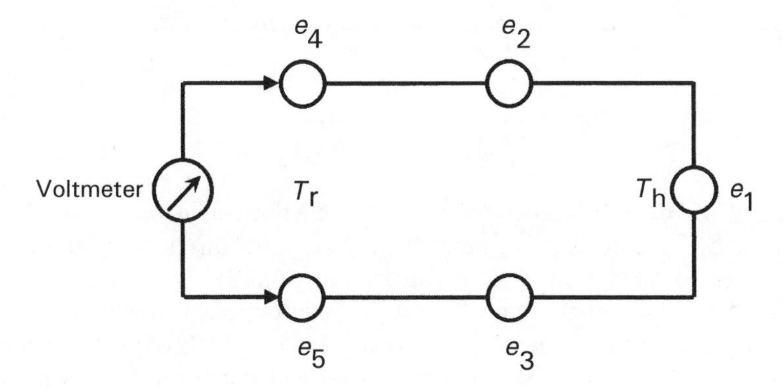

Figure 13.3 *Equivalent circuit of thermocouple with extension leads.*

Table 13.1 *Thermocouple table for iron–constantan (type J) thermocouple*

Temperature (°C)	Output (mV)	Temperature (°C)	Output (mV)	Temperature (°C)	Output (mV)	Temperature (°C)	Output (mV)
−210	−8.096	150	8.008	510	27.949	870	49.989
−200	−7.890	160	8.560	520	28.511	880	50.621
−190	−7.659	170	9.113	530	29.075	890	51.249
−180	−7.402	180	9.667	540	29.642	900	51.875
−170	−7.122	190	10.222	550	30.210	910	52.496
−160	−6.821	200	10.777	560	30.782	920	53.115
−150	−6.499	210	11.332	570	31.356	930	53.729
−140	−6.159	220	11.887	580	31.933	940	54.341
−130	−5.801	230	12.442	590	32.513	950	54.949
−120	−5.426	240	12.998	600	33.096	960	55.553
−110	−5.036	250	13.553	610	33.683	970	56.154
−100	−4.632	260	14.108	620	34.273	980	56.753
−90	−4.215	270	14.663	630	34.867	990	57.349
−80	−3.785	280	15.217	640	35.464	1000	57.942
−70	−3.344	290	15.771	650	36.066	1010	58.533
−60	−2.892	300	16.325	660	36.671	1020	59.121
−50	−2.431	310	16.879	670	37.280	1030	59.708
−40	−1.960	320	17.432	680	37.893	1040	60.293
−30	−1.481	330	17.984	690	38.510	1050	60.877
−20	−0.995	340	18.537	700	39.130	1060	61.458
−10	−0.501	350	19.089	710	39.754	1070	62.040
0	0.000	360	19.640	720	40.382	1080	62.619
10	0.507	370	20.192	730	41.013	1090	63.199
20	1.019	380	20.743	740	41.647	1100	63.777
30	1.536	390	21.295	750	42.283	1110	64.355
40	2.058	400	21.846	760	42.922	1120	64.933
50	2.585	410	22.397	770	43.563	1130	65.510
60	3.115	420	22.949	780	44.207	1140	66.087
70	3.649	430	23.501	790	44.852	1150	66.664
80	4.186	440	24.054	800	45.498	1160	67.240
90	4.725	450	24.607	810	46.144	1170	67.815
100	5.268	460	25.161	820	46.790	1180	68.389
110	5.812	470	25.716	830	47.434	1190	68.963
120	6.359	480	26.272	840	48.076	1200	69.536
130	6.907	490	26.829	850	48.717		
140	7.457	500	27.388	860	49.354		

$$E_{(T_h,T_0)} = E_{(T_h,T_r)} + E_{(T_r,T_0)} \tag{13.1}$$

where $E_{(T_h,T_0)}$ is the e.m.f. generated with the junctions at temperatures T_h and T_0 ($T_0 = 0\,°C$), $E_{(T_h,T_r)}$ is the e.m.f. generated with the junctions at temperatures T_h and T_r (T_r = nonzero reference junction temperature), and $E_{(T_r,T_0)}$ is the e.m.f. generated with the junctions at temperatures T_r and T_0. For example, if the reference junction of an iron–constantan thermocouple is maintained at 30 °C, and the output e.m.f. measured is 13.681 mV, the measured temperature at the hot junction can be calculated from (13.1) as follows:

$E_{(T_h,T_r)}$ = measured output e.m.f. with reference junction at 30 °C = 13.681 mV
$E_{(T_r,T_0)}$ = output e.m.f. with junction temperatures at 30 °C and 0 °C = 1.536 mV (using Table 13.1)
Hence, $E_{(T_h,T_0)} = E_{(T_h,T_r)} + E_{(T_r,T_0)} = 13.681 + 1.536 = 15.217\,\text{mV}$.

Applying Table 13.1, the e.m.f. of 15.217 mV indicates a hot junction temperature of 280 °C.

Thermocouples only have a relatively low output of a few millivolts, and signal processing must be applied that provides amplification with high gain but low noise. They are also delicate instruments, which must be treated carefully if their specified operating characteristics are to be reproduced. One major source of error is induced strain in the hot junction, which reduces the e.m.f. output, and precautions are normally taken to minimise this, by mounting the thermocouple horizontally rather than vertically. In some operating environments, no further protection is necessary to obtain satisfactory performance from the instrument. However, thermocouples are prone to contamination by various metals, and protection is often necessary to minimise this. Such contamination alters the thermoelectric behaviour of the device, such that its characteristic varies from that published in standard tables. Contamination also makes thermocouples brittle and shortens their life.

Protection takes the form of enclosing the thermocouple in a sheath. Some common sheath materials and their maximum operating temperatures are shown in Table 13.2. A thermocouple naturally has a first-order type of step response characteristic, but the time constant is usually negligible when it is used unprotected. However, when enclosed in a sheath, the time constant of the thermocouple and sheath combination is significant. The size of the thermocouple, and hence the diameter required for the sheath, has a large effect on the importance of this. The time constant of a thermocouple in a 1-mm diameter sheath is only 0.15 s, and this has little practical effect in most measurement situations, whereas a larger sheath of 6 mm diameter gives a time constant of 3.9 s, which cannot be ignored so easily.

The diameter of wire used to construct thermocouples is usually in the range between 0.4 mm and 2 mm. The larger diameters are used where ruggedness and long life are required, although these advantages are gained at the expense of increasing the measurement time constant. In the case of noble-metal thermocouples, the use of large diameter wire incurs a substantial cost penalty. Some special applications

Table 13.2 *Common sheath materials for thermocouples*

Material	Maximum operating temperature (°C)	Material	Maximum operating temperature (°C)
Mild steel	900	Recrystallised alumina	1850
Nickel–chromium	900	Beryllia	2300
Fused silica	1000	Magnesia	2400
Special steel	1100	Zirconia	2400
Mullite	1700	Thoria	2600

Note: The maximum operating temperatures quoted assume oxidising or neutral atmospheres. For operation in reducing atmospheres, the maximum allowable temperature is usually reduced.

have a requirement for a very fast response time in the measurement of temperature, and, in such cases, wire diameters as small as 0.1 μm (0.1 microns) can be used.

The mode of construction of thermocouples means that their characteristics can be incorrect even when they are new, due to faults in either the homogeneity of the thermocouple materials or in the construction of the device. Therefore, calibration checks should be carried out on all-new thermocouples before they are put into use. Thereafter, the rate of change of thermoelectric characteristics with time is entirely dependent upon the operating environment and the degree of exposure to it. Particularly relevant factors in the environment are the type and concentration of trace-metal elements and the temperature (the rate of contamination of thermocouple materials with trace elements of metals is a function of temperature). A suitable calibration frequency can therefore only be defined by practical experimentation, and this must be reviewed whenever the operating environment and conditions of use change.

Finally, before leaving the subject of thermocouples, mention must also be made of the *thermopile*, since this often appears in manufacturers' catalogues. A thermopile is simply a temperature-measuring device that consists of several thermocouples connected together in series, such that all the reference junctions are at the same cold temperature, and all the hot junctions are exposed to the temperature being measured. The effect of connecting n thermocouples together in series is to increase the measurement sensitivity by a factor of n. A typical thermopile manufactured by connecting together 25 chromel–constantan thermocouples gives a measurement resolution of 0.001 °C. As the thermopile is essentially a multiple thermocouple, all the previous discussion about thermocouples applies equally well to thermopiles, and calibration requirements are also identical.

13.1.2 Varying-resistance devices

Varying-resistance devices rely on the physical principle that the resistance of materials varies with temperature. The instruments working on this principle are known as either resistance thermometers or thermistors, according to whether the material used for their construction is a metal or a semiconductor material.

Resistance thermometers

Resistance thermometers, also known as resistance temperature devices (RTDs), use the principle that the resistance of metals varies with temperature, according to the relationship:

$$R = R_0(1 + a_1T + a_2T^2 + a_3T^3 + \cdots + a_nT^n)$$

This equation is inconvenient, because it is nonlinear. However, for the metals used as resistance thermometers, the terms in a_2T^2 and higher powers of T are negligible over a limited temperature range, and so the equation is approximately linear:

$$R \approx R_0(1 + a_1T)$$

The characteristics of metals used in resistance thermometers are shown in Figure 13.4(a). For platinum, the resistance–temperature relationship is linear within ±0.4% over the temperature range between –200 °C and +40 °C. Even at +1000 °C, the quoted maximum inaccuracy figure is only ±1.2%. Platinum is also chemically inert in most environments, and is therefore usually the first choice for resistance thermometers over the wide measuring range from –270 °C to +650 °C (platinum can actually be used up to 1000 °C, but tungsten is often preferred above 650 °C). Nickel and copper resistance thermometers are sometimes used where cost is important, since platinum is very expensive. They are cheaper but less accurate, and can be used from –200 °C to +260 °C (copper), and from –200 °C to +430 °C (nickel). However, these two metals are very susceptible to oxidation and corrosion, which limits their accuracy and life. Another metal, tungsten, is also used in some circumstances, particularly for high-temperature measurements. Its measuring range is –270 °C to +1100 °C.

Two types of resistance thermometer exist. The first consists of a coil of wire, usually protected inside a glass or ceramic sheath. The second kind are thin-film sensors, which are made by depositing metal on to an aluminium oxide substrate and then etching the metal layer to produce a resistance element. Thin-film devices can be as small as a few square millimetres in area. The different devices available have resistance elements ranging from 10 Ω right up to 25 kΩ. The devices with high resistance have several operational advantages. The first of these is that any connection resistances within the circuit become negligible in their effect. The second advantage is that the relatively high-voltage output produced by these instruments means that

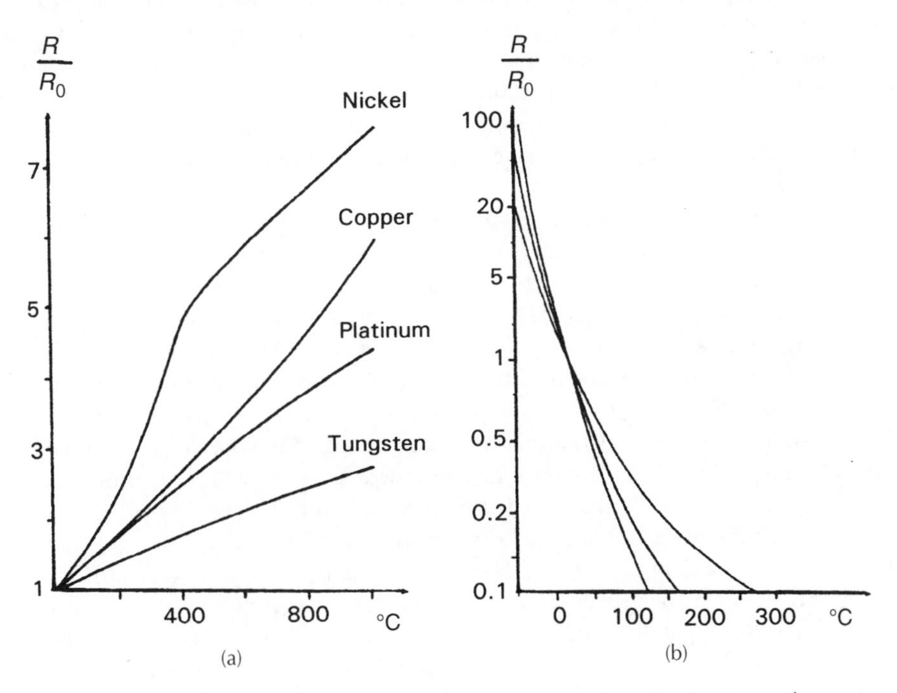

Figure 13.4 *Varying-resistance devices: (a) characteristics of metals used in resistance thermometers; (b) characteristics of thermistors.*

any induced e.m.f.'s produced by thermoelectric behaviour at the connection junction are negligible.

The normal method of measuring resistance is to use a d.c. bridge. The excitation voltage of the bridge has to be chosen very carefully, because, although a high value is desirable for achieving high measurement sensitivity, the self-heating effect of high currents flowing in the temperature transducer creates an error by increasing the temperature of the device, and so changing the resistance value.

In the case of noncorrosive and nonconducting environments, resistance thermometers are used without protection. In all other applications, they are protected inside a sheath. As in the case of thermocouples, such protection affects the speed of response of the system to rapid changes in temperature. A typical time constant for a sheathed platinum resistance thermometer is 0.4 seconds.

Thermistors

Thermistors are manufactured from beads of semiconductor material prepared from oxides of chromium, cobalt, copper, iron, manganese, nickel and titanium. Such semiconductor materials normally have a negative temperature coefficient, where the resistance decreases as the temperature increases according to:

$$R = R_0 \exp[\beta(1/T - 1/T_0)]$$

This relationship is very nonlinear, and it is impossible to make a linear approximation, even over a small temperature range. The negative temperature coefficient is also fundamentally different to that in resistance thermometers, where the temperature coefficient is positive. In fact, the temperature coefficient is so large that the resistance/temperature characteristic is normally drawn on a logarithmic scale, and the resistance is expressed as a ratio relative to the resistance at a nominal temperature, as shown in Figure 13.4(b).

Many types of thermistor are available, which vary according to the temperature coefficient and the nominal resistance. In each case, tables showing the resistance/temperature characteristic are provided by manufacturers. The nominal resistance is usually quoted at 25 °C, and the typical resistance at this temperature varies between 2000 Ω and 30 000 Ω in different devices. Devices with high resistance have the advantage that the effect of the connection-lead resistance is minimised. Any one device usually only has a small measuring range of between 50 and 100 °C. However, particular types of thermistor can be obtained to cover the full range from −200 °C up to +1000 °C, although types designed to operate at temperatures above +150 °C are usually hermetically sealed in glass to avoid damage.

The major advantages of thermistors are their low cost, small size and tolerance of shock and vibration. The size advantage means that the time constant of thermistors operated in sheaths is small. However, the size reduction also decreases its heat dissipation capability, and so makes the self-heating effect greater. In consequence, thermistors have to be operated at generally lower current levels than resistance thermometers, and so the measurement sensitivity afforded is less. Like the resistance thermometer, the resistance of a thermistor is usually measured by a d.c. bridge circuit.

It should also be mentioned that other thermistors exist that are made from heavily doped compositions of barium, lead and strontium titanates, with other additives[2]. These are advertised as being positive temperature coefficient devices, but in fact they have a small negative temperature coefficient up to a certain temperature where there is quite an abrupt change to a positive coefficient. Such devices are typically used as switches to detect things like over-current in motors, but they are rarely used as temperature measurement sensors.

13.1.3 Radiation thermometers

Radiation thermometers use the physical principle that all bodies emit electromagnetic radiation as a function of their temperature, according to the equation: $E = KT^4$, where E is the total rate of radiation emission per second, and T is the temperature of the body in degrees Kelvin. Therefore, measuring the magnitude of this energy emission allows the temperature of the body to be determined. The power spectral density of the emission varies with temperature, in the manner shown in Figure 13.5. The major part of the frequency spectrum lies within the band of wavelengths between 0.3 μm and 40 μm, which corresponds with the visible (0.3–0.72 μm) and infrared (0.72–1000 μm) ranges. Choice of the best method of measuring the emitted radiation depends on the temperature of the body. At low temperatures, the peak of the power spectral density function lies in the infrared region, whereas at higher temperatures it moves towards the visible part of the spectrum.

The term 'radiation thermometer' actually describes a whole class of devices that between them cover the temperature range from −100 °C up to +10 000 °C.

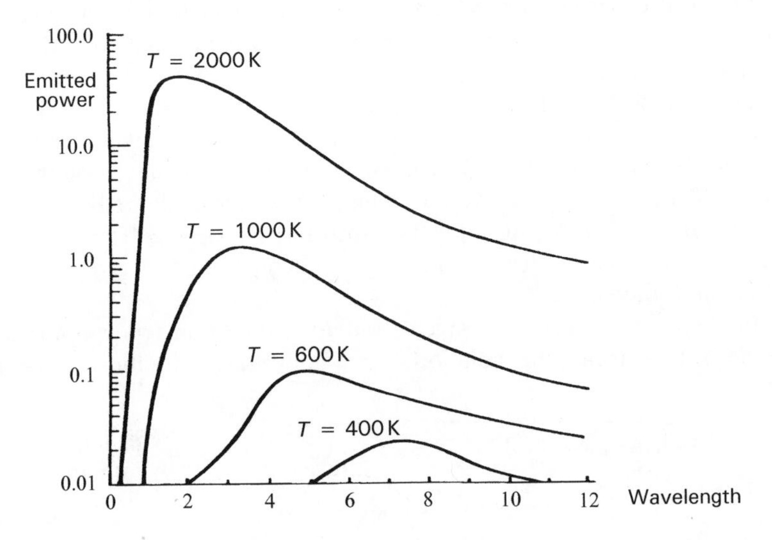

Figure 13.5 Power spectral density of radiated energy emission at various temperatures.

Measurement uncertainty as low as ±0.05% is claimed for some devices. The oldest device available is the *optical pyrometer*, but this requires manual operation, and so it is excluded from the discussion below because it is unsuitable for unmanned process parameter monitoring. However, most other types of radiation pyrometer are suitable for automatic monitoring systems. A typical radiation pyrometer consists of a tube that is pointed towards the energy source, as shown in Figure 13.6. A lens system within the tube focuses the energy emitted by the source to a focal point where an energy detector is positioned. The radiation detector is either a thermal detector, which measures the temperature rise in a black body at the focal point of the optical system, or a photon detector. Thermal detectors respond equally to all wavelengths in the frequency spectrum, whereas photon detectors respond selectively to a particular narrow band of wavelengths within the range 0.5 μm–1.2 μm. Devices using thermal detectors are usually called broadband radiation thermometers, whereas devices using photon detectors are normally called narrowband ones. In general, the accuracy of broadband instruments deteriorates significantly over time, and an error of 10 °C is common after 1–2 years' operation. Narrowband instruments are much more stable, and many only have a 1 °C error after 10 years.

Thermopiles, resistance thermometers and thermistors are all used as thermal detectors in different versions of these instruments. These typically have time constants of several milliseconds, because of the time taken for the black body to heat up and the detector to respond to the temperature change. Photon detectors are usually of the photoconductive or photovoltaic type. Both of these respond very much faster to temperature changes than thermal detectors, because they involve atomic processes, and typical measurement time constants are a few microseconds.

Apart from differences in the technique used to measure radiation, radiation pyrometers also differ in the range of energy wavelengths that they are designed to measure, and hence in the temperature range measured. One further difference is the material used to construct the energy-focusing lens. Outside the visible part of the spectrum, glass becomes almost opaque to infrared wavelengths, and other lens materials such as arsenic trisulphide are used.

An important feature of radiation thermometers is that they measure the temperature of a body, without being in contact with it. This makes them especially suitable for applications like food processing, since there is no possibility of contamination. They can also measure very high temperatures that are beyond the capabilities of contact instruments like thermocouples, resistance thermometers and thermistors. Furthermore, they can measure the temperature of moving bodies like steel bars in a rolling mill.

Despite these advantages, the usage of radiation thermometers is not straightforward. One problem is that the radiation from a body varies with the composition and

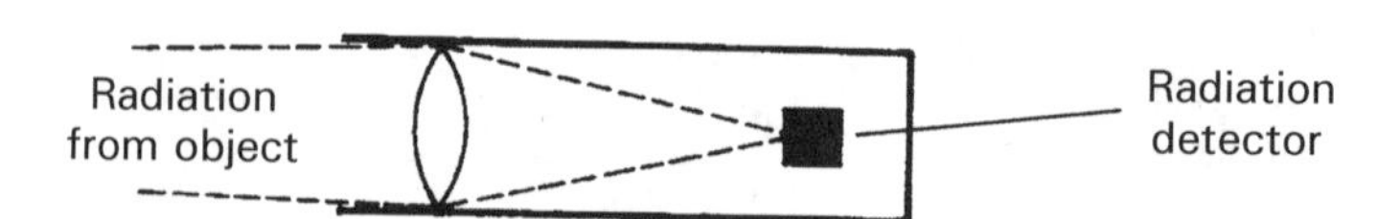

Figure 13.6 Structure of a radiation pyrometer. From Morris (1997) Measurement and Calibration Requirements, © John Wiley & Sons, Ltd. Reproduced with permission.

surface condition of the body as well as with temperature. This dependence on surface condition is quantified by a parameter known as the *emissivity* of the body. The use of radiation thermometers is further complicated by absorption and scattering of the energy between the emitting body and the radiation detector. Energy is scattered by atmospheric dust and water droplets and absorbed by carbon dioxide, ozone and water vapour molecules. Therefore, radiation thermometers normally have to be carefully calibrated for each separate application. However, this requirement is avoided in the two-colour pyrometer.

The *two-colour pyrometer* (alternatively known as a *ratio pyrometer*) is a system constructed as shown in Figure 13.7. Radiation from the body is split equally into two parts, which are applied to separate narrowband filters. The outputs from the filters consist of radiation within two narrow bands of wavelength λ_1 and λ_2. Detectors sensitive to these frequencies produce output voltages V_1 and V_2 respectively. The ratio of these outputs (V_1/V_2), can be shown[3] to be a function of temperature, and to be independent of the emissivity, provided that the two wavelengths λ_1 and λ_2 are close together. Unfortunately, this assumption does not entirely hold in practice, and therefore the accuracy of the two-colour pyrometer may not be as good as claimed. Also two-colour pyrometers typically cost 50–100% more than other types of pyrometer. However, the instrument is still very useful when the target is obscured by fumes or dust, which is a common problem in many applications.

13.1.4 Semiconductor sensors

Semiconductor devices in the form of either integrated circuit transistors or diodes have the advantage of small size, and they provide relatively cheap temperature measurement. However, they only give limited measurement accuracy, and they have the disadvantage of needing an external power supply to the sensor. They are usually supplied in a stainless-steel protective sheath.

Integrated circuit transistors have better linearity than either thermocouples or resistance thermometers. They only cost a few pounds and produce an output proportional to the absolute temperature. Different types are configured to give an

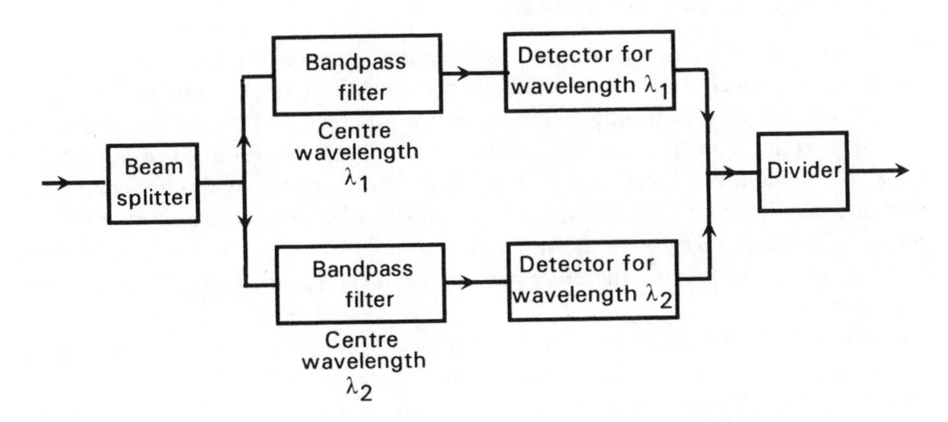

Figure 13.7 Two-colour pyrometer.

output in the form of either a varying current (typically 1 μA/K) or varying voltage (typically 10 mV/K). Unfortunately, they only have a small measurement range from −50 °C to +150 °C, and their typical inaccuracy is ±3%, which limits their range of application.

Diodes also have good output linearity, and their typical inaccuracy is only ±0.5%. The output is in the form of a variation in the forward voltage across the device, which varies linearly with temperature. Silicon diodes cover the temperature range from −50 to +300 °C, and Germanium ones from −270 to +40 °C.

13.1.5 Intelligent temperature transmitters

Intelligent temperature transmitters are effectively microprocessor-based signal-processing packages that are designed for use with transducers that have either a d.c. voltage output in the mV range or an output in the form of a resistance change. They are therefore suitable for use in conjunction with thermocouples, thermopiles, resistance thermometers, thermistors and radiation pyrometers. The transmitters usually include adjustable damping, noise rejection, self-adjustment for zero and sensitivity drifts and expanded measurement range, and measurement uncertainty levels down to ±0.05% of full scale are often specified. They also have nonvolatile memories where all constants used in correcting output values for environmental inputs, etc., are stored, thus enabling the instrument to survive power failures without losing such information.

Intelligent transmitters cost significantly more than nonintelligent sensors, and justification purely on the grounds of their superior accuracy is hard to make. However, their expanded measurement range brings immediate savings, because of the reduction in the number of spare instruments needed to cover a number of measurement ranges. Their capability for self-diagnosis and self-adjustment also means that they require attention much less frequently, giving additional savings in maintenance costs. However, although the signal-processing element within an intelligent transmitter is largely self-calibrating, the appropriate calibration routines described earlier have to be applied to the temperature sensor.

13.1.6 Choice between temperature transducers

The most commonly used device in industry for temperature measurement is the base-metal thermocouple. This is relatively cheap, with prices varying widely from a few pounds upwards according to the thermocouple type and sheath material used. Typical inaccuracy is ±0.5% of full scale over the temperature range of −250 °C to +1200 °C. Noble-metal thermocouples are much more expensive, but are chemically inert and can measure temperatures up to 2300 °C with an inaccuracy of ±0.2% of full scale. However, all types of thermocouple have a low-level output voltage, making them prone to noise and therefore unsuitable for measuring small temperature differences.

Resistance thermometers are also in common use within the temperature range of −270 °C to +650 °C, with a measurement inaccuracy of ±0.5%. Whilst they have a smaller temperature range than thermocouples, they are more stable and can measure small temperature differences. The platinum resistance thermometer is generally regarded as offering the best ratio of price to performance for measurement between −200 °C and +500 °C, with prices starting from £15.

Thermistors are another relatively common class of devices. They are small and cheap, with a typical cost of around £5. They give a fast output response to temperature changes, with good measurement sensitivity, but their measurement range is quite limited.

Semiconductor devices have better linearity than thermocouples and resistance thermometers, and are therefore a viable alternative in many applications. Integrated circuit transistor sensors are particularly cheap (from £10 each), although their accuracy is relatively poor and they have a very limited measurement range (−50 °C to +150 °C). Diode sensors are much more accurate, and have a wider temperature range (−270 °C to +200 °C), although they are also more expensive (typical costs vary from £50 to £500).

A major virtue of radiation thermometers is their noncontact, noninvasive mode of measurement. Costs vary from £250 up to £3000, according to type. Although calibration for the emissivity of the measured object often poses difficulties, this is not required in the two-colour pyrometer. Various forms of radiation pyrometer are used over the temperature range between −20 °C and +1800 °C and can give measurement inaccuracies as low as ±0.05%. One particular merit of narrowband radiation pyrometers is their ability to measure fast temperature transients of a duration as short as 10 μs. No other instrument can measure transients anywhere near as fast as this.

13.1.7 Calibration of temperature measuring devices

A number of different temperature sensors can be used as a working standard to calibrate process instruments. Provided that the instrument is kept in good conditions and reserved solely for calibration duties, the list of suitable devices for use as a working standard at this calibration level includes mercury-in-glass thermometers, base-metal thermocouples (type *K*), noble-metal thermocouples (types *B*, *R* and *S*), platinum resistance thermometers and radiation pyrometers. However, as process instruments are often calibrated *in situ* in their normal operating environment, it is essential that the instrument used as a working standard for calibration can withstand whatever hostile conditions may be present in the environment, and will not become contaminated.

At the next calibration level, the working standard instruments themselves are calibrated using special, purpose-designed equipment that provides controlled reference temperatures. Such equipment is commercially available from many sources. For calibration of all temperature transducers other than radiation thermometers, a furnace consisting of an electrically heated ceramic tube is commonly used. The temperature of such a furnace can typically be controlled within limits of ±2 °C over the range from 20 °C to 1600 °C. Up to 630 °C, the platinum resistance thermometer is commonly used as a reference standard. Above that temperature, a type *R* (platinum/13% rhodium–platinum) thermocouple is usually employed. Type *K* (chromel–alumel) thermocouples are also used as an alternative reference standard for temperature calibration up to 1000 °C.

For calibration of working-standard radiation thermometers, a radiation source that approximates as closely as possible to the behaviour of a black body is required. Some form of optical bench is required, so that the instruments being calibrated can be held firmly and aligned accurately. The actual value of the emissivity of the source must also be measured by a surface pyrometer. This contains a hemispherical, gold-plated surface, which is supported on a telescopic arm that allows it to be put into contact with the hot surface. The radiation emitted from a small hole in the hemisphere is independent of the surface emissivity of the measured body and is equal to that which would be emitted by the body if its emissivity value were 100. This radiation is measured by a thermopile with its cold junction at a controlled temperature. A black hemisphere is also provided with the instrument that can be inserted to cover the gold surface. This allows the instrument to measure the normal radiation emission from the hot body, and so it allows the surface emissivity to be calculated by comparing the two radiation measurements.

The simplest form of radiation source is a hot plate heated by an electrical element, in which the plate surface has a typical emissivity of 0.85 and the temperature can be controlled within limits of ±1 °C over the range from 0 °C to 650 °C. Type *R* noble-metal thermocouples embedded in the plate are normally used as the reference instrument. A heat source with a much better emissivity is provided by a black-body cavity that is lined with a refractory material and heated by an electrical element. This can be constructed in various alternative forms according to the temperature range of the radiation thermometers to be calibrated, though a common feature is a blackened conical cavity with a cone angle of about 15°. This gives a typical emissivity of 0.998 and temperature control within limits of ±0.5 °C up to 1200 °C and ±1 °C up to 1600 °C. Type *R* noble-metal thermocouples embedded in the plate are used as the reference instrument in the range 200–1200 °C and type B thermocouples (30% rhodium–platinum/6% rhodium–platinum) in the range 600–1600 °C. As an alternative to thermocouples, radiation thermometers can also be used as a standard within ±0.5 °C over the temperature range from 400 °C to 1250 °C. For calibration at temperatures above 1600 °C, a carbon cavity furnace is used. This consists of a graphite tube with a conical radiation cavity at its end, where temperatures up to 2600 °C can be maintained with an accuracy of ±5 °C. Narrowband radiation thermometers are used as the reference standard instrument.

For calibration at low temperatures (20 °C to 200 °C), a black-body cavity immersed in a liquid bath is used, which provides a constant temperature (±0.5 °C). The typical emissivity of a cavity heated in this way is 0.995. Water is suitable for the bath in the temperature range of 20–90 °C and a silicone fluid in the range of 80–200 °C. Below 20 °C, a stirred water bath is used to provide a constant reference temperature, and the same equipment can be used for temperatures up to 100 °C. Similar stirred liquid baths containing oil or salts (potassium/sodium nitrate mixtures) can be used to provide reference temperatures up to 600 °C. At these low temperatures, a platinum resistance thermometer is often used as the reference instrument.

At the highest level of calibration, reference temperatures are defined in terms of the freezing and boiling points of substances, where the transition between solid, liquid and gaseous states is sharply defined. The International Practical Temperature Scale (IPTS) is defined on this basis and consists of the six *primary fixed points* shown in Table 13.3.

Table 13.3 Primary fixed points of temperature

Fixed point	Temperature	Fixed point	Temperature
Triple point of equilibrium hydrogen	−259.34 °C	Freezing point of zinc	419.58 °C
Boiling point of oxygen	−182.962 °C	Freezing point of silver	961.93 °C
Boiling point of water	100.0 °C	Freezing point of gold	1064.43 °C

Table 13.4 Some examples of secondary fixed points of temperature

Fixed point	Temperature	Fixed point	Temperature
Freezing point of tin	231.968 °C	Freezing point of rhodium	1963 °C
Freezing point of antimony	630.74 °C	Freezing point of iridium	2447 °C
Freezing point of nickel	1455 °C	Freezing point of tungsten	3387 °C

The six primary fixed points are widely spaced and have an upper limit of 1064.43 °C. To provide higher reference temperatures, and also to provide additional intermediate reference points, a number of secondary fixed points are defined in terms of the freezing points of certain metals. Some examples are provided in Table 13.4.

The procedure for calibrating instruments against these absolute reference standards of temperature at the primary and secondary fixed freezing points of metals involves heating an ingot of pure metal (better than 99.99% pure) beyond its melting point in an electric resistance furnace, and allowing it to cool. The metal is heated inside a graphite crucible with a close-fitting lid, to protect it against oxidation. Alternatively, to calibrate radiation thermometers, the pure metal is heated in a black-body furnace inside a conical-ended, cylindrical radiation cavity. If the temperature of the metal is monitored as it cools, an arrest period is observed in its cooling curve at the freezing point. Up to 1100 °C, a measurement uncertainty of less than ±0.5 °C is achievable.

It should be noted that this calibration technique only provides standard reference temperatures at the fixed melting points of the metals used. However, this is still a very useful check on the accuracy of reference standard instruments, and, once their accuracy has been verified at such points within their measuring range, much greater confidence is created in their accuracy at intermediate temperatures.

13.2 Pressure Measurement

Pressure measurement is principally needed to avoid the development of high pressures in industrial processes that might lead to environmental pollution. A common effect of high pressure is the rupture of pipes that are carrying fluids, which can cause pollution in several ways. Firstly, any liquid products that are released can find their ways into watercourses. Secondly, any gaseous products released can cause atmospheric pollution. Thirdly, the release of fluids can lead to combustion processes or explosions that pollute the atmosphere. As explained for the case of temperature measurement, pressure measurement systems for environmental purposes generally need to operate in an automatic fashion and record the measurements obtained, to demonstrate that the EMS is operating satisfactorily. However, apart from the environmental protection purpose, pressure measurements and the responses made with the intention of avoiding unintended pressure variations also have significant economic benefits, since pressure variations away from their target values invariably have an adverse effect on costs.

Pressure measuring instruments normally measure gauge pressure or differential pressure rather than absolute pressure, because the measurement of absolute pressure is very difficult. The exception to this is a form of U-tube manometer in which one end of the tube is sealed and evacuated, although this only measures absolute pressure with limited accuracy. The instruments available for measuring gauge pressure can be divided into four categories, according to the range of pressure that they are designed to measure. These four pressure ranges are:

(1) high pressures greater than 7000 bar (6910 atmospheres);
(2) mid-range pressures in the span between 1.013 bar and 7000 bar (1 to 6910 atmospheres);
(3) low pressures between 0.001 mbar and 1.013 bar (the mean atmospheric pressure);
(4) very low pressures less than 0.001 mbar.

For environmental protection purposes, there is rarely a requirement to measure pressures outside the mid-range described above. Therefore, this chapter only covers instruments that are appropriate to measurements in this mid-range from atmospheric pressure up to 7000 bar. For measurements above 7000 bar, it is normal to use an instrument consisting of a coil of manganin wire enclosed in sealed, paraffin-oil filled, flexible bellows. At pressures below 1 atmosphere (1.013 bar) down to 0.001 mbar, special versions of the types of instrument used to measure in the mid-range of pressures can be used, although special instruments also exist, such as the thermocouple gauge, Pirani gauge and thermistor gauge. These latter three devices can also measure very low pressures less than 0.001 mbar, but a special instrument called the ionisation gauge is often used for measurement in this range. Further information on instruments used for measuring pressures outside the mid-range of 1 atmosphere to 7000 bar can be obtained from various sources, such as Reference 4.

Apart from only covering devices suitable for mid-range pressure measurements, this chapter also limits discussion to devices that are suitable for unmanned, automatic pressure monitoring functions. Thus, manometers (which are indicators that require humans to read and interpret the output reading) are not included in the list of measurement devices, although the U-tube manometer and micromanometer are mentioned in the later section covering calibration.

13.2.1 Measurement of mid-range pressures (1.013 bar–7000 bar)

Mid-range pressures are the ones measured most commonly in industrial applications. Suitable instruments for this range are the diaphragm, Bourdon tube and bellows-type instrument.

Diaphragm

The diaphragm is a common measurement device, and is shown schematically in Figure 13.8. Applied pressure causes displacement of the diaphragm, and this movement is measured by a displacement transducer. Both gauge pressure and differential pressure can be measured by different versions of diaphragm-type instruments. The usual range of gauge pressure measured is 1 to 10 bar, but special versions of diaphragm devices can measure low pressures down to 0.001 mbar and high pressures up to 2000 bar. Differential pressures up to 2.5 bar can also be measured, and in this case the two pressures are applied to either side of the diaphragm, and the displacement of the diaphragm corresponds with the pressure difference. The typical magnitude of diaphragm displacement is 0.1 mm, which is well suited to a strain gauge type of measuring transducer. The diaphragm itself can be plastic, metal alloy, stainless steel or ceramic. Plastic diaphragms are cheapest, but metal diaphragms give better accuracy. Stainless steel is normally used in high-temperature or corrosive environments. Ceramic diaphragms are resistant even to strong acids and alkalis, and are used when the operating environment is particularly harsh.

If strain gauges are used to measure the diaphragm displacement, it is normal to use four gauges in a bridge configuration, with the excitation voltage applied across two opposite points of the bridge. The output voltage measured across the other two points of the bridge is then a function of the resistance change due to the strain in the diaphragm. This arrangement automatically provides compensation for environmental temperature changes. Older pressure transducers of this type used metallic strain gauges, bonded to a diaphragm typically made of stainless steel. Apart from manufacturing difficulties arising from the problem of bonding the gauges, metallic strain gauges have a low gauge factor. This means that the strain gauge bridge only has a low output that has to be amplified by an expensive direct current amplifier. The development of semiconductor (piezoresistive) strain gauges provided a solution to the low output problem, as they have gauge factors up to one hundred times greater than metallic gauges. However the difficulty of bonding gauges to the diaphragm remained, and a new problem emerged regarding the highly nonlinear characteristic of the strain/output relationship.

The problem of strain gauge bonding was solved with the emergence of monolithic piezoresistive pressure transducers, and these are now the most commonly used

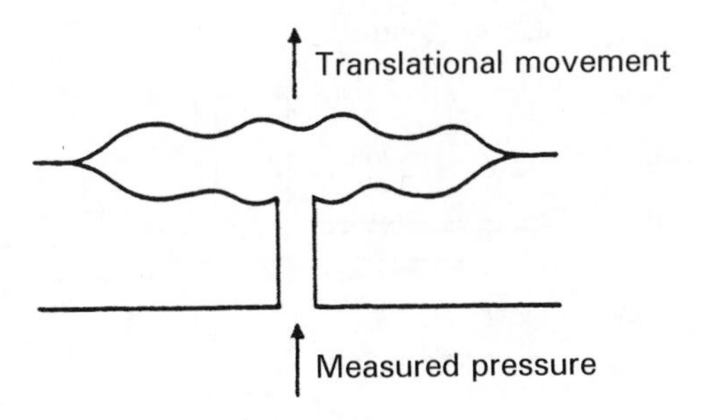

Figure 13.8 *Diaphragm pressure sensor. From Morris (1997)* Measurement and Calibration Requirements, *© John Wiley & Sons, Ltd. Reproduced with permission.*

types of diaphragm pressure transducer. The monolithic cell consists of a diaphragm made of a silicon sheet into which resistors are diffused during the manufacturing process. Besides avoiding the difficulty with bonding, such monolithic silicon measuring cells have the advantage of being very cheap to manufacture in large quantities, and their sensitivity is three times better than that of unbonded strain gauges. Another advantage is that they can be made extremely small, down to 0.75 mm in diameter, and in this form they find application in specialist areas, such as on the tip of medical catheters. Although the inconvenience of a nonlinear characteristic remains, this is normally overcome by applying signal processing to linearise the output signal.

As an alternative to strain-gauge type displacement measurement, capacitive or inductive transducers are sometimes used. Capacitive types are often known as *capacitive pressure sensors*. As a further option, developments in optical fibres have been exploited in the *Fotonic Sensor* (Figure 13.9), which is a diaphragm-type device in which the displacement is measured by optoelectronic means. Light travels from a light source down an optical fibre, is reflected back from the diaphragm and travels back along a second fibre to a photodetector. There is a characteristic relationship between the light reflected and the distance from the fibre ends to the diaphragm, thus making the amount of reflected light dependent upon the measured pressure.

Bellows

The bellows sensor, illustrated in Figure 13.10, is another elastic-element type device. It operates on a very similar principle to the diaphragm, although its use is much less common, since it has a high manufacturing cost and it is prone to failure. Pressure changes within the bellows produce translational motion of the end of the bellows, which can be measured by capacitive, inductive (LVDT) or potentiometric transducers, according to the range of movement produced. The principal attribute of a bellows sensor is that its sensitivity is greater than that of a diaphragm. Typical measurement uncertainty is ±0.5%. Bellows type instruments can typically measure pressures up to 150 bar. However, special versions are also available that are designed to measure low pressures down to 0.1 mbar.

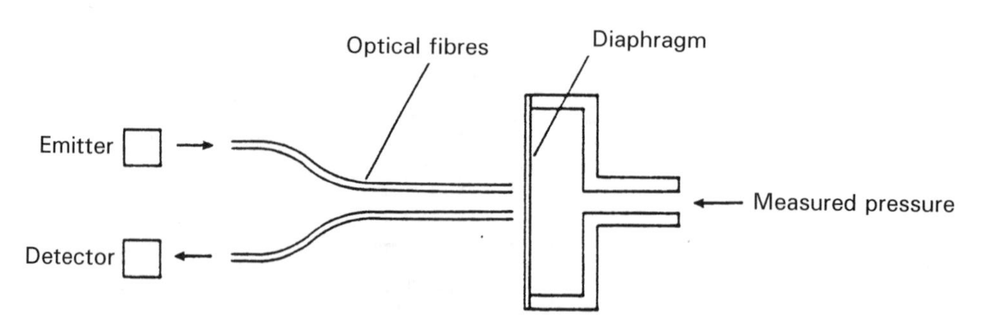

Figure 13.9 *Fotonic sensor. From Morris (1997)* Measurement and Calibration Requirements, © *John Wiley & Sons, Ltd. Reproduced with permission.*

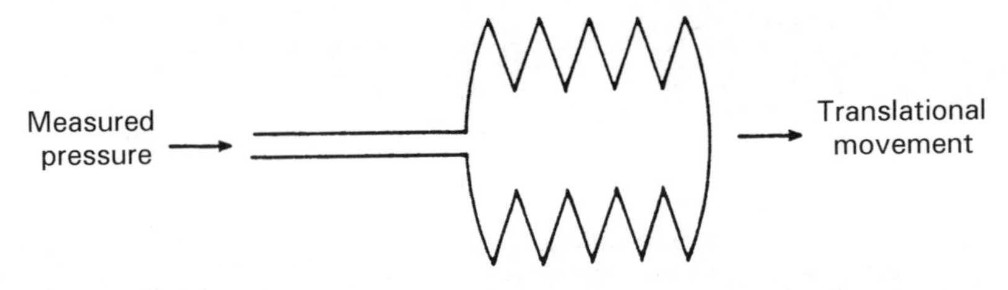

Figure 13.10 *Bellows pressure sensor. From Morris (1997)* Measurement and Calibration Requirements, *© John Wiley & Sons, Ltd. Reproduced with permission.*

Bourdon tube

The Bourdon tube is another common industrial measuring instrument that is used for measuring the pressure of both gaseous and liquid fluids. It consists of a specially shaped piece of oval-section flexible tube that is fixed at one end and free to move at the other. When pressure is applied at the fixed end of the tube, the oval cross-section becomes more circular. As the cross-section of the tube tends towards a circular shape, a deflection of the closed, free end of the tube is caused. When used as a pressure-indicator, this motion is translated into the movement of a pointer against a scale. Alternatively, if an electrical output is required, the displacement is measured by some form of displacement transducer, which is commonly a potentiometer or LVDT (linear variable differential transformer), or less often a capacitive sensor or optical sensor.

The three common shapes of Bourdon tube are shown in Figure 13.11(a), (b) and (c). The maximum possible deflection of the free end of the tube is proportional to the angle subtended by the arc through which the tube is bent. For a C-type tube, the maximum value for this arc is somewhat less than 360°. Where greater measurement sensitivity and resolution are required, spiral and helical tubes are used, in which the possible magnitude of the arc subtended is limited only by a practical limit on how many turns it is convenient to have in the helix or spiral. However, this increased measurement performance is only gained at the expense of a substantial increase in manufacturing difficulty and cost compared with C-type tubes, and is also associated with a large decrease in the maximum pressure that can be measured.

C-type tubes are available for measuring pressures up to 6000 bar. A typical C-type tube of 25 mm radius has a maximum displacement travel of 4 mm, giving a moderate level of measurement resolution. Measurement inaccuracy is typically quoted as ±1% of full-scale deflection. Similar accuracy is available from helical and spiral types, but whilst the measurement resolution is higher, the maximum pressure measurable is only 700 bar. Special versions of the Bourdon tube are also available for measuring low pressures in the range down to 10 mbar.

13.2.2 Intelligent pressure transducers

Adding microprocessor power to any of the pressure transducers discussed earlier brings about substantial improvements in their characteristics. Measurement sensitivity improvement, extended measurement range, compensation for hysteresis and

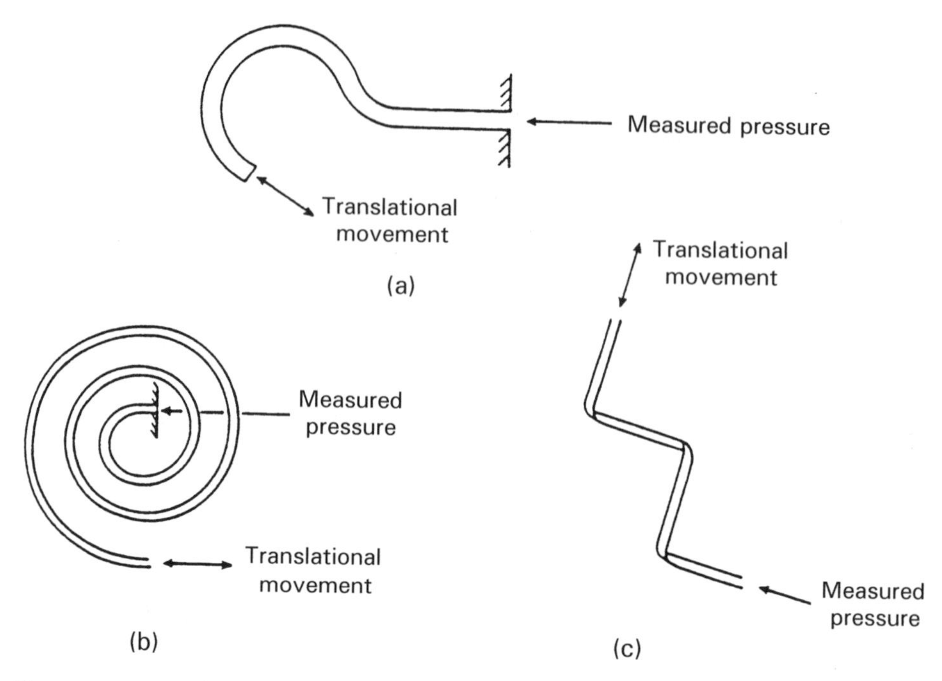

Figure 13.11 Bourdon tubes: (a) C-type; (b) spiral type; (c) helical type. From Morris (1997) Measurement and Calibration Requirements, © John Wiley & Sons, Ltd. Reproduced with permission.

other nonlinearities, and correction for ambient temperature and pressure changes are just some of the facilities offered by intelligent pressure transducers. For instance, inaccuracies of only ±0.1% can be achieved with diaphragm-type piezoresistive-bridge silicon devices.

13.2.3 Calibration of pressure-measuring devices

All of the instruments that are commonly used to measure mid-range pressures, like the diaphragm, Bourdon tube and bellows, contain an elastic element whose movement is measured by a displacement transducer. Both the elastic element and the displacement transducer are mechanical in nature, and therefore such instruments suffer changes in characteristics for a number of reasons. One factor is the characteristics of the operating environment and the degree to which the instrument is exposed to it. Another reason is the amount of mishandling it receives. These parameters are entirely dependent upon the particular application in which the instrument is used, and the frequency with which it is used and exposed to the operating environment. Therefore, a suitable calibration frequency can only be determined on an experimental basis.

Pressure calibration requires the output reading of the instrument being calibrated to be compared with the output reading of a standard reference instrument when the same pressure is applied to both. This necessitates designing a suitable leakproof seal to connect the pressure-measuring chambers of the two instruments. The calibration of pressure transducers used for process measurements often has to be carried out *in situ* in order to avoid serious production delays. Such devices are often remote from the nearest calibration laboratory, and to transport them there for calibration would take an unacceptably long time. Because of this, portable reference instruments have been developed for calibration at this level in the calibration chain. These use a standard air supply connected to an accurate pressure-regulator to provide a range of reference pressures. An inaccuracy of ±0.025% is achieved when calibrating mid-range pressures in this manner. Calibration at higher levels in the calibration chain must of course be carried out in a proper calibration laboratory, maintained in the correct manner.

U-tube manometers, dead-weight gauges and barometers can all be used as a standard for calibration of instruments measuring mid-range pressures. The vibrating cylinder gauge also provides a very accurate reference standard over part of this pressure range. As this chapter has only covered devices designed to measure mid-range pressures, it would not be appropriate to describe the calibration in detail of instruments that are primarily used to measure pressures above and below this range. However, for the sake of completeness, it is appropriate to mention the names of devices that are used as a calibration standard at high and low pressures, and to direct the reader to consult other sources (e.g. Reference 4) if further information is required. Above 20 bar, a gold–chrome alloy resistance instrument is normally used as a reference standard. For low pressures in the range of 10^{-1} to 10^{-3} mbar, both the McLeod gauge and various forms of micromanometer are used as a reference standard. At even lower pressures below 10^{-3} mbar, a pressure dividing technique is normally used to establish calibration.

U-tube manometer

A U-tube manometer consists of a glass tube shaped into a letter U and filled with a liquid, which is usually water when the device is being used for calibration purposes. In its usual form, as shown in Figure 13.12, the gauge pressure of the fluid (p) is related to the difference (h) between the levels of fluid in the two halves of the tube: p (gauge pressure) = $h \times d$, where d is the specific gravity of the fluid.

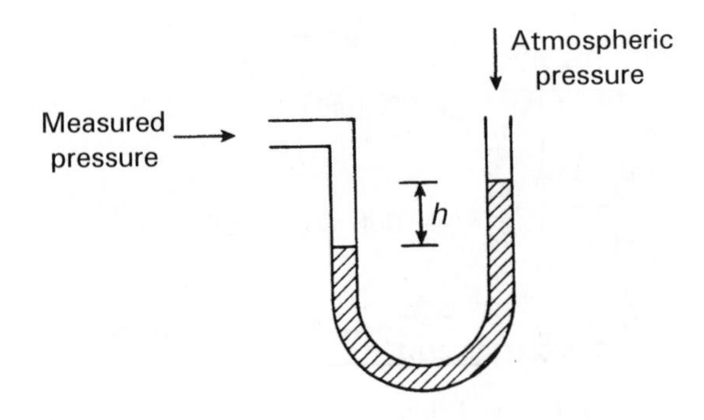

Figure 13.12 *U-tube manometer.*

Although a U-tube manometer is a deflection-type rather than a null-type of instrument, it is used as a calibration standard because its characteristics are very stable, although small errors can develop in the longer term, because of volumetric changes in the glass. However, when used for calibration duties, special precautions are necessary. Firstly, the vertical alignment of the U-tube must be set up carefully. Secondly, particular care must be taken to ensure that there are no temperature gradients across the two halves of the tube, since these would cause local variations in the specific weight of the manometer fluid, resulting in measurement errors. Correction must also be made for the local value of g (acceleration due to gravity).

Dead-weight gauge (pressure balance)

The dead-weight gauge or pressure balance, as shown in Figure 13.13, consists of a piston that moves inside a cylinder. It is a null-reading type of measuring instrument, in which weights are added to the piston platform until a mark on the piston is adjacent to a fixed reference mark on the cylinder, at which time the downward force of the weights on top of the piston is balanced by the pressure exerted by the fluid beneath the piston. The fluid pressure is therefore calculated in terms of the weight added to the platform and the known area of the piston.

Special precautions are necessary in the manufacture and use of dead-weight gauges. Friction between the piston and cylinder must be reduced to a very low level to minimise measurement error. This is accomplished firstly by machining the cylinder to a slightly greater diameter than the piston, and secondly by designing the piston and cylinder so that they can be turned relative to one another. Unfortunately, a small amount of fluid flows past the seals between the piston and cylinder due to the small gap, which produces a viscous shear force that partly balances the weight on the platform. In theory, the magnitude of this shear force can be calculated and compensated. However, in practice, the piston deforms under pressure and alters the piston/cylinder gap, and so the shear force calculation and correction can only be approximate. In spite of these difficulties, the instrument gives a typical measurement inaccuracy of only ±0.01%. It is normally used for calibrating pressures in the range of 20 mbar up to 20 bar. However, special versions can measure pressures down to 0.1 mbar or up to 7000 bar.

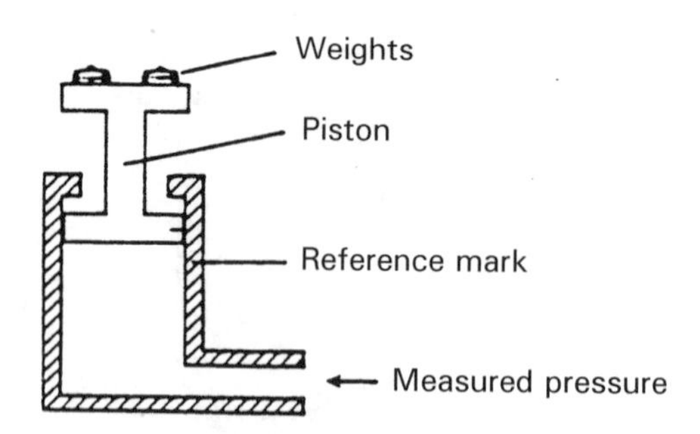

Figure 13.13 *Dead-weight gauge. From Morris (1997)* Measurement and Calibration Requirements, *© John Wiley & Sons, Ltd. Reproduced with permission.*

Barometers

The type of barometer that is most commonly used for calibration duties is the Fortin barometer. This is a highly accurate instrument that provides measurement inaccuracy levels of between ±0.03% of full-scale reading and ±0.001% of full-scale reading, depending on the measurement range. To achieve such levels of accuracy, the instrument has to be used under very carefully controlled conditions of lighting, temperature and vertical alignment. It must also be manufactured to exacting standards, and is therefore very expensive to buy. Corrections have to be made to the output reading, according to the ambient temperature, local value of gravity and atmospheric pressure. Because of its expense and the difficulties in using it, the barometer is not normally used for calibration other than as a primary reference standard at the top of the calibration chain.

Vibrating cylinder gauge

The vibrating cylinder gauge, shown in Figure 13.14, acts as a reference standard instrument for calibrating pressure measurements up to 3.5 bar. It consists of a cylinder in which vibrations at the resonant frequency are excited by a current-carrying coil. The pressure-dependent oscillation frequency is monitored by a pick-up coil, and this frequency measurement is converted to a voltage signal by a microprocessor and signal-conditioning circuitry contained within the package. By evacuating the space on the outer side of the cylinder, the instrument is able to measure the absolute pressure of the fluid inside the cylinder. Measurement errors are less than 0.005% over the absolute pressure range up to 3.5 bar.

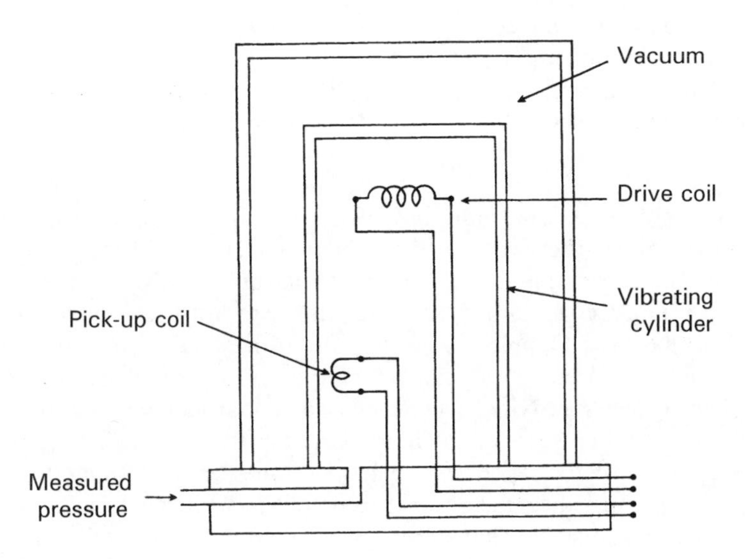

Figure 13.14 *Vibrating cylinder gauge. From Morris (1997)* Measurement and Calibration Requirements, *© John Wiley & Sons, Ltd. Reproduced with permission.*

13.3 Flow Measurement

In general, flow measurement can involve the measurement of both mass flow rate (mass flowing per unit time) and volume flow rate (volume flowing per unit time). However, there is rarely a need to measure mass flow rate for environmental purposes, and therefore this section is only concerned with volume flow rate. Volume flow rate is needed for various purposes. Firstly, when waste materials are discharged into a watercourse, there is a need to measure the volume flow rate of both the waste and the watercourse, in order to ensure that the subsequent concentration of pollutants from the waste in the watercourse does not exceed target values. Secondly, when gaseous waste products are being discharged into the atmosphere, knowledge of the volume flow rate is needed, so that the resultant concentration of pollutants in the atmosphere can be calculated, given particular dispersion characteristics and wind conditions. Thirdly, flow rates of chemicals and fuels in industrial processes can affect the rate of reaction and even the products of reactions, and thus it is important that flow rates are measured and kept at their target values. Fourthly, adequate coolant water flow is important in many applications, since lack of sufficient flow can cause temperature rises and consequent environmental damage. Therefore, continuous monitoring of flow rate is needed to detect the onset of problems before serious problems have arisen. The instruments suitable for measuring volume flow rate as part of an EMS can be divided into the following classes:

(1) differential pressure meters;
(2) positive displacement meters;
(3) turbine flowmeters;
(4) electromagnetic flowmeters;
(5) vortex shedding flowmeters;
(6) open channel flowmeters;
(7) ultrasonic flowmeters;
(8) intelligent flowmeters (versions of all of the above types).

Devices that only give a visual indication of pressure, like the variable area meter, are not covered, since these have to be read by a human operator and are therefore unsuitable for automatic flow monitoring schemes that are usually needed for environmental protection. Certain other flowmeters that are not widely used, such as the target meter and laser-Doppler flowmeter, are also excluded from the discussion. However, this still leaves a large number of types to choose from. The factors to be considered when specifying a flowmeter for a particular application includes the temperature and pressure of the fluid, its density, viscosity, chemical properties and abrasiveness, whether it contains particles, whether it is a liquid or gas, etc. The required performance factors of accuracy, measurement range, acceptable pressure drop, output signal characteristics, reliability and service life must also be assessed.

13.3.1 Differential pressure meters

Differential pressure meters involve the insertion of some device into a fluid-carrying pipe that causes an obstruction. In consequence, the velocity of the fluid through the restriction increases and the pressure decreases. The volume flow rate is then proportional to the square root of the pressure difference across the obstruction. The manner in which this pressure difference is measured is important. Measuring the two pressures with different instruments and calculating the difference between the two measurements is not satisfactory, because of the large measurement error that can arise when the pressure difference is small. Therefore, the normal procedure is to use a diaphragm-based, differential pressure transducer.

All applications of this method of flow measurement assume that flow conditions upstream of the obstruction device are in steady-state, and a certain minimum length of straight run of pipe ahead of the flow measurement point is specified to ensure this. The minimum lengths required for various pipe diameters are specified in British Standards tables (and also in alternative but equivalent national standards used in other countries), but a useful rule of thumb is to specify a length that is ten times the pipe diameter. If physical restrictions make this impossible to achieve, special flow smoothing vanes can be inserted immediately ahead of the measurement point.

Flow-restriction type instruments are popular because they have no moving parts and are therefore robust, reliable and easy to maintain. However, one disadvantage is that the obstruction causes a permanent loss of pressure in the flowing fluid. The magnitude of this loss depends on the type of obstruction element used, but it is sometimes necessary to recover the lost pressure by an auxiliary pump further down the flow line if the pressure loss is large. The obstruction device used in the majority of applications is the *orifice plate*, which consists of a metal disc with a hole. Figure 13.15(a) shows the flow profile across the orifice plate, and Figure 13.15(b) shows the typical pattern of pressure variation. The minimum cross-sectional area of flow does not occur within the obstruction but rather at a point downstream. The point of minimum pressure coincides with this point of minimum cross-section of flow. A small rise in pressure is also apparent immediately before the obstruction. Therefore, it requires some skill to position the instrument measuring P_2 exactly at the point of minimum pressure and also to measure pressure P_1 at a point upstream of the point where the pressure starts to rise before the obstruction.

In the absence of any heat-transfer mechanisms, and assuming frictionless flow of an incompressible fluid through the pipe, the theoretical volume flow rate of the fluid, Q, is given by:

$$Q = \frac{A_2\sqrt{[2(P_1 - P_2)/d]}}{\sqrt{\left[1-(A_2/A_1)^2\right]}} \tag{13.2}$$

where A_1 and P_1 are the cross-sectional area and pressure of the fluid flow before the obstruction, A_2 and P_2 are the cross-sectional area and pressure of the fluid flow at the narrowest point of the flow beyond the obstruction, and d is the fluid density. However, equation (13.2) is never applicable in practice, for several reasons. Firstly,

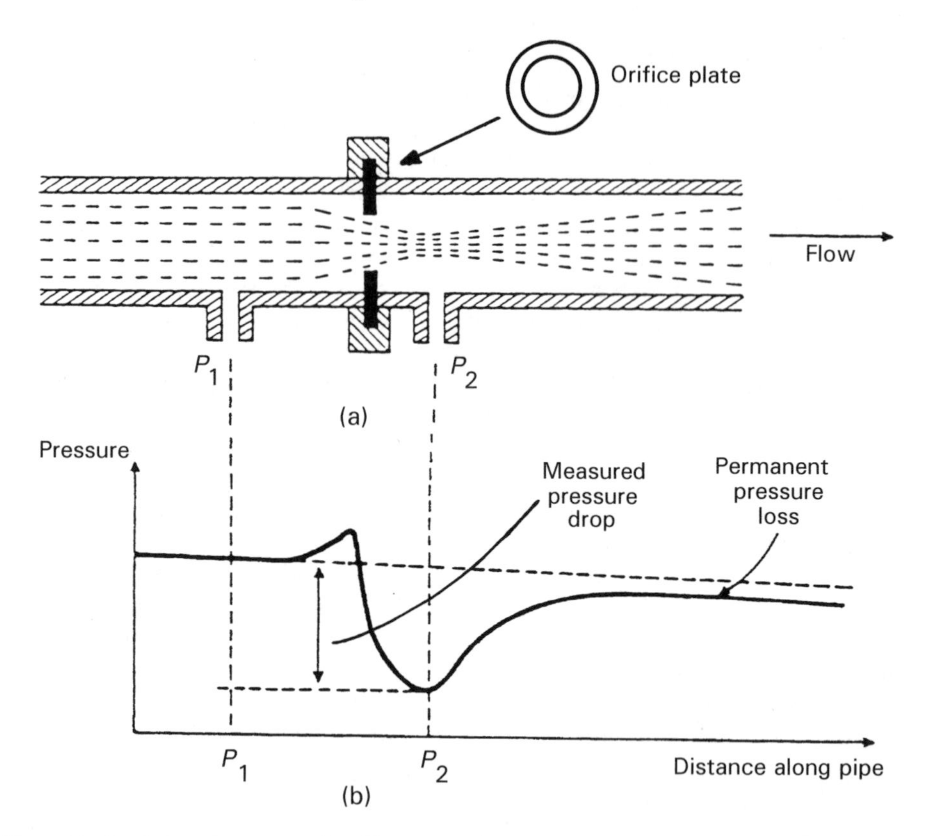

Figure 13.15 *Orifice plate: (a) profile of flow across orifice plate; (b) pattern of pressure variation along pipe.*

frictionless flow is never achieved, although, in the case of turbulent flow through smooth pipes, friction is low and it can be adequately accounted for by a variable called the *Reynold's number*, which is a measurable function of the flow velocity and the viscous friction. The other reasons for the nonapplicability of equation (13.2) are that the initial cross-sectional area of the fluid flow is less than the diameter of the pipe carrying it, and the minimum cross-sectional area of the fluid is less than the diameter of the obstruction. Therefore, neither A_1 nor A_2 can be measured. These problems are taken account of by modifying equation (13.2) to the following:

$$Q = \frac{C_D A_2' \sqrt{[2(P_1 - P_2)/d]}}{\sqrt{[1-(A_2'/A_1')^2]}} \tag{13.3}$$

where A_1' and A_2' are the pipe diameters before and at the obstruction, and C_D is a constant, known as the *discharge coefficient*, which accounts for the Reynold's number and the difference between the pipe and flow diameters. Before equation (13.3) can be evaluated, the discharge coefficient must be calculated. As this varies between each measurement situation, it would appear that the discharge coefficient must be determined by practical experimentation in each case. However, provided that certain conditions are met, standard tables can be used to obtain the value of the discharge coefficient appropriate to the pipe diameter and fluid involved.

It is particularly important in applications of flow restriction methods to choose an instrument whose range is appropriate to the magnitudes of flow rate being measured. This requirement arises because of the square-root type of relationship between the pressure difference and the flow rate, which means that, as the pressure difference decreases, the error in flow-rate measurement can become very large. In consequence, restriction-type flowmeters are only suitable for measuring flow rates between 30% and 100% of the instrument range.

The orifice plate is widely used, because it is simple, cheap and available in a wide range of sizes. However, its typical inaccuracy is ±2%, and the permanent pressure loss in the flow is between 50% and 90% of the pressure difference (P_1–P_2) in magnitude. Other problems with the orifice plate are a gradual change in the discharge coefficient over a period of time as the sharp edges of the hole wear away, and a tendency for any particles in the flowing fluid to stick behind the hole and gradually build up and reduce its diameter. The latter problem can be minimised by using an orifice plate with an eccentric hole close to the bottom of the pipe, so that particles are swept through and do not build up behind the plate. A very similar problem arises if there are any bubbles of vapour or gas in the flowing fluid when liquid flow is involved. These also tend to build up behind an orifice plate and distort the pattern of flow. This difficulty can be avoided by mounting the orifice plate in a vertical run of pipe.

As an alternative to an orifice plate, other obstruction devices that have a relatively smooth internal shape can be used. These have various names such as the *Venturi*, *Dall flow tube* and *flow nozzle*. These vary in the amount of precision engineering required and hence price, but all are much more expensive than an orifice plate. The advantage gained is better measurement accuracy and a reduction in the permanent pressure loss in the measured system to around 10–15% of the pressure-difference (P_1–P_2). The smooth internal shape of these devices also means that they are unaffected by solid particles or gaseous bubbles in the flowing fluid, with low maintenance requirements and a long working life. An example of a Venturi is shown in Figure 13.16(a).

The *Pitot tube*, shown in Figure 13.16(b) is a further differential pressure device that operates on somewhat different principles. It measures the local flow velocity at a particular point within a pipe, rather than the average flow velocity as measured by other obstruction devices. The Pitot tube brings to rest that part of the fluid that impinges on it, and the loss of kinetic energy is converted to an increase in pressure inside the tube. The flow velocity can be calculated from the formula:

$$v = C\sqrt{[2g(P_1 - P_2)]}$$

The Pitot tube coefficient C is a factor that corrects for the fact that not all fluid incident on the end of the tube will be brought to rest: a proportion will slip around it, according to the design of the tube. The volume flow rate can then be calculated by multiplying v by the cross-sectional area of the flow pipe. If a single Pitot tube is

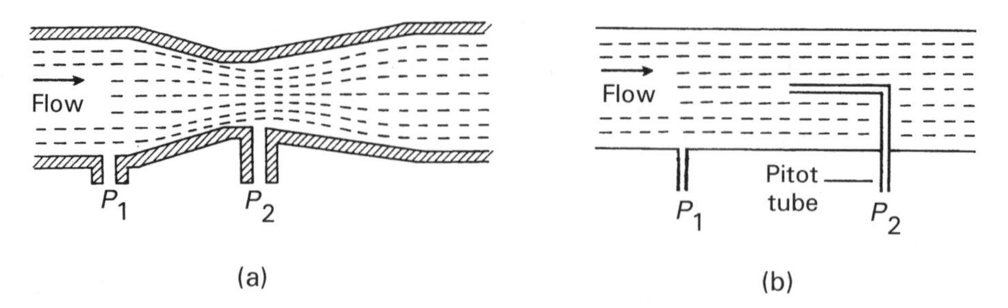

Figure 13.16 *Other differential pressure flowmeters: (a) the Venturi (b) the Pitot tube.*

used, the average fluid flow rate is only measured accurately if the flow profile across the pipe is very uniform. However, if this condition is not met, multiple Pitot tubes that are spatially distributed across the pipe cross-section (sometimes called an *Annubar*) can be used, with the mean flow rate being calculated from the average tube pressure measured. The averaging Pitot tube is particularly useful for measuring steam flow as well as other liquids and gases.

Pitot tubes have several advantages. They are cheap, cause negligible pressure loss in the flow and can be installed very simply by pushing them down a small hole drilled in the flow-carrying pipe. Measurement uncertainty down to ±1% is achievable with carefully designed Pitot tubes.

13.3.2 Positive displacement flowmeters

Positive displacement flowmeters account for about 10% of the total number of flowmeters used in industry. They can measure the flow of clean, noncorrosive liquids and gases, and they are particularly useful for measuring highly viscous fluids where the volume flow rate is low. They consist of a cylindrical unit that rotates inside a cylindrical chamber. The rotating unit has mechanical divisions that create compartments of known volume, although many different mechanical arrangements exist to achieve this that are named accordingly, for example the *rotary piston meter*. As the measured fluid flows through the instrument, the compartments are repeatedly filled and emptied, and the flow rate is calculated in terms of the rate of rotation. All versions of positive displacement meter are low-friction, low-maintenance and long-life devices, although they do impose a small permanent pressure loss on the flowing fluid. Cost varies widely according to measurement accuracy given. The cheapest versions have an inaccuracy of about ±1.5%. However, inaccuracy as low as ±0.2% can be achieved in very expensive versions.

13.3.3 Turbine meters (inferential meters)

A turbine or inferential flowmeter consists of a multibladed wheel mounted in a pipe along an axis parallel to the direction of fluid flow, as shown in Figure 13.17(a). The flow of fluid past the wheel causes it to rotate at a rate that is proportional to the volume flow rate of the fluid. This rate of rotation is measured by constructing

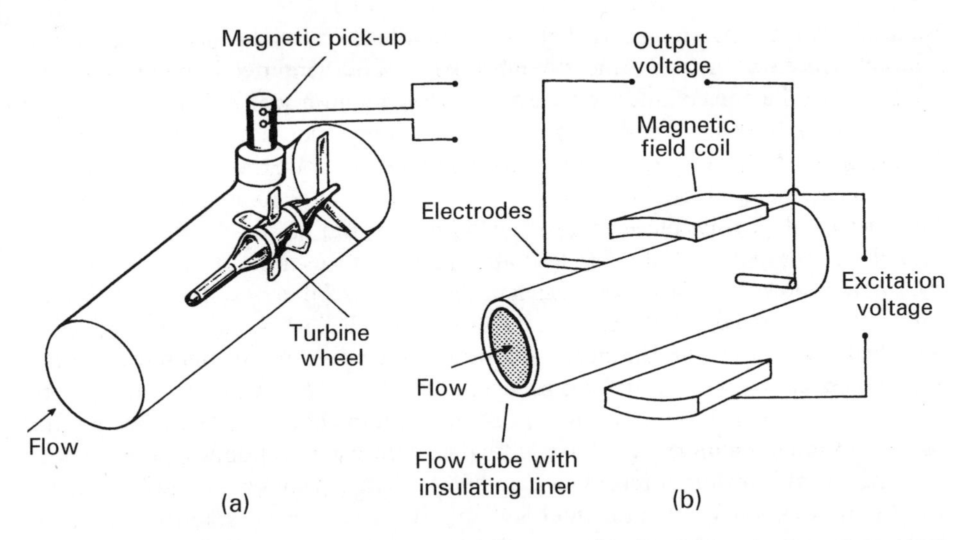

Figure 13.17 *Other types of flowmeter: (a) turbine meter; (b) electromagnetic meter. From Morris (1997)* Measurement and Calibration Requirements, *© John Wiley & Sons, Ltd. Reproduced with permission.*

the flowmeter, such that it behaves as a variable reluctance tachogenerator. This is achieved by fabricating the turbine blades from a ferromagnetic material and placing a permanent magnet and coil inside the meter housing. A voltage pulse is induced in the coil as each blade on the turbine wheel moves past it, and, by using a pulse counter, the pulse frequency and flow rate are calculated. Turbine meters cause some pressure loss in the flowing fluid. In addition, they are badly affected by any particulate matter in the flowing fluid, which causes bearing wear.

Turbine meters have a similar market share to positive displacement meters. Like positive displacement meters, cost varies widely, from cheap versions with ±2% inaccuracy to expensive versions with ±0.1% inaccuracy. For any given level of accuracy, their cost is similar to that of positive displacement meters. Turbine meters are smaller and lighter than the latter, and are preferred for low-viscosity, high-flow measurements. However, positive-displacement meters are superior in conditions of high viscosity and low flow rate.

13.3.4 Electromagnetic flowmeters

Electromagnetic flowmeters are limited to measuring the volume flow rate of electrically conductive fluids. However, despite this application limitation and their relatively high cost, magnetic flowmeters now account for about 15% of the new flowmeters sold. As they have no moving parts, reliability is high and maintenance needs are low. Typical measurement uncertainty is ±1.0%, but ±0.2% can be achieved in very expensive versions.

The instrument, shown in Figure 13.17(b), consists of a stainless-steel cylindrical tube, fitted with an insulating liner, that carries the measured fluid. A magnetic field is created in the tube by placing mains-energised field coils either side of it, and the voltage induced in the fluid is measured by two electrodes inserted into opposite sides of the tube. The ends of these electrodes are usually flush with the inner surface of the cylinder. The electrodes are constructed from a material that is unaffected by the flowing fluid. By Faraday's law of electromagnetic induction, the voltage, E, induced across a length, L, of the flowing fluid moving at velocity, v, in a magnetic field of flux density, B, is given by:

$$E = B \times L \times v \tag{13.4}$$

L is the distance between the electrodes, which is the diameter of the tube, and B is a known constant. Hence, measurement of the voltage E induced across the electrodes allows the flow velocity v to be calculated from equation (13.4). Having thus calculated v, it is a simple matter to multiply v by the cross-sectional area of the tube to obtain a value for the volume flow rate. The typical voltage signal measured across the electrodes is 1 mV when the fluid flow rate is 1 m/s.

The internal diameter of a magnetic flowmeter is normally the same as that of the rest of the flow-carrying pipework in the system. Therefore, there is very little obstruction to the fluid flow, and, consequently, negligible pressure loss associated with measurement. Like other forms of flowmeter, a minimum length of straight pipework is required immediately prior to the point of flow measurement, in order to guarantee the accuracy of measurement, although a length equal to five pipe diameters is usually sufficient.

Whilst the flowing fluid must be electrically conductive to an extent, some recent versions can measure the flow of fluids where the conductivity is as low as 1 μ Siemen/cm^3. The method can measure a wide variety of liquids, including corrosive ones, and it is particularly useful for measuring the flow of slurries in which the liquid phase is electrically conductive. One operational problem is that the insulating lining is subject to damage when abrasive fluids are handled, and this can give the instrument a limited life.

13.3.5 Vortex-shedding flowmeters

The operating principle of vortex-shedding flowmeters is based on the natural phenomenon of vortex shedding, created by placing an unstreamlined obstacle (known as a bluff body) in a fluid-carrying pipe, as indicated in Figure 13.18(a). When fluid flows past the obstacle, boundary layers of viscous, slow-moving fluid are formed along the outer surface. Because the obstacle is not streamlined, the flow cannot follow the contours of the body on the downstream side, and the separate layers become detached and roll into eddies or vortices in the low-pressure region behind the obstacle. The shedding frequency of these alternately shed vortices is proportional to the fluid velocity past the body. Various thermal, magnetic, ultrasonic and capacitive vortex detection techniques are employed in different instruments.

Such instruments have no moving parts, require little maintenance, operate over a wide flow range and have low power consumption. They are particularly suitable for measuring steam flow, as well as general liquid and gas flows, and a common

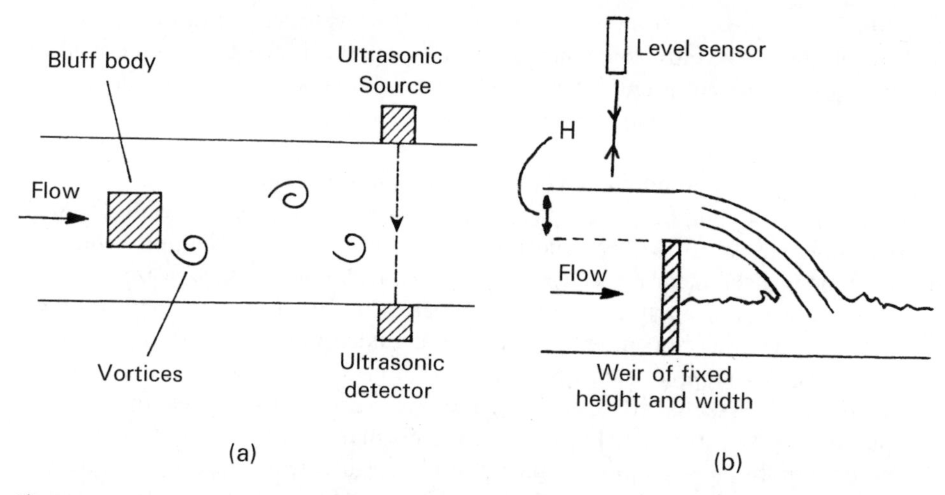

Figure 13.18 *More types of flowmeter: (a) vortex-shedding flowmete; (b) open-channel flowmeter. From Morris (1997)* Measurement and Calibration Requirements, *© John Wiley & Sons, Ltd. Reproduced with permission.*

inaccuracy figure quoted is ±1% of the full-scale reading measuring liquid flow and ±3% measuring gas or steam flow, although this can be seriously downgraded in the presence of flow disturbances upstream of the measurement point, and a straight run of pipe before the measurement point of 50 pipe diameters is recommended.

13.3.6 Open channel flowmeters

Open channel flowmeters measure the flow of liquids in open channels, and are particularly relevant to measuring the flow of water in rivers as part of environmental management schemes. The normal procedure is to build a weir or flume of constant width across the flow and measure the height of liquid immediately before the weir or flume with an ultrasonic or radar level sensor, as shown in Figure 13.18(b). The volume flow rate can then be calculated from this measured height.

As an alternative to building a weir or flume, magnetic flowmeters up to 180 mm wide are available that can be placed across the channel to measure the flow velocity, providing that the flowing liquid is conductive. If the channel is wider than 180 mm, two or more magnetic meters can be placed side by side. Apart from measuring the flow velocity in this way, the height of the flowing liquid must also be measured, and the width of the channel must also be known in order to calculate the volume flow rate.

13.3.7 Ultrasonic flowmeters

The ultrasonic technique is an entirely noninvasive method of volume flow rate measurement, because the instrument clamps on externally to existing pipework, rather than being inserted as an integral part of the flow line. As the procedure of breaking into a pipeline to insert a flowmeter can be as expensive as the cost of the flowmeter itself, the ultrasonic flowmeter has significant cost advantages. Also, because the instrument is entirely isolated from the flowing fluid, it can measure all manner of hazardous fluids, including corrosive, poisonous, radioactive, flammable and explosive ones. This has safety advantages, since the person installing the flowmeter will thus avoid coming into contact with such hazardous fluids. Also, any contamination of the fluid being measured (e.g. food substances and drugs) is avoided.

Two different types of ultrasonic flowmeter exist, which use Doppler shift and transit time measurement respectively. These two technologies have distinct characteristics and areas of application, and many situations exist where one form is very suitable and the other not suitable. As a general guide, transit-time types are best for clean liquids, and Doppler-shift types are best for gritty, aerated liquids. However, in both cases, it is essential to ensure that there is a stable flow profile ahead of the measurement point. It is usual to increase the normal specification of the minimum length of straight pipe-run prior to the point of measurement, expressed as a number of pipe diameters, from a figure of 10 up to 20, or, in some cases, even 50 diameters.

Doppler-shift ultrasonic flowmeter

The principle of operation of the Doppler-shift flowmeter is shown in Figure 13.19(a). A fundamental requirement is the presence of scattering elements within the flowing fluid, which deflect the ultrasonic energy output from the transmitter such that it enters the receiver. These can be provided by either solid particles, gas bubbles or eddies in the flowing fluid. The scattering elements cause a frequency shift between the transmitted and reflected ultrasonic energy, and measurement of this shift enables the fluid velocity to be inferred.

The instrument essentially consists of an ultrasonic transmitter–receiver pair clamped on to the outside wall of the fluid-carrying vessel. The transmitter emits a train of short sinusoidal bursts of energy at a frequency between 0.5 MHz and 20 MHz. The flow velocity, v, is given by:

$$v = \frac{c(f_t - f_r)}{2 f_t \cos\theta}$$

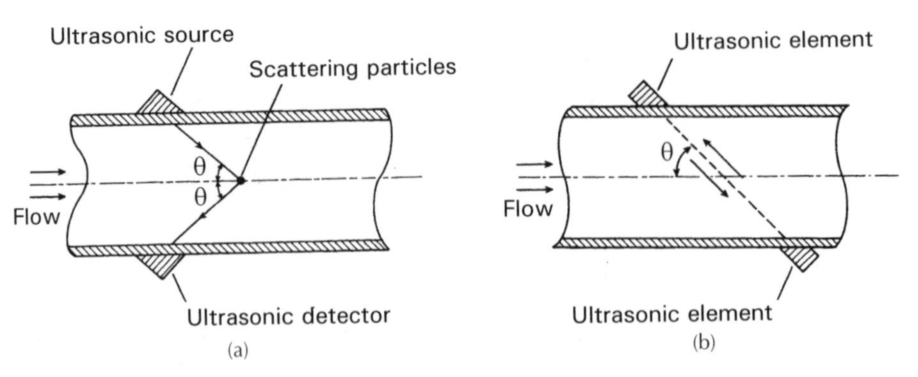

Figure 13.19 *Ultrasonic flowmeters: (a) Doppler-shift type, (b) transit time type. From Morris (1997)* Measurement and Calibration Requirements, *© John Wiley & Sons, Ltd. Reproduced with permission.*

where f_t and f_r are the frequencies of the transmitted and received ultrasonic waves respectively, c is the velocity of sound in the fluid being measured, and θ is the angle that the incident and reflected energy waves make with the axis of flow in the pipe. Volume flow rate is then readily calculated by multiplying the measured flow velocity by the cross-sectional area of the fluid-carrying pipe.

The measurement accuracy obtained depends on many factors, such as the flow profile, the constancy of pipe-wall thickness, the number, size and spatial distribution of scatterers, and the accuracy with which the speed of sound in the fluid is known. Consequently, in order to achieve accurate measurements, the instrument has to be carefully calibrated for each particular flow measurement application.

Transit-time ultrasonic flowmeter

The transit-time ultrasonic flowmeter is designed to measure the volume flow rate in clean liquids or gases. It consists of a pair of ultrasonic transducers mounted along an axis aligned at an angle with respect to the fluid-flow axis, as shown in Figure 13.19(b). Each transducer consists of a transmitter–receiver pair, with the transmitter emitting ultrasonic energy that travels across to the receiver on the opposite side of the pipe. These ultrasonic elements are normally piezoelectric oscillators of the same type as used in Doppler-shift flowmeters. Fluid flowing in the pipe causes a time difference between the transit times of the beams travelling upstream and downstream, and measurement of this difference allows the flow velocity to be calculated. The typical magnitude of this time difference is 100 ns in a total transit time of 100 μs, and high-precision electronics are therefore needed to measure it. There are three distinct ways of measuring the time shift. These are direct measurement, conversion to a phase change and conversion to a frequency change. The third of these options is particularly attractive, as it obviates the need to measure the speed of sound in the measured fluid that the first two methods require.

Transit-time flowmeters are of more general use than Doppler-shift flowmeters, particularly where the pipe diameter involved is large, so that the transit time is sufficiently large to be measured with reasonable accuracy. Measurement uncertainty can be reduced to ±0.5% in some instruments. In general, transit-time meters cost more than Doppler-shift meters, because of the greater complexity of the electronics needed to make accurate transit-time measurements.

13.3.8 Intelligent flowmeters

All the usual benefits associated with intelligent instruments are obtained in various versions of intelligent flowmeters that are available. These have various features, such as extended measurement range, improved measuring accuracy, compensation for thermal expansion of meter components and temperature-induced viscosity changes, correction for variations in flow pressure, self-diagnosis and self-adjustment capability. There is now also a trend towards total-flow computers that can process inputs from almost any type of transducer. Such devices allow user-input of parameters like specific gravity, fluid density, viscosity, pipe diameters, thermal expansion coefficients, discharge coefficients, etc. Auxiliary inputs from temperature transducers are also catered for. After processing the raw flow transducer output with this additional data, flow computers are able to produce measurements of flow to a very high degree of accuracy.

13.3.9 Calibration of flowmeters

Calibration of volume-flow-measuring instruments is a relatively expensive procedure, even when only a moderate level of accuracy is demanded. Where high accuracy is required, the cost of calibration can be very great indeed. Therefore, it is particularly important to establish exactly what accuracy level is needed, so that the calibration system instituted does not cost more than necessary. In some cases, such as handling valuable fluids, or where there are legal requirements, as in petrol pumps, high accuracy levels (e.g. error $\leq 0.1\%$) are justified. In other situations, such as measuring additives to the main stream in a process plant, only low levels of accuracy are needed (e.g. error $\approx 5\%$).

Normal practice is to calibrate flowmeters on-site as far as possible. This ensures that calibration is performed in the actual flow conditions, which are difficult or impossible to reproduce exactly in a laboratory. This is necessary because the accuracy of flow measurement is greatly affected by the flow conditions and characteristics of the flowing fluid. On account of this, it is also standard practice to repeat flow calibration checks until the same reading is obtained in two consecutive tests. The equipment and procedures used for calibration depend on whether gaseous or liquid flows are being measured. Therefore, separate sections are devoted to each of these cases. It must also be stressed that all calibration procedures refer only to flows of single-phase fluids (i.e. liquids or gases).

Calibration equipment and procedures for liquid flow

The *calibrated tank* is probably the simplest piece of equipment available for calibration. It consists of a cylindrical vessel, as shown in Figure 13.20(a), with conical ends that facilitate draining and cleaning of the tank. A sight tube with a graduated scale is placed alongside the final, upper, cylindrical part of the tank, and this allows the volume of liquid in the tank to be measured accurately. Flow-rate calibration is performed by measuring the time taken, starting from an empty tank, for a given volume of liquid to flow into the vessel. Because the calibration procedure starts and ends in zero-flow conditions, it is not suitable for calibrating instruments that are affected by flow acceleration and deceleration characteristics, such as differential pressure meters, turbine flowmeters and vortex-shedding flowmeters. The technique is further limited to the calibration of low-viscosity liquid flows, so that the tank can be fully drained between each use, although lining the tank with an epoxy coating can allow the system to cope with somewhat higher viscosities.

Pipe provers are another type of calibration device. The bidirectional type, shown in Figure 13.20(b), consists of a U-shaped tube of metal of accurately known cross-section. The purpose of the U-bend is to give a long flow path within a compact spatial volume. Alternative versions with more than one U-bend also exist to cater for situations where an even longer flow path is required. Inside the tube is a hollow, inflatable sphere that is filled with water until its diameter is about 2% larger than

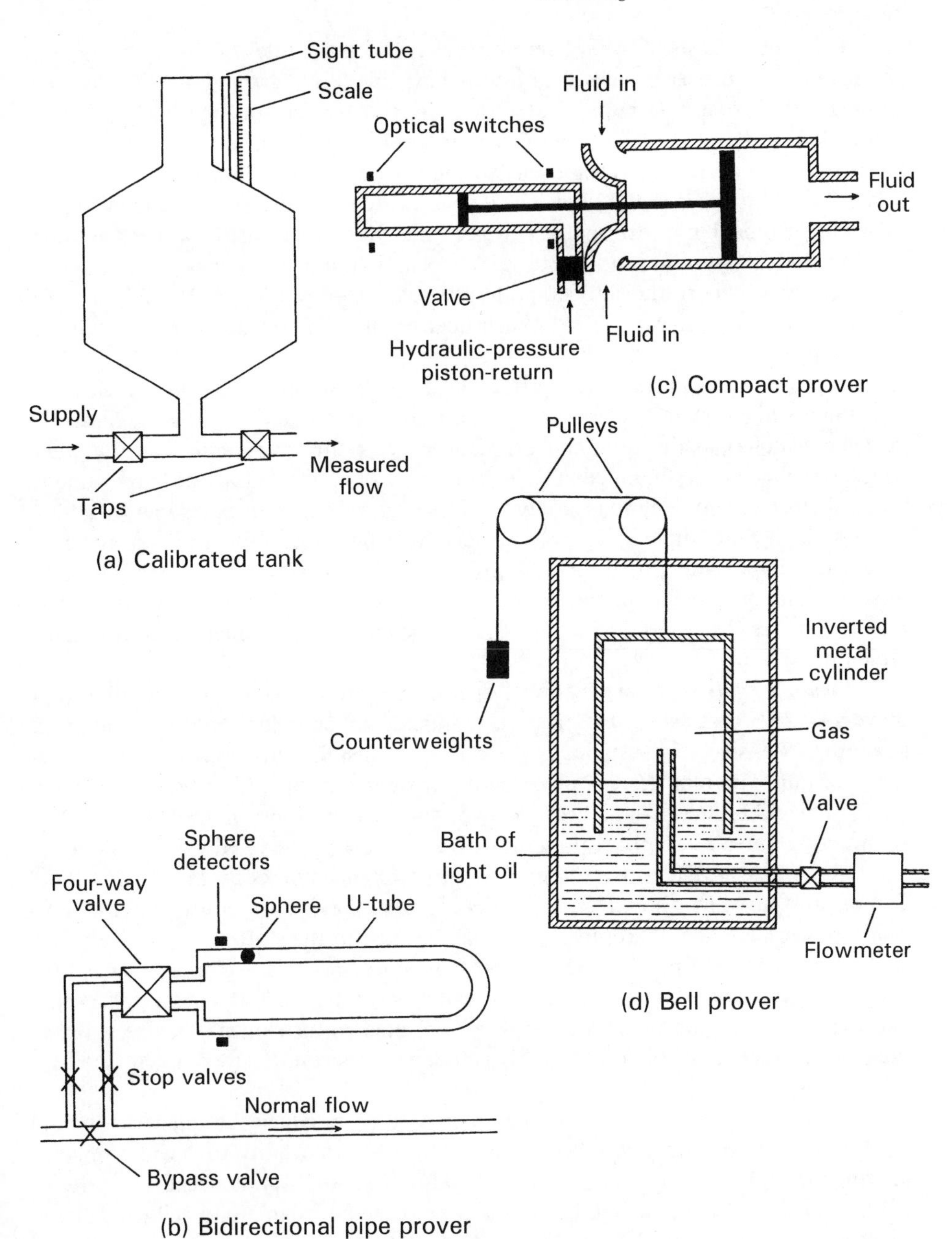

Figure 13.20 *Flow instrument calibration: (a) calibrated tank; (b) bidirectional type of pipe prover; (c) compact prover; (d) bell prover. From Morris (1997)* Measurement and Calibration Requirements, *© John Wiley & Sons, Ltd. Reproduced with permission.*

that of the tube. As such, the sphere forms a seal with the sides of the tube, and acts as a piston. The prover is connected into the existing fluid-carrying pipe network via tappings either side of a bypass valve. A four-way valve at the start of the U-tube allows fluid to be directed in either direction around it. Calibration is performed by diverting flow into the prover and measuring the time taken for the sphere to travel between two detectors in the tube. The detectors are normally of an electromechanical, plunger type. Unidirectional versions of the above also exist in which fluid only flows in one direction around the tube. A special handling valve has to be provided to return the sphere to the starting point after each calibration, but the absence of a four-way flow control valve makes such devices significantly cheaper than bidirectional types.

Pipe provers are particularly suited to the calibration of pressure-measuring instruments like turbine meters that have a pulse type of output. In such cases, the detector switches in the tube can be made to gate the instrument's output pulse counter. This enables the ancillary electronics to be calibrated at the same time as the basic instrument. The inaccuracy level of such provers can be as low as ±0.1%, even for high fluid-viscosity levels and very high flow rates. Even higher accuracy is provided by an alternative form of prover that consists of a long, straight metal tube containing a metal piston. However, such devices are more expensive than the other types discussed above, and their large space requirements can also cause difficulties.

The *compact prover* has an identical operating principle to that of the other pipe provers described above, but it occupies a much smaller spatial volume. It is therefore used extensively in situations where there is insufficient room to use a larger prover. Many different designs of compact prover exist, operating both in the unidirectional and bidirectional modes, and one such design is shown in Figure 13.20(c). Common features of compact provers are an accurately machined cylinder containing a metal piston that is driven between two reference marks by the flowing fluid. The instants at which the reference marks are passed are detected by switches. Provision has to be made for returning the piston back to the starting point after each calibration, and a hydraulic system is commonly used for this. Again, measuring the piston traverse time is made easier if the switches can be made to gate a pulse train, and therefore compact provers are also most suited to instruments having a pulse-type output such as turbine meters. Measurement uncertainty levels down to ±0.1% are possible.

A *gravimetric* calibration method can also be used in which the quantity of fluid flowing in a given time is weighed. Apart from its applicability to a wide range of instruments, this technique is not limited to low-viscosity fluids, because any residual fluid in the tank before calibration will be detected by the load cells and therefore compensated for. In the simplest implementation of this system, fluid is allowed to flow for a measured length of time into a tank resting on load cells. As before, the stop–start mode of fluid flow makes this method unsuitable for calibrating differential pressure, turbine and vortex-shedding flowmeters. It is also unsuitable for measuring high flow rates, because of the difficulty in bringing the fluid to rest. These restrictions can be overcome by directing the flowing fluid into the tank via diverter valves, although, if this is done, the timing system must be carefully synchronised with the operation of the diverter valves.

High-quality versions of normal flow measuring instruments can also be used for calibration. *Positive displacement meters* can give measurement inaccuracy levels down to ±0.2%. *Turbine meters* can calibrate with a measurement uncertainty down to ±0.2%, and are used particularly for calibrating high flow-rates through large-bore pipes. Sometimes even a *certified orifice plate* can be used for calibration if the typical measurement uncertainty of ±1% is acceptable.

Calibration equipment and procedures for gaseous flow

Calibration of gaseous flows poses considerable difficulties compared with calibrating liquid flows. These problems include the lower density of gases, their compressibility and the difficulty in establishing a suitable liquid/air interface, as utilised in many liquid flow measurement systems. In consequence, the main methods of calibrating gaseous flows, as described below, are small in number. Certain other specialised techniques, including the gravimetric method and the pressure–volume–temperature method, are also available, but the equipment required is so expensive that it is usually only available in National Standards Laboratories.

The *bell prover* consists of a hollow, inverted, metal cylinder suspended over a bath containing light oil, as shown in Figure 13.20(d). The air volume in the cylinder above the oil is connected, via a tube and a valve, to the flowmeter being calibrated. An airflow through the meter is created by allowing the cylinder to fall downwards into the bath, thus displacing the air contained within it. The flow rate, which is measured by timing the rate of fall of the cylinder, can be adjusted by changing the value of counterweights attached via a low-friction pulley system to the cylinder. This is essentially laboratory-only equipment, and therefore on-site calibration is not possible.

A *positive displacement flowmeter* can also be used for the calibration of gaseous flows, with inaccuracy levels down to ±0.2%.

13.4 Level Measurement

The main reason for level measurement in an EMS is to ensure that storage tanks do not overflow, although, in many applications, high measurement accuracy is not needed, since only a rough indication of level is needed. A variety of suitable level-measuring instruments are covered in the paragraphs below. These can all measure liquid level, and some can also measure the level of solids that are in the form of powders or small particles. Some level-measuring instruments have been excluded from the discussion, for various reasons. Firstly, instruments that require a human to read the output are excluded, because automatic, continuous measurement is needed, so that the rate of rise of level in a tank can be calculated and action taken before overflow occurs. Secondly, very expensive devices, such as those using radiation and laser methods, are excluded. Thirdly, devices that are not in common use, such as thermal and magnetostrictive sensors, are also excluded.

13.4.1 Float and displacer systems

Float systems, in which the position of a float on the surface of a liquid is measured by means of a suitable transducer, are simple, cheap and reasonably accurate, but usually have high maintenance requirements. One type of float system sensor was shown earlier, in Figure 3.3. An alternative arrangement is shown in Figure 13.21(a), where a ball floats on the liquid surface and movement of a rod attached to the ball is detected by a linear variable displacement transducer (LVDT) or alternative displacement transducer. The *float and tape gauge* (or *tank gauge*) is an alternative float system in which the float has a tape attached to it that passes round a pulley situated vertically above the float. The other end of the tape is attached to either a counterweight or a negative-rate counterspring. The amount of rotation of the pulley, measured by either a synchro or a potentiometer, is then proportional to the liquid level. *Displacer level sensors* work on a slightly different principle, where the force exerted by the fluid on the displacer is balanced by the downward force of a spring, as shown in Figure 13.21(b). The equilibrium position is measured by either an LVDT or other displacement transducer.

13.4.2 Hydrostatic system (pressure-measuring device)

A hydrostatic system measures the hydrostatic pressure exerted by a liquid, which is directly proportional to its depth and hence to the level of its surface. This is a relatively simple and cheap method of measurement. In the case of open-topped vessels (or covered ones that are vented to the atmosphere), the level is usually measured by inserting an appropriate pressure transducer at the bottom of the vessel, as shown in Figure 13.22(a). The liquid level, h, is then related to the measured pressure, P, according to: $h = P/\rho g$, where ρ is the liquid density and g is the acceleration due to gravity. In the case of liquids contained in sealed vessels, the liquid level can be measured by measuring the differential pressure between the top and bottom of the vessel. One source of error in this method can be imprecise knowledge of the liquid density. This can be a particular problem in the case of liquid solutions and mixtures

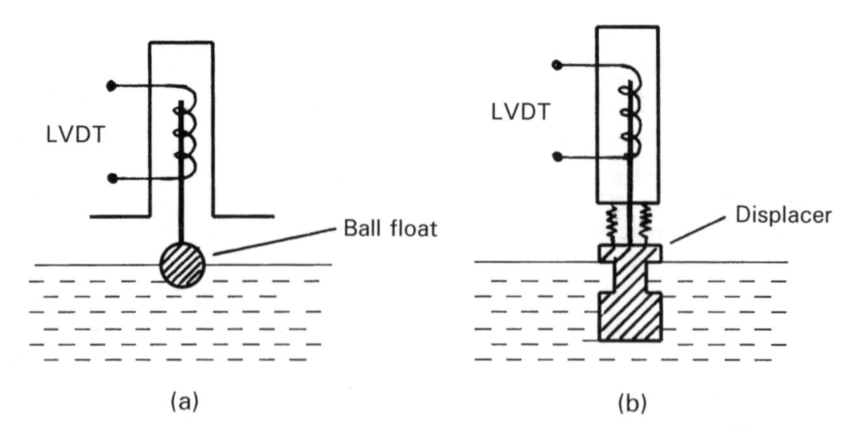

Figure 13.21 Level measurement: (a) ball float sensor; (b) displacer sensor.

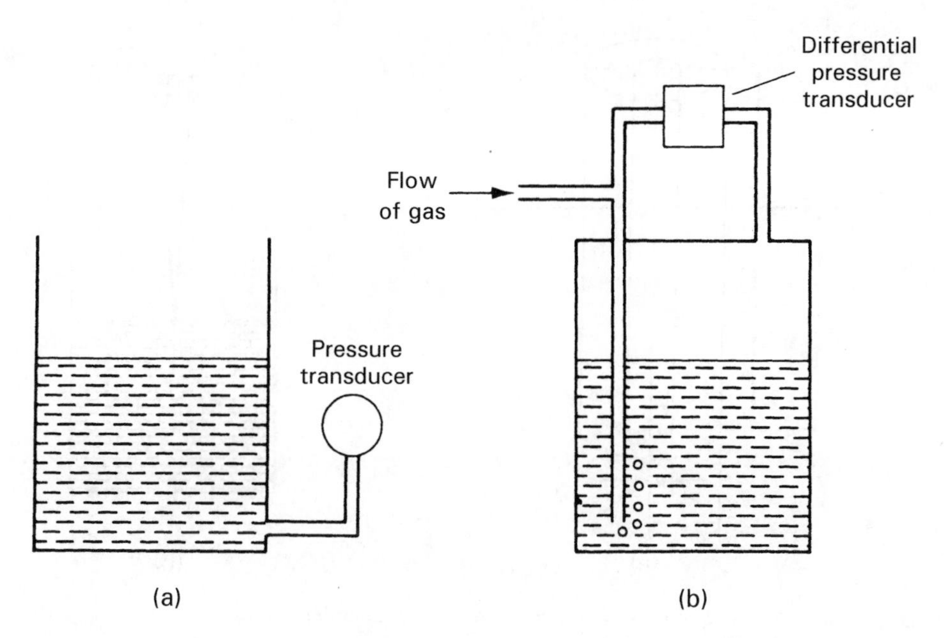

Figure 13.22 *Hydrostatic level measurement: (a) open-topped vessel; (b) bubbler unit.*

(especially hydrocarbons), and in some cases only an estimate of density is available. Even with single liquids, the density is subject to variation with temperature, and therefore temperature measurement may be required if very accurate level measurements are needed. A further problem is that the point of entry into the vessel for the pressure transducer may become a source of leakage.

The bubbler unit shown in Figure 13.22(b) is an alternative hydrostatic system that does not need a pressure tapping, and so avoids this potential cause of leakage. The bubbler unit uses a dip pipe that reaches to the bottom of the tank and is purged free of liquid by a steady flow of gas through it. The rate of flow is adjusted until gas bubbles are just seen to emerge from the end of the tube. The pressure in the tube, measured by a pressure transducer, is then equal to the liquid pressure at the bottom of the tank. It is important that the gas used is inert with respect to the liquid in the vessel. Nitrogen, or sometimes just air, is suitable in most cases. Gas consumption is low, and a cylinder of nitrogen may typically last for six months. The method is suitable for measuring the liquid pressure at the bottom of both open and sealed tanks.

13.4.3 Capacitive devices

Capacitive level sensors (alternatively known as admittance or RF sensors) are widely used for measuring the level of both liquids and solids in powdered or granular form. Two versions are used according to whether the measured substance is conducting or not. For nonconducting ($<0.1\,\mu$ Siemen/cm^3) substances, two bare-metal capacitor plates in the form of concentric cylinders are immersed in the substance, as shown in Figure 13.23(a). The substance behaves as a dielectric between the plates,

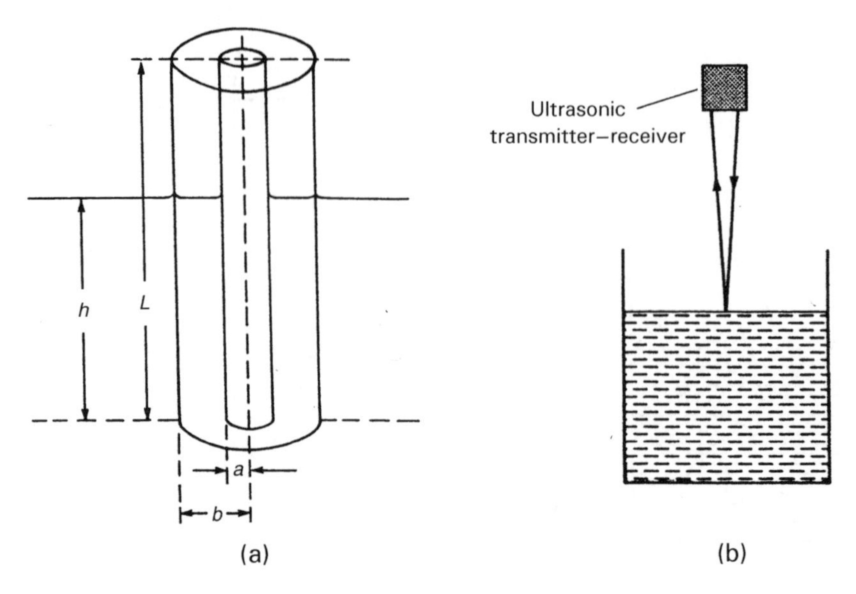

Figure 13.23 More level sensors: (a) capacitive sensor; (b) ultrasonic sensor.

according to the depth of the substance. For concentric cylinder plates of radius a and b ($b > a$), and total height L, the depth of the substance h is related to the measured capacitance C by:

$$h = \frac{C \log_e(b/a) - 2\pi\varepsilon_o}{2\pi\varepsilon_o(\varepsilon - 1)} \tag{13.5}$$

where ε is the relative permittivity of the measured substance and ε_o is the permittivity of free space. The capacitance is measured by one of the methods discussed in the next chapter.

In the case of conducting substances, exactly the same measurement techniques are applied, but the capacitor plates are encapsulated in an insulating material. The relationship between C and h in equation (13.5) then has to be modified to allow for the dielectric effect of the insulator.

Capacitive sensors provide good accuracy and can cope with difficult conditions, such as measuring liquid metals (high temperatures), liquid gases (low temperatures), corrosive liquids (acids, etc.), high-pressure processes and conditions where the liquid surface is turbulent or frothy. The main limitation is that they become inaccurate if the measured substance is prone to contamination by agents that change the dielectric constant. Ingress of moisture into powders is one such example of this.

13.4.4 Ultrasonic level gauge

The principle of the ultrasonic level gauge is illustrated in Figure 13.23(b). Energy from an ultrasonic source above the liquid is reflected back from the liquid surface into an ultrasonic energy detector. Measurement of the time of flight allows the liquid level to be inferred. Different versions allow the sensor to be placed anywhere between 0.1 m and 15 m above the liquid surface. In alternative versions, the ultrasonic source is placed at the bottom of the vessel containing the liquid, and the time of flight between emission, reflection off the liquid surface and detection back at the bottom of the vessel is measured.

Ultrasonic level sensors provide simple measurement at relatively low cost. They are especially useful in measuring the position of the interface between two immiscible liquids contained in the same vessel, or measuring the sludge or precipitate level at the bottom of a liquid-filled tank. In either case, the method employed is to fix the ultrasonic transmitter–receiver transducer at a known height in the upper liquid, as shown in Figure 13.24. This establishes the level of the liquid/liquid or liquid/sludge level in absolute terms.

When using ultrasonic instruments, it is essential that proper compensation is made for the working temperature if this differs from the calibration temperature. The speed of ultrasound through air varies with temperature at the rate of 0.607 m/s per °C. The speed of ultrasound also has a small sensitivity to humidity, air pressure and carbon dioxide concentration, but these factors are usually insignificant.

13.4.5 Radar (microwave) level sensor

Radar sensors direct a constant-amplitude, frequency-modulated microwave signal at the liquid surface. A receiver measures the phase difference between the reflected signal and the original signal transmitted directly through air to it, as shown in Figure 13.25. This measured phase-difference is linearly proportional to the liquid level. The system is similar in principle to ultrasonic level measurement, but has the important advantage that the transmission time of radar through air is almost totally unaffected by ambient temperature and pressure fluctuations. Radar level sensors are relatively expensive, but are in widespread use since their high cost is often justified by their maintenance-free operation, ability to work in difficult conditions such as measurement in closed tanks, measurement of slurries, measurement where the liquid surface

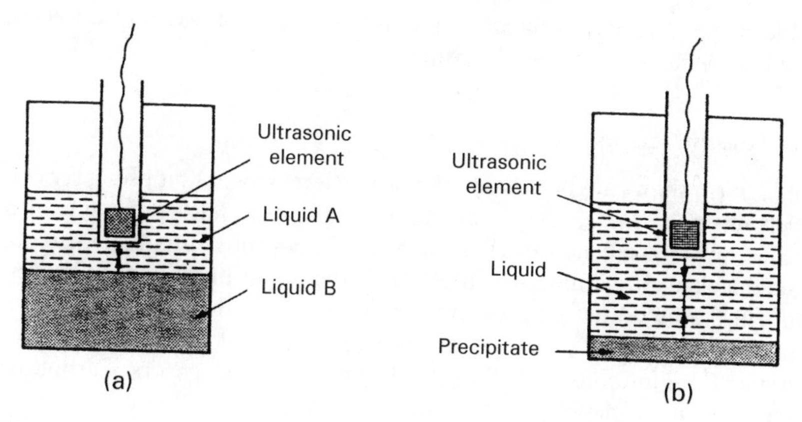

Figure 13.24 Measuring interfaces: (a) liquid/liquid interface; (b) liquid/precipitate interface.

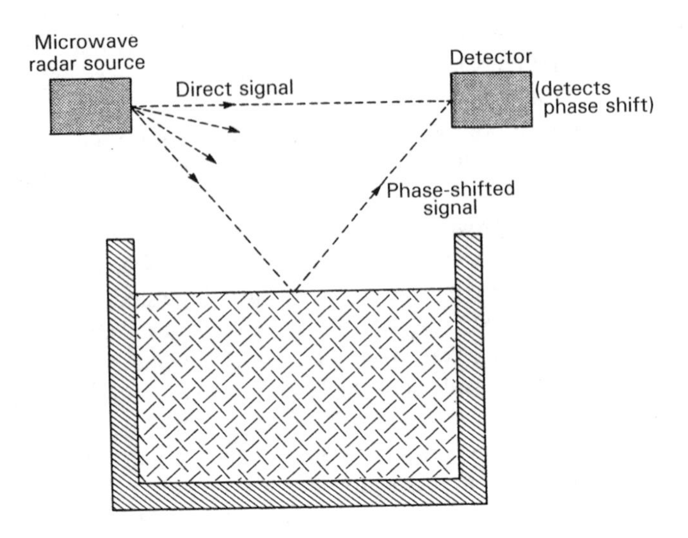

Figure 13.25 Radar level detector.

is turbulent or covered in foam, and measurement in the presence of obstructions such as mist and steam condensate. However, because the microwave frequency is within the band used for radio communications, strict conditions on amplitude levels have to be satisfied, and appropriate licences have to be obtained.

13.4.6 Vibrating level sensor

The principle of the vibrating level sensor is illustrated in Figure 13.26. The instrument consists of two piezoelectric oscillators fixed to the inside of a hollow tube that generate flexural vibrations in the tube at its resonant frequency. The resonant frequency of the tube varies according to the depth of its immersion in the liquid. A phase-locked loop circuit is used to track these changes in resonant frequency and adjust the excitation frequency applied to the tube by the piezoelectric oscillators. Liquid level measurement is therefore obtained in terms of the output frequency of the oscillator when the tube is resonating.

13.4.7 Resistive sensors

Figure 13.27(a) shows a typical form of resistive level sensor that uses a series of hot-wire elements placed at regular intervals along a vertical line up the side of a tank. The heat transfer coefficient of such elements differs substantially depending upon whether the element is immersed in air or in the liquid in the tank. Consequently, elements in the liquid have a different temperature and therefore a different resistance to those in air. This method of level measurement is a simple one, but the measurement resolution is limited to the distance between sensors. Carbon resistors are sometimes used in place of hot-wire elements. In an alternative arrangement

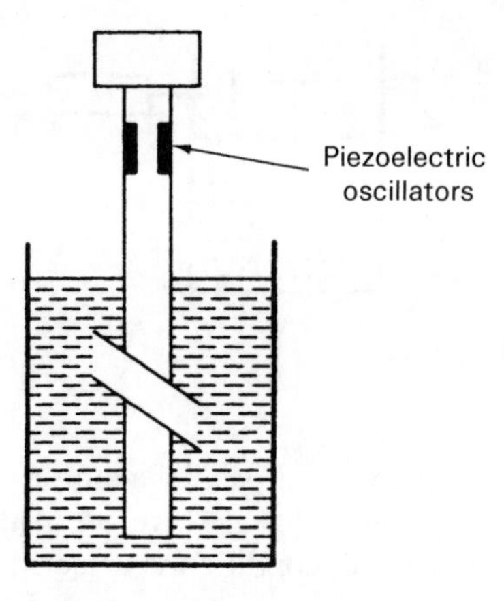

Figure 13.26 Vibrating level sensor.

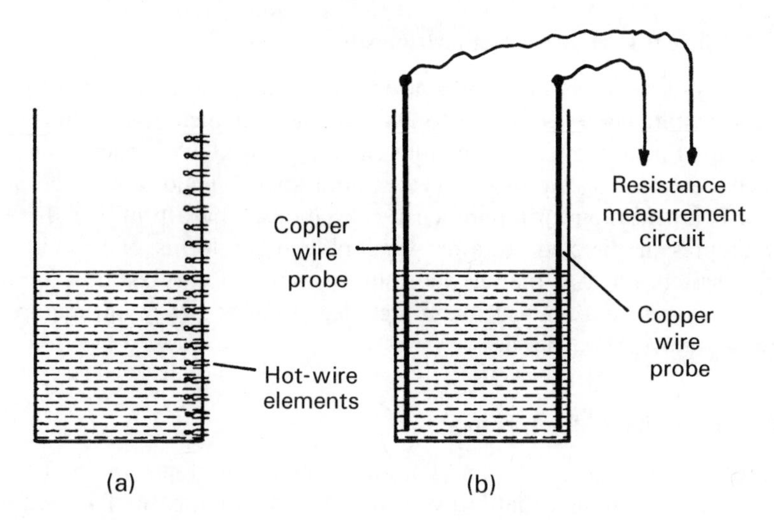

Figure 13.27 Resistive sensors: (a) hot-wire elements; (b) electrode type.

shown in Figure 13.27(b), two electrodes are placed along the sides of the tank. The resistance between the electrodes is then measured. This varies according to the relative amount of air and liquid between the electrodes, and so allows the liquid depth to be calculated. This system must be calibrated according to the conductivity of the liquid in the tank.

13.4.8 Paddle-wheel level sensors

Paddle-wheel sensors are primarily used as a switch to detect when powdered, granular, pelletised or similar solid material reaches a certain level in a container. As

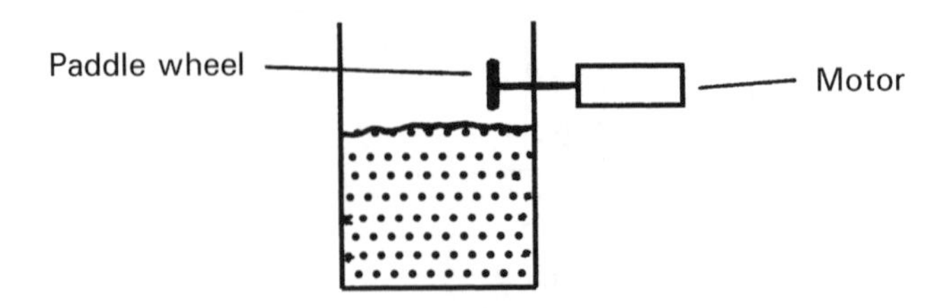

Figure 13.28 Paddle wheel level sensor.

shown in Figure 13.28, a paddle wheel driven by a motor rotates freely until the level of solids in the container reaches a level where the paddle wheel makes contact with the surface and stops. The stall condition is detected and power removed from the motor. The sensor is reset once the excess material in the container has been removed.

13.4.9 Intelligent level-measuring instruments

Only certain types of level-measuring device are suitable for use as intelligent instruments. Hydrostatic systems are obvious candidates for inclusion within intelligent level-measuring instruments, and versions claiming only ±0.05% inaccuracy are available. Such instruments can also carry out additional functions, such as providing automatic compensation for liquid density variations. Apart from this, there is little activity at present directed towards developing other forms of intelligent level-measuring systems. This is mainly the result of market forces, as most applications of liquid measurement do not demand high degrees of accuracy, but require instead just a rough indication of level.

13.4.10 Calibration of level sensors

Calibration of ultrasonic, radar and vibrating-element level sensors can be achieved by using a calibration tank that has vertical sides and a flat base of known area. By adding known volumes of liquid to the tank, a set of reference levels can be created on which the level sensor can be tested. For other types of level sensor that are an integral part of a vessel containing either liquids or solids, adequate calibration can often be achieved by using a dipstick. A dipstick is simply a graduated metal rod that is lowered into the vessel until it touches the bottom, at which time the level is determined by reading the position of the material surface against the graduated scale. Alternatively, level sensors that are an integral part of the tank being measured can be calibrated by first emptying the tank and then progressively adding known volumes of liquid to it.

Ideally, calibration should be performed with the measured vessel containing the material that it usually holds. However, provided that it can be shown that no measurement error is created if alternative materials are put in the tank, it is often less-expensive, more convenient and safer in the case of hazardous materials to use just water in the tank for calibration purposes.

References

1. Brookes, C. (1985) Nicrosil–nisil thermocouples, *Journal of Measurement and Control*, **18(7)**, 245–8.
2. Thermometrics Corporation, 2002 (*www.thermometrics.com*).
3. Dixon, J. (1987) Industrial radiation thermometry, *Journal of Measurement and Control*, **20(6)**, 11–16.
4. Morris, A.S. (1997) *Measurement and Calibration Requirements for Quality Assurance to ISO 9000* (John Wiley & Sons, Chichester, UK).

Appendix 1

Summary of ISO 14000 Series Standards

ISO 14001 (2002): Environmental Management Systems: Specification with Guidance for Use – This sets out the requirements for establishing and maintaining an environmental management system that assures a company of conformance with the environmental objectives that it sets for itself and also with any relevant legislation. If desired, this can be audited and certified by national accrediting bodies.

ISO 14004 (1996): Environmental Management Systems: General Guidelines on Principles, Systems and Supporting Techniques – This gives general guidance and practical advice on implementing or improving an EMS, and is supported by a number of examples.

ISO 14010 (1996): Guidelines for Environmental Auditing: General Principles – This sets out the general principles governing the audit process of an EMS, and sets out the necessary basic conditions that have to be satisfied before an audit can be carried out, such as the availability of adequate documentation, the provision of sufficient resources to support the audit, and the willing cooperation and assistance of the company whose EMS is being audited. Guidance on the structure of the report giving the audit findings is also provided. This standard is soon to be replaced by ISO 19011.

ISO 14011 (1996): Guidelines for Environmental Auditing: Audit Procedures: Auditing of Environmental Management Systems – This explains the usual objectives of an EMS audit to confirm that the EMS has been properly implemented and is meeting its objectives, is managed satisfactorily and is reviewed regularly to try to achieve continual improvement in its performance. The duties of lead and other auditors are set down, and also those of the company whose EMS is being audited. A summary of the necessary steps in executing the audit is also presented, and recommendations are made about the structure, content and distribution of the audit report. This standard is soon to be replaced by ISO 19011.

ISO 14000 Environmental Management Standards: Engineering and Financial Aspects. Alan S. Morris.
 ISBN 0-470-85128-7

ISO 14012 (1996): Guidelines for Environmental Auditing: Qualification Criteria for Environmental auditors – This standard sets out the educational qualifications required, personal skills needed and the level of training and experience necessary for auditors who are appointed by Standards Bodies to carry out audits and certify an EMS. This standard is soon to be replaced by ISO 19011.

ISO 14015 (2001): Environmental Management Systems: Environmental Assessment of Sites and Organisations – This standard gives guidance on the conduct of an environmental assessment of a site or organisation. The usual circumstances in which such an assessment is carried out is when a company is considering purchasing another company, to ensure that the purchase would not involve it in large costs in rectifying nonconformance of the site with good environmental management principles, or with present or likely future environmental legislation. Like ISO 14001, it is applicable to a wide range of companies of all types, in all industries and of all sizes. The assessment will cover all present and proposed future operations, but will often have to consider past operations that may have caused ongoing environmental damage, such as contaminated land.

ISO 14020 (2001): Environmental Labels and Declarations: General principles – This gives guidelines on the use of labels to indicate the environmental impact of products. It requires that labels should be accurate and verifiable. Such labelling is provided for the benefit of environmentally conscious customers, who can decide whether or not to purchase a product, according to its environmental performance.

ISO 14021 (2001): Environmental Labels and Declarations: Self-declared Environmental Claims (Type II Environmental Labelling) – This sets common standards for the use of particular environmental labels and symbols. It specifies the requirements that must be satisfied for the application of various labels such as 'degradable' and 'recyclable'.

ISO 14024 (2001): Environmental Labels and Declarations: Type I Environmental Labels: Principles and Procedures – This extends ISO 14020 by providing a framework for issuing licenses to permit the use of particular environmental labels on products, and to assess the continued compliance of the products at regular intervals.

ISO 14031 (2000): Environmental Management: Environmental Performance Evaluation Guidelines – This is provided as a management tool to help companies to assess whether its environmental performance is meeting the targets set. The document is particularly useful to companies that have not implemented a formal EMS, since it includes guidance on identifying the environmental impact of operations and setting targets for reduction in environmental damage.

ISO 14032 (2000): Environmental Management: Examples of Environmental Performance Evaluation – This document provides 17 case studies across a wide range of industries, showing how ISO 14031 has been implemented.

ISO 14040 (1997): Environmental Management: Life Cycle Assessment: Principles and Framework – This standard defines a suitable framework for conducting life-cycle assessments for products in terms of their environmental impact. Recommendations are made about setting environmental targets and defining relevant indicators of environmental performance for all stages in a product's life cycle, including raw material acquisition, product manufacture, product use and product disposal at the end of its useful life. ISO 14040 defines the four phases in life-cycle assessment as defining goals, inventory analysis, impact assessment and interpretation. Further guidance on these four phases are provided in ISO 14041–ISO 14043.

ISO 14041 (1998): Environmental Management: Life Cycle Assessment: Goal and Scope Definition and Inventory Analysis – This extends ISO 14040 by providing further guidance on defining the goals of life-cycle assessment and collecting data to assess environmental performance against the indicators set.

ISO 14042 (2000): Environmental Management: Life Cycle Assessment: Life Cycle Impact Assessment – This provides guidance on how the performance indicator data defined according to ISO 14041 should be used to assess the environmental impact of a product and identify opportunities to improve the product to reduce its environmental impact.

ISO 14043 (2000): Environmental Management: Life Cycle Assessment: Life Cycle Interpretation – This provides guidance on how the impact assessments made according to ISO 14042 should be summarised, interpreted and discussed with respect to the environmental goals set according to ISO 14041.

ISO 14048 (2002): Environmental Management: Life Cycle Assessment: Data Documentation Format – This provides a standard framework for the documentation of the performance indicators as specified in ISO14041 and provides appropriate standards for data collection, quality and analysis.

ISO 14049 (2000): Environmental Management: Life Cycle Assessment: Examples of Application of ISO14041 to Goal and Scope Definition and Inventory Analysis – This provides a number of examples indicating how the various requirements of ISO 14041 might be implemented.

ISO 14050 (2002): Environmental Management Vocabulary – This defines all concepts, words and phrases used in ISO 14000 standards, to ensure that a common international understanding is achieved.

ISO 14061 (1998): Information to Assist Forestry Organisations in the Use of Environmental Management System Standards ISO 14001 and ISO 14004 – This provides information to assist forestry organisations in the use of ISO 14001 and ISO 14004. As well as guidelines specific to the forestry industry, appropriate reference material is provided and a number of case studies are included.

ISO 19011 (2002): Guidelines for Quality and/or Environmental Management Systems Auditing – This mainly copies the guidance on system auditing contained in ISO 14010–14012, but does so without specific reference to environmental systems, so that it is applicable either to quality assurance systems (ISO 9000) or to environmental management systems. This single document supersedes ISO 14010, ISO 14011 and ISO 14012 (and also corresponding documents in the ISO 9000 series).

Appendix 2

Typical Structure of an Environmental Management System Manual

The following example structure is based on real environmental management system (EMS) documentation, but the company name and all site names are fictitious. Any resemblance to the names of real companies or sites is unintended.

Because of space limitations, there is not room to give an example of a complete manual, and therefore this appendix is only able to give direction about the content and structure that an EMS manual should have. In practice, the detail necessary in a full manual would necessarily extend to many more pages. Nevertheless, the guidelines given here, together with the explanation of the content and details necessary, as covered in Chapter 2 and elsewhere, should assist companies in writing a manual that is appropriate to the scale of their operations.

A comment should also be made about the structure and order of presentation of the material in the following example. The structure adopted in the example is felt by the author to be a sensible order in which to present the necessary information. However, whilst ISO 14001 sets out requirements in terms of what should be documented, it does not specify any particular order in which the information should be given. Companies may therefore find it useful to use the following structure as a guideline, but should feel free to adapt it as necessary to fit their own particular circumstances.

ISO 14000 Environmental Management Standards: Engineering and Financial Aspects. Alan S. Morris.
 ISBN 0-470-85128-7

MORRIS FABRICATIONS LIMITED

ENVIRONMENTAL MANAGEMENT SYSTEM MANUAL

****

THIS MANUAL IS A CONTROLLED DOCUMENT

IT MUST BE KEPT IN A SAFE BUT ACCESSIBLE LOCATION

IT MUST BE MAINTAINED IN GOOD CONDITION

IT MUST BE UPDATED WITH REVISED PAGES WHICH WILL BE CIRCULATED FROM TIME TO TIME, IN ACCORDANCE WITH THE INSTRUCTIONS PROVIDED WITH THE UPDATE PAGES

****

COPY NUMBER: 4

AUTHORISED HOLDER: Works Manager – West Cross Site

Cover Page Revision number: 07 Date: 08 Mar 2003

LIST OF REVISION NUMBERS AND DATES OF CURRENT MANUAL SECTIONS

Section number	Revision number	Date	Section number	Revision number	Date
1	1	16 Jan 2002	¦	¦	¦
2	7	08 Mar 2003	¦	¦	¦
3	7	08 Mar 2003	¦	¦	¦
¦	¦	¦	¦	¦	¦
¦	¦	¦	¦	¦	¦
¦	¦	¦	¦	¦	¦

It is strongly recommended that page numbers within sections should be numbered consecutively in the style: page 1 of 3, page 2 of 3, page 3 of 3, etc. Besides the page number, each page should also be labelled with the revision number and date of the section that it is part of. This will assist in ensuring that all manual pages are the latest revision.

Where a company has two or more separate sites that have different characteristics and activities, EMS details and targets may vary from site to site. In this case, it is advisable to have a separate set of pages in the EMS manual for each site.

Revision numbers and dates of current section numbers Page 1 of 1

Revision number: 07 Date: 08 Mar 2003

CONTENTS

Section number

1 Statement of environmental policy.
2 Management structure for EMS.
3 Design of EMS:
   How environmental impact of operations is assessed; How information on up-to-date legal and regulatory requirements is obtained; How environmental targets are set in terms of cost/benefits equation; How the legal/regulatory requirements are satisfied and the actual environmental targets set.
4 EMS elements:
   Person responsible; Purpose; Detail of hardware designs; Operational/maintenance instructions; Emergency procedures to respond to unexpected incidents; Interaction with other elements of EMS; Mechanisms for communication of matters relevant to EMS; Elements where activities of suppliers and customers are relevant.
5 List of approved suppliers.
6 Training programmes and schedule.
7 Training records.
8 Measurement and monitoring requirements:
   Process variables to be monitored; Measurement frequencies; Measurement techniques; Type of measuring instruments to be used; Special precautions necessary; Details of signal processing, conversion, transmission and recording procedures; Instrument calibration procedures; Training requirements for personnel involved in measurement and calibration; Training records.
9 EMS performance records:
   System inspection and test procedures; System inspection and test results; Running costs; Length of time for which records should be kept.
10 Incident reports:
   Records of incidents that led to pollution including date of incident, cause of incident, remedial action taken in response, and modifications made to EMS to prevent re-occurrence of incident.
11 Internal audits: procedures and records of audits.
12 EMS reviews:
   Person responsible for reviews; Frequency of reviews; Review procedure; Changes made to EMS as a result of review.
13 Control of EMS manual:
   Person responsible; Procedure for amendments to manual.
14 Distribution list: Authorised holders and locations of environmental management system (EMS) manual.
15 Procedure for amending manual.

Contents page Page 1 of 1 Revision number: 07 Date: 08 Mar 2003

1 Statement of environmental policy

Morris Fabrications is committed to a clean and healthy environment, and operates an environmental management system that is certified to ISO 14001 and is designed to minimise the environmental impact of our operations. We will identify all possible environmental impacts of our operations and comply fully with all relevant environmental legislation and regulations. We will also reduce pollution beyond our legal obligation, wherever it is practical and economic to do so.

We believe that having a sound environmental policy benefits our customers and also the public at large. Our aim is to provide customers with safe and reliable products, and to minimise environmental impacts both in their manufacture and subsequent use. We also accept a responsibility to influence environmental protection in all ways possible, and we therefore offer a free advice service in good environmental practice to all our customers and suppliers.

To fulfil these commitments, we will:

- Consider environmental impacts in the manufacture, subsequent use and final disposal of our products, and make design changes as necessary to minimise these impacts.
- Seek to reduce the amount of waste generated in our manufacturing operations.
- Use recycled materials in our products wherever possible.
- Use energy as efficiently as possible.
- Install and operate special equipment, as necessary, to reduce emissions to the environment.
- Avoid pollution by ensuring that storage tanks do not overflow, and by monitoring pipes carrying hazardous fluids to detect leakage.
- Specify procedures to be followed that will minimise pollution if emergency situations develop.
- Provide appropriate training to employees.
- Ensure that all employees understand the aims and objectives of our environmental management system (EMS) and are fully committed to it.
- Measure all relevant parameters to ensure that our EMS is operating as intended and is meeting its stated environmental protection objectives.
- Review the performance of the EMS at regular intervals.
- Maintain records of all monitored parameters and performance measurements.
- Work continuously to improve environmental performance and update the EMS as necessary.

Section 1 Page 1 of 1 Revision number: 01 Date: 16 Jan 2002

(Start new page)

2 Management structure for EMS

Person with overall responsibility for EMS: Company EMS director.

Person responsible for EMS at Fiveoaks Site: Works Manager, Fiveoaks Site.

Person responsible for EMS at West Cross Site: Operations Manager, West Cross Site.

Person responsible for EMS at North Hill site: Production Manager, North Hill Site.

|
|
|
|

Section 2 Page 1 of 1 Revision number: 07 Date: 08 Mar 2003

(Start new page)

3 Design of EMS

This section should explain how the environmental impact of operations is assessed, how information on up-to-date legal and regulatory requirements is obtained, and how environmental targets are set in terms of cost/benefits equation. The legal and regulatory requirements and the actual environmental targets set should be documented within this section.

|
|
|
|

Section 3 Page 1 of 1 Revision number: 07 Date: 08 Mar 2003

---

4 EMS elements

This section should list all elements in the EMS, including all procedures designed to minimise pollution in normal operations and all emergency procedures to respond to unexpected incidents. The following information should be given for each element:

person responsible;
purpose;
detail of hardware designs and operating/maintenance instructions;
interaction with other elements of EMS;
mechanisms for any internal or external communication necessary for element.

In addition, all EMS elements where activities of suppliers and customers are relevant should be listed, and procedures for monitoring and influencing these activities should be stated.

|
|
|
|

Section 4 Page 1 of *x* Revision number: 07 Date: 08 Mar 2003

(Start new page)

5 List of approved suppliers

In every instance where substandard quality and reliability in material and equipment bought from suppliers can have an adverse environmental impact, a list of approved suppliers should be listed. This list will obviously include materials like chemicals and fuel that are used in industrial processes. However, it should also extend to equipment such as filters and other devices designed to remove pollutants, as well as emergency-response equipment.

|
|

Section 5 Page 1 of x Revision number: 06 Date: 14 Jan 2003

(Start new page)

6 Training programmes and schedule

Training programme	Personnel involved	Frequency
Basic awareness of EMS	All	every 12 months
Waste disposal	All personnel involved in operations that generate waste	every six months
|	|	|
|	|	|

Details of training programmes can be found in the separate Training Manual.

|

Section 6 Page 1 of x Revision number: 02 Date: 23 Mar 2002

(Start new page)

7 Training records

In the case of a small company, it is feasible to record within the EMS manual the details and dates of all training programmes undergone by company personnel. In the case of larger companies, it is more appropriate to maintain details of training programmes and training records in a separate Training Manual. In a large company with several sites, it might be sensible to have separate training manuals at each site. However, if training procedures and records are not included within the EMS manual, it is essential that the EMS manual explains where such training procedures and records can be inspected.

|

Section 7 Page 1 of x Revision number: 07 Date: 08 Mar 2003

(Start new page)

8 Measurement and monitoring requirements

This section should list all process variables and parameters associated with the operations of a company that have to be monitored as part of the EMS.

For each measurement requirement, the following information should be given:

- Required measurement frequency.
- Measurement technique to be used, and any special precautions necessary, such as environmental control.
- Type of measuring instrument to be used.
- Person responsible for each measuring instrument.
- Instructions on the proper way of using instruments.
- Necessary training courses for personnel using instruments.
- Instrument calibration procedures, including required calibration frequency, standard instruments to be used, handling of standard instruments and traceability of calibration to national reference standards, method of recording of calibration results, procedure to be followed if an instrument is found to be outside calibration limits, including the method of marking it to prevent use until faults are corrected, training programmes for personnel involved in calibration duties.
- Procedure for reviewing calibration procedures.
- Records of calibration procedure reviews.
- Details of signal processing, conversion, transmission and recording procedures

|
|

Section 8 Page 1 of *x* Revision number: 05 Date: 19 Nov 2002

(Start new page)

9 EMS performance records

This section should record reviews and tests to show whether the EMS is operating as intended and the cost of operation. This information should be divided between the following sections:

- system inspection and test procedures;
- system inspection and test results, including equipment maintenance records;
- operational costs of EMS;
- length of time that performance records should be kept.

|
|

Section 9 Page 1 of *x* Revision number: 03 Date: 10 May 2002

(Start new page)

10 Records of incidents that led to pollution

All incidents that lead to pollution of the environment should be recorded in this section. For each, the action taken to prevent recurrences of the incident should be recorded. For each incident, the following information should be given:

- Date of incident;
- Cause of incident;
- Remedial action taken;
- Modifications made to EMS to prevent re-occurrence of incident.

|
|

Section 10 Page 1 of *x* Revision number: 04 Date: 06 Sep 2002

(Start new page)

11 Internal audits: procedures and records of audits

This section should document the procedures followed in carrying out a formal internal audit of the EMS. These procedures should examine thoroughly whether the EMS conforms to the requirements of ISO 14001, has been properly implemented, is properly maintained, is achieving the targets set, is reviewed at regular intervals and is accompanied by a full set of up-to-date documentation. All audit results must be documented within this section.

|
|

Section 11 Page 1 of *x* Revision number: 06 Date: 14 Jan 2003

(Start new page)

12 EMS Reviews

This section should document procedures for reviewing and revising the EMS in response to EMS performance records, records of past incidents that have led to pollution and audit results (from both internal and external audits). This information should be divided into the following sections:

- person responsible for reviews;
- frequency of reviews and review procedure;
- result of reviews;
- changes made to EMS as a result of review.

|
|

Section 12 Page 1 of *x* Revision number: 04 Date: 06 Sep 2002

(Start new page)

13 Control of EMS manual

This section should document:

- the person responsible for producing, maintaining and revising the EMS manual;
- the procedure for making amendments to the manual.

|
|

Section 13 Page 1 of *x* Revision number: 03 Date: 10 May 2002

(Start new page)

14 Distribution list: authorised holders and locations of environmental management system (EMS) manual

AUTHORISED HOLDERS OF MANUAL

Copy number	Job title of holder	Location where kept
1	Engineering Management System Director	EMS Director's office
2	Works Manager – Fiveoaks Site	Fiveoaks Site library
3	EMS manager – Fiveoaks Site	EMS Manager's office
4	Works Manager – West Cross site	Works Manager's office
5	EMS manager – West Cross site	EMS Manager's office
6	Works Manager – North Hill site	Works Manager's office
7	EMS manager – North Hill site	EMS Manager's office

NO FURTHER COPIES OF THIS MANUAL SHALL BE MADE, UNLESS THESE ARE AUTHORISED BY THE QUALITY ASSURANCE DIRECTOR AND DETAILS OF SUCH FURTHER COPIES AND THEIR AUTHORISED HOLDERS ARE ENTERED ON THIS PAGE.

|
|

Section 14 Page 1 of 1 Revision number: 07 Date: 08 Mar 2003

(Start new page)

15 Procedure for amending manual

Any suggestions for amendments to this manual, and the environmental management procedures described therein, should be addressed to the Engineering Management System Director. These will be considered, and any positive response will be in the form of an official revision of the manual.

NO ALTERATION WHATSOEVER MUST BE MADE TO THIS MANUAL, EXCEPT FOR SUCH OFFICIAL REVISIONS MADE IN ACCORDANCE WITH THE FOLLOWING PROCEDURE.

Procedure for amendment of manual:

1. Amendments as necessary will be issued by the Engineering Management System Director.
2. Each amendment package issued will consist of a set of pages, together with instructions about how these are to be integrated into the existing manual. These instructions will typically require some totally new pages to be added and some existing pages to be replaced with newer versions.
3. The amendment package will include a page that gives the correct revision levels of all pages in the manual. After updating the document, the manual holder should check all pages carefully to ensure that they are all of the correct revision level. After use, this page should be filed at the front of the manual.
4. All sheets removed from the manual in accordance with these instructions must be destroyed.
5. When the amendments to the manual have been completed, the 'Amendment Receipt Slip' provided with the amendment package should be signed and dated and returned to the Engineering Management System Director.

|

Section 15 Page number 1 of *x* Revision number: 03 Date: 16 Jan 2002

Index

ISO 14000 Environmental Management Standards: Engineering and Financial Aspects. Alan S. Morris.
 ISBN 0-470-85128-7